D1222529

MACROMOLECULES AND BEHAVIOR

Second Edition

MACROMOLECULES AND BEHAVIOR

Second Edition

EDITED BY

John Gaito

YORK UNIVERSITY
TORONTO, ONTARIO, CANADA

APPLETON-CENTURY-CROFTS
Educational Division
MEREDITH CORPORATION
New York

7101-1

Library of Congress Catalog Card Number: 76-150497

PRINTED IN THE UNITED STATES OF AMERICA
390-34301-3

Contributors

JOSEPH ALTMAN
Purdue University, Lafayette, Indiana

STANLEY H. APPEL
Duke University, Durham, North Carolina

SAMUEL H. BARONDES
University of California (San Diego), La Jolla, California

EDWARD L. BENNETT
University of California (Berkeley), Berkeley, California

JAMES BONNER
California Institute of Technology, Pasadena, California

MARIAN C. DIAMOND
University of California (Berkeley), Berkeley, California

J. A. DEUTSCH
University of California (San Diego), La Jolla, California

JOHN GAITO
York University, Toronto, Ontario, Canada

EDWARD GLASSMAN
University of North Carolina, Chapel Hill, North Carolina

HOLGER HYDÉN
Institute of Neurobiology, Faculty of Medicine, University of Göteborg, Göteborg, Sweden

ZAFAR IQBAL
All-India Institute of Medical Science, New Delhi, India

BRANISLAV D. JANKOVIĆ
Microbiological Institute, University of Belgrade; Department of Experimental Immunology, Institute of Immunology and Virology, and Immunology Unit, Institute for Biological Research, Belgrade, Yugoslavia

EDWARD KOENIG
State University of New York (Buffalo), Buffalo, New York

PAUL W. LANGE
Institute of Neurobiology, Faculty of Medicine, University of Göteborg, Göteborg, Sweden

SIDNEY OCHS
Indiana University Medical Center, Indianapolis, Indiana

LEONID Z. PEVZNER
Pavlov Institute of Physiology of the Academy of Sciences of the U.S.S.R., Leningrad, U.S.S.R.

KARL H. PRIBRAM
Stanford University, Stanford, California

MARK R. ROSENZWEIG
University of California (Berkeley), Berkeley, California

LARRY R. SQUIRE
University of California (San Diego), La Jolla, California

G. P. TALWAR
All-India Institute of Medical Science, New Delhi, India

GEORGES UNGAR
Baylor College of Medicine, Houston, Texas

JOHN E. WILSON
University of North Carolina, Chapel Hill, North Carolina

Contents

Preface

In the five years since the publication of the First Edition, a tremendous amount of research has been conducted attempting to relate macromolecules and behavior. New approaches to the overall problem have developed. Several chapters of the First Edition were not continued in this edition; most of those which were included have been revised thoroughly. A number of chapters by other contributors were invited to help round out the volume. New chapters are those by Glassman and Wilson; Squire and Barondes; Ungar; Janković; Hydén and Lange; Koenig; Appel; Rosenzweig, Bennett, and Diamond; and Deutsch. This edition should provide a representative view of the important research and theorizing which is occurring in this area.

This volume is built around two important topics: Products of DNA Activation (Section II) and Macromolecules and Intracellular, Intercellular, and Synaptic Events (Section III). The Introduction (Section I) leads to these topics, and Models of Memory (Section IV) attempts to provide an integration of these topics with brain function.

The preparation of this volume was facilitated by grants from the National Research Council (Canada) and by Contract No. Nonr 4935(00) with the Physiological Psychology Division of the Office of Naval Research (U.S.A.).

October 1971 *John Gaito*

Preface to the First Edition

An important problem for many biological scientists is that of determining the mechanisms whereby organisms record and store information relative to their life experiences. For a number of years scientists have suggested that neurological changes occur during learning. Also there has been the expectation that some molecular modification would underlie these neurological changes. The molecular modification would be the mechanism for symbolizing the experimental events. Any one of a number of molecular events might be involved in the experiential process.

The rapid and outstanding developments in molecular biology in recent years have had a stimulating effect on the thinking and experimental research of a number of behavioral scientists and have led to hypotheses concerning the involvement of DNA and RNA (and other macromolecules) in learning and memory. On April 20 to 22, 1964, a conference was held at Kansas State University under the sponsorship of the Office of Naval Research to discuss the role of macromolecules in complex behavior. In attendance were molecular biologists, neurobiologists, and psychologists. The conference provided an opportunity to critically analyze research and ideas in this area and to suggest new ideas and research efforts.

A number of the participants were interested in publishing a revised and current version of their presentation. Other individuals who were unable to participate in the conference also agreed to contribute. This volume is the outcome of that effort. The editor wishes to thank Drs. Gilbert Tolhurst and Joseph Saunders of the Office of Naval Research for their interest and support of the conference, without which this volume might not have materialized.

MACROMOLECULES AND BEHAVIOR

Second Edition

Section I

INTRODUCTION

Chapter 1
INTRODUCTION

JOHN GAITO

York University, Toronto, Ontario, Canada

There are numerous macromolecules that one might refer to behavior. The ones that have been of chief concern to many investigators are the nucleic acids and proteins. Lipids have received lesser emphasis. This volume will be concerned with four types of macromolecules: deoxyribonucleic acid (DNA), ribonucleic acid (RNA), proteins, and lipids; however, chief emphasis will be on the first three.

Up to approximately 25 years ago it was believed that proteins were the genetic material. Although the nucleic acids had been discovered a century ago, their exact function was unknown. Thus genetic theory was built around protein molecules.

In the area of learning and memory, protein also was given a prominent role. For example, Gerard (1953) viewed changes in nerve proteins as synaptic events which facilitate learning and maintain memory. Katz and Halstead (1950) and Halstead (1951) developed an elaborate set of hypotheses attributing an important role to nucleoproteins. These molecules were hypothesized to have the ability to act as templates on which replica molecules were formed. At first the neurons of the brain were supposed to contain random configurations of protein. Stimulation of neural tissue by impulses during learning events caused the randomly oriented molecules to assume a specific configuration. These nucleoproteins then became templates. These workers believed that these templates were like those of germ cells in representing native endowment but differed from the latter in arising from external stimulation. The ordering of the protein templates could take place in various components of the cell and its processes, including the synapse. However, the reorganized protein replicas ultimately resided in the neural membranes where they participated as "traces."

Recent evidence in molecular genetics has indicated that DNA is the genetic material. Thus protein now has been given a secondary—although important—role to that of DNA for genetic functioning.

This work in molecular genetics has been acclaimed by some individuals as the greatest biological discoveries in the 20th century. The advances in this area have been extremely rapid, with researchers from multiple disciplines actively at work. The transfer of genetic information during protein synthesis which involves DNA, a number of RNAs, and amino acids, has been well analyzed. RNA codes have been proposed for each amino acid. Even the sequence of a number of short RNA molecules has been determined. An air of excitement and enthusiasm seems to permeate researchers with the expectation that man is getting closer and closer to nature's basic secrets.

In that the linear sequence of bases in DNA provides the information which specifies the genetic potentialities of an organism, it was immediately evident to a number of behavioral scientists that the linear sequence of bases in neural DNA or in one of the neural RNA fractions (or other mechanisms involving these macromolecules) might subserve the function of an experiential code, i.e., maintain the information contained in memory. Thus the investigations concerned with genetic coding have had a profound effect on a number of behavioral scientists who are concerned with simple and complex behavior and has led to the development of an area which might be appropriately entitled *molecular psychobiology*, an integration of the ideas and methods of psychology and related behavioral sciences with those from molecular biology. This area includes some of the subject matter of biochemistry, biophysics, virology, bacteriology, genetics, cytology, psychology, neurology, neurophysiology, neurochemistry, and neurohistology.

The more simple aspects of behavior—such as stimulation and reduction of stimulation—have not been so exciting and interest-catching to many individuals as have been learning aspects. However, these simple aspects are basic for the more complex behavior; some discussion of these are found in Sections 2 and 3. In general one finds that there are increases in RNA, protein, and lipids during stimulation and a decrement during decreased stimulation. Overstimulation tends also to bring about a decrement.

A number of individuals attempted to relate DNA, RNA, and/or proteins to learning, e.g., Hydén (1959, 1961), Gaito (1961), Dingman and Sporn (1961). Holger Hydén has been the pioneer in proposing and evaluating the hypothesis that RNA is of basic importance in learning. In a number of articles, he has indicated that base ratio changes occur in glial RNA and nuclear RNA of neurons (but not in neuronal cytoplasmic RNA) during the performance of two learning tasks: rats balancing on a wire to reach food and rats forced to use the nonpreferred hand to reach food. The results of these experiments were summarized by Hydén and Lange (1965). Base ratio analyses indicated that during the early portion (three to five days) of the tasks, adenine and uracil predominated in the change in RNA. Later, increased amounts of guanine and cytosine were present. Hydén assumed that these base ratio changes indicated that the RNAs synthesized during learning tasks differed from those synthesized during control conditions.

The meaning of the base ratio changes is not clear. Two major alternatives are suggested. A qualitative change may be occurring with polymer RNA being modified to produce a new molecular species or foreign RNA may permeate certain cells such that the RNA population contains new types of RNA. On the other hand, quantitative changes may be involved, i.e., the relative amounts of synthesis and/or degradation of the species of transfer, ribosomal, and messenger RNAs may be changing. The latter possibility appears to be the most likely one.

Based on results such as the above, Hydén developed some interesting theoretical notions. At first (1959, 1961), he advocated a qualitative-change approach (an instructive model). He hypothesized that memory involves a change in the sequence of bases in the RNA molecule through frequency modulation; one or more bases are exchanged with the surrounding cytoplasmic materials. The new base at this space is now stable under the influence of the modulated frequency. The new pattern remains and constitutes the specification of the RNA in the nerve cell. Since the sequence of the bases in the template RNA is now changed, new protein formed through the mediation of the RNA will also be specified. Stimulation causes a rapid dissociation of the specified protein and the combination of the dissociated products with a complementary molecule or an energy activation of the specified dissociated protein. Through a rapid combination of the dissociated protein with a complementary molecule, an activation occurs of the transmitter substance, and the postsynaptic structure is excited.

In this model, a nerve cell responds differentially depending on whether the pattern of impulses it receives is novel or familiar, as well as on the pattern itself. No protein will have the correct configuration if the incoming pattern of impulses is new; therefore, no dissociation of the protein can occur. The electrical pattern must first shape a new RNA molecule, which in turn shapes a protein molecule that can dissociate. The protein molecule fragments then react with a complementary molecule, causing the triggering of a substance (excitatory or inhibitory) across the synapse. If, on the other hand, the incoming impulse is familiar, protein molecules will already be present that are capable of dissociating rapidly.

A similar type of model, but one concerned with DNA base changes, was advocated by Gaito (1961). He suggested that changes might occur at the attachment of the two strands of DNA, with adenine at one locus changing to guanine and the associated pyrimidine changing from thymine to cytosine. Deletion, addition, and rearrangement of bases were considered as other means of changing the code. These changes would provide a basis for modification of the genetic potential in nerve cells by means of external stimulation during learning, which would be a somatic mutation not transmitted to the offspring. Other possible biochemical mechanisms were also suggested which could be involved in memory function, i.e., changes in RNA or amino acid sequences.

Dingman and Sporn (1961) hypothesized that RNA changes were the basis for memory. However, they suggested that the linear sequence of bases (primary structure) was only one possible means of coding experiential events; changes in the helical structure (secondary structure) and overall configuration (tertiary structure) also could be the basis for memory.

Early models emphasizing unique contributions of RNA during learning were suggested also by Cameron (1963), Gerard (1963), and by McConnell (1962).

In recent years there has been a shift away from the emphasis on the base sequence of nucleic acids (instructive models) to aspects involving the interaction of two or more types of molecules as in enzyme induction and DNA-histone complexing ideas (selective models). Associated with this shift is the increasing concern with multiple macromolecules such as lipids and neural circuitry, as suggested by Dingman and Sporn (1964) and by Barondes (1965).

Smith (1962) maintained that the enzyme induction model might be suitable for learning events. He suggested that the inducer substance may be acetylcholine (ACh) which is released at the synapse during stimulation. The proteins then induced via RNA synthesis would be choline acetylase (ChA) and acetylcholinesterase (AChE). These events were presumed to increase the amounts of ACh at the synapse and increase the probability that stimulation in neural units would activate other neural units, thus leading to greater potential for adaptive behavior. Briggs and Kitto (1962) and Goldberg (1964) expressed somewhat similar views.

A later approach to learning by Hydén (Hydén, 1967; Hydén and Lange, 1965) is a selective model: DNA sites in glia and in neurons are stimulated such that specific RNA species are synthesized. During the early part of the learning process, the RNA that is synthesized is high in adenine and uracil, thus being DNAlike in composition. This RNA is formed during the establishment of functional synapses for the new behavior, and this stage represents short-term memory. During the later portion of the learning process an RNA rich in guanine and cytosine (similar to ribosomal RNA) is formed. This stage is supposed to constitute the fixation of long-term memory with a high synthesis of transmitters at the synapse. Hydén (1967) suggested also that higher-type learning, e.g., insightful learning, could involve "an instructive mechanism, a permanent change of information rich macromolecules during the life cycle" (p. 340). He did not specify the exact mechanism, however.

Other similar approaches are the DNA derepression model of Bonner (1966) and the DNA activation model of Gaito (1966). Bonner suggested that in neurons there is a gene, or a few particular genes, which are repressed, but which can be derepressed by certain substances as a result of electrical stimulation of specific portions of nerve cells, viz., dendrites. Once derepressed, the gene makes more RNA and ultimately more enzyme. The enzyme is then involved in the chemical reaction which makes more of the substance that carries the effect of the dendritic input to the repressed DNA site. These genes, once derepressed, remain derepressed permanently. Such derepression would account for the increased rate of RNA and protein synthesis which is reported in learning experiments. (See Chapter 16, present volume.)

The DNA activation model hypothesized that, during learning events, stimulation of specific nerve cells in a particular portion of the brain causes a modification in the DNA complex such that DNA is activated to synthesize RNA. Four types of RNA are synthesized: messenger(s) for synaptic and nonsynaptic protein; messenger(s) for ribosomal protein; messenger(s) for enzymes to synthesize lipids; and ribosomal RNA. The protein proceeds to synaptic junctions to increase the surface area, to prepare for the development of connections with other nerve cells, and to make postsynaptic neurons more susceptible to excitation. The lipids are incorporated along with proteins into membranes at the synapse and elsewhere. The ribosomal RNA and protein aggregate to make new ribosomes. The synaptic changes link neural circuits together to make more probable the inclusion of certain cells in the electrochemical stimulation that results from the physical energies impinging upon the receptors of the organism when external stimulation occurs. This stimulation allows the DNA complex in a greater number of nerve cells to be potentially available for modification so that further RNA and protein synthesis can occur. Since they function as vehicles for protein synthesis, the increased ribosomal content (in aggregates as polysomes) would provide for a more rapid rate of protein synthesis during later stimulation.

Another selective model is that by Flexner and Flexner (1966), who assumed that specific messenger RNAs are synthesized during learning events.

Recently a model was suggested which incorporated both instructive and selective mechanisms. Griffith and Mahler (1969) speculated that base changes (via methylation or demethylation) in control DNA sites could determine the number of times that an induced messenger RNA would be used in protein synthesis. Each messenger RNA was assumed to have a number of RNA tickets adjacent to the coding portion. Only the messenger RNAs with base changes in one or more RNA tickets, as a result of DNA base changes, would function in protein synthesis. A protein or polypeptide would be synthesized for each modified RNA ticket. This model is reminiscent of the earlier Gaito (1961) model, but it incorporates the selective model which is widely accepted for molecular biological phenomena. At this time, however, the model must be considered highly speculative, as is the case with any of the models. Research within the next few years should provide a better basis for evaluation of these models.

An interesting aspect of some models in this area (e.g., by Hydén), as well as in neuropsychology (e.g., Galambos, 1961; Pribram, 1966) is the increased importance attached to glial cells in complex behavior. Although the suggestions are unique and of great interest, the evidence cited is of circumstantial nature. (However, one should refer to Chapters 15 and 17 of the present volume.) This is true of the Hydén results, which suggest a relationship between glial and neural RNA during certain behaviors. To show the causal relationship that he is assuming occurs, he would need, e.g., to produce change in glial RNA and note the effect on neural RNA.

One assumption in all of the above models is that RNA and/or protein changes mediate the behavioral changes during learning events, i.e., that RNA-protein modifications have a primary effect in memory. However, it is possible that the RNA-protein changes are secondary effects of the behavioral events (Pevzner, 1966) or that the two are parallel, noninteracting systems. Unfortunately, although there is considerable evidence to indicate a relationship between RNA-protein changes and behavior (Pevzner, 1966), there is no research which clearly and conclusively differentiates these three possibilities. Thus, although the data and models are abundant relative to the role of macromolecules in behavior, the exact function of these molecules in behavior is not clear; however, the time is ripe for basic contributions in this area.

References

Barondes, S. H. (1965). Nature, 205:18.

Bonner, J. (1966). In: Gaito, J., ed., Macromolecules and Behavior, 1st ed. New York: Appleton-Century-Crofts.

Briggs, M. H., and Kitto, G. B. (1962). Psychol. Rev., 69:537.

Cameron, D. E. (1963). Brit. J. Psychiat., 109:325.

Dingman, W., and Sporn, M. B. (1961). J. Psychiat. Res., 1:1. ——— and Sporn, M. B. (1964). Science, 144:26.

Flexner, L. B., and Flexner, J. B. (1966). Proc. Nat. Acad. Sci. U.S.A., 55: 369.

Gaito, J. (1961). Psychol. Rev., 66:288. ——— (1966). Molecular Psychobiology. Springfield, Ill.: Thomas.

Galambos, R. (1961). Proc. Nat. Acad. Sci. U.S.A., 47:129.

Gerard, R. W. (1953). Sci. Amer., 189:118. ——— (1963). J. Verbal Learning Verbal Behavior, 2:22.

Goldberg, A. L. (1964). Science, 144:1529.

Griffith, J. S., and Mahler, H. R. (1969). Nature, 223:580.

Halstead, W. C. (1951). In: Jeffress, L. A., ed., Cerebral Mechanisms in Behavior. New York: Wiley.

Hydén, H. (1959). In: Biochemistry of the Central Nervous System, Vol. III, Proc. 4th Internat. Congr. Biochem. London: Pergamon Press. ——— (1961). Sci. Amer., 205:62. ——— (1967). Proc. Amer. Phil. Soc., 111: 326. ——— and Lange, P. W. (1965). Proc. Nat. Acad. Sci. U.S.A., 53: 946.

Katz, J., and Halstead, W. C. (1950). Comp. Psychol. Monogr., 20, 103:1.

McConnell, J. V. (1962). J. Neuropsychiat., 3:42.

Pevzner, L. B. (1966). In: Gaito, J., ed., Macromolecules and Behavior, 1st ed. New York: Appleton-Century-Crofts.

Pribram, K. (1966). In: Gaito, J., ed., Macromolecules and Behavior, 1st ed. New York: Appleton-Century-Crofts.

Smith, C. E. (1962). Science, 138:889.

Section II

PRODUCTS OF DNA ACTIVATION

This section is concerned with macromolecular events during and subsequent to the assumed activation of DNA during behavior. If macromolecules contribute to behavior, they do so in a dynamic cell within the neurological milieu. Thus one needs to know how macromolecules relate to the electrical events of the membrane which encases the nerve cell. Part of Chapter 2 (Talwar and Iqbal) is concerned with these electrical aspects. The remainder of Chapter 2 and Chapters 3 and 4 (Glassman and Wilson; Gaito) describe research efforts in which a direct approach was used (behavior is varied as the independent variable and chemical changes are evaluated as the dependent variable); the first is concerned with sensory aspects whereas the others deal with learning phenomena. The next two chapters (Squire and Barondes; Ungar) discuss research that utilized indirect procedures (the independent and dependent variables are the reverse of that in the direct procedure). Squire and Barondes use RNA and protein synthesis inhibitors to affect behavior, and Ungar is concerned with the "transfer experiment" paradigm. Chapter 7 (Janković) describes an immunoneurological approach to understanding the relationship between macromolecules and behavior, and the contribution of S-100 proteins in a simple learning task are evaluated in the final chapter (Hydén and Lange) describing this approach.

Chapter 2

MACROMOLECULES—FUNCTIONAL AND BIOCHEMICAL CORRELATES

G. P. TALWAR and ZAFAR IQBAL

All-India Institute of Medical Science,
New Delhi, India

The role of small molecules, in particular the monovalent and divalent cations and neurohumors, has been well recognized in the central nervous system. The former contribute to the development of resting potentials across the membranes. They also impart, along with structural components, the properties of excitability to these cells. Their dynamics are inherently associated with the generation of action potentials. Neurohumors as a class have an important function in transmission of messages, excitatory or inhibitory, from one cell to the other. Several such molecules have been identified and chemically characterized. The fact that many compounds with pronounced psychoactive effects have structures closely resembling the neurotransmitters points to the great relevance of these "micro" molecules to the brain function. Some of these issues have recently been discussed elsewhere (Talwar and Singh, 1970) and would not come under the purview of this chapter.

Nerve cells, in particular the cells of limbic system and hypothalamus, also make and secrete substantial amounts and varieties of intermediate size molecules that are mostly polypeptides. These polypeptides act either as hormones (oxytocin, vasopressin) or as "release factors" for the secretion of anterior pituitary hormones. They have, therefore, a role in the integration of nervous system and endocrines. It is clear that the part played by the central nervous system in homeostasis is exercised through these *intermediate-size* molecules. They do not, however, qualify on grounds of strict definition to be among macromolecules and hence will be omitted in the present discussion even though they are functionally important.

Macromolecules constitute the structural elements of the central nervous system. The brain is a highly complex and intricately organized organ. Structure and organization are essential to the function of most cells, but these qualities are of primordial consideration in the central nervous system. The pioneer of neurochemistry, Thudichum, analyzed the chemical components of the brain almost 100 years ago. His treatise is an early classic on the subject. However, with the development of finer techniques of extraction and separation over the past years, as well as because of advances in protein chemistry and enzymology, much more is known on the type of compounds present in the brain. But very little is understood of their organization and precise intermolecular relations in situ. A part of this chapter will consider some macromolecules pertinent to the ontogenesis of spontaneous and evoked electrical activity in the developing brain.

Learning and memory have inherently an element of sensory input. Metabolic events taking place in response to sensory stimulation are not fully known. This chapter will discuss the synthesis and turnover of proteins in cortical areas of the brain when animals are exposed to visual stimuli. We will also summarize salient facts and fallacies about brain RNA and its possible role in brain function.

Macromolecules and Ontogenesis of Electrical Activity

The manifestation of spontaneous and evoked electrical activity is a characteristic trait of a functionally developed nervous system. The stage at which these attributes appear in a developing nervous system varies from animal to animal. In the chick embryo brain, there is practically no manifestation of electrical activity till Day 4, when occasional slow electroencephalographic waves with intermittent silent intervals are recorded (Sharma et al., 1964). These occasional slow waves gradually become regular and increase in amplitude with age. By Day 13 fairly well-marked, low-voltage fast activity appears that becomes prominent in periods after this age. The capacity for evoked response also emerges after Day 12. Application of strychnine produces spikes in chick embryo brain only on and after Day 12 (Garcia-Austt, 1954). Lately our laboratory has been interested in evaluating the biochemical changes taking place in the chick brain between Days 4 and 12. Of particular interest are the changes related to the distribution of cations and membrane linked components.

Capacity for Retention of Potassium

Most cells have a higher intracellular concentration of potassium as compared to the extracellular environment. This capacity is particularly marked in the nerve tissue. The intracellular concentration of potassium in the brain is computed to be about 30-fold higher than the concentration of potassium in an extracellular medium.

There is a sharp rise in the K^+ concentration of the chick embryo brain between Days 6 and 10. It rises from 331 μEq on Day 6 to 635 μEq/g dry weight of the tissue or from 16.7 ± 0.7 to 44.7 ± 1.4 μEq/mg DNA on Day 10 (Iqbal et al., unpublished data). This rise is obviously related to the increase in protein content of the brain. The cations would no doubt be substantially bound by anions such as glutamate, aspartate, gangliosides, glycolipids, mucopolysaccharides, and others (Folch et al., 1957; Hayden et al., 1961). The avidity of the lipids is particularly high for binding of divalent cations. The univalent cations are bound only at high concentrations (Breyer, 1965). It has been estimated that 50-fold higher concentrations of Na^+ or K^+ are required to displace half of Ca^{++} bound to acidic lipids of stomach or spinal cord (Wooley and Campbell, 1962). Even from cerebrosides, which are good K^+ binders, 16 μmoles of K^+ were needed to displace half of 4.5 μmoles of Ca^{++}.

There are reasons to believe that acidic lipids may not be a unique class of compounds involved in the retention of potassium in the brain. Tracer studies show that about 20 percent of the brain potassium does not exchange with the environmental potassium (Brinley, 1963), which would imply that either this fraction of the cation is held in nonionic bonds and is not freely exchangeable, or that this fraction of the total potassium is sequestered behind a potassium-impermeable barrier. The cation bound to lipids is essentially exchangeable. This is an indirect argument for a "non-lipid" form of this fraction of potassium. Another observation pointing to the need for further search of alternate potassium-binding molecules in the nerve cells is the report by Katzman and Wilson (1961). These workers found that the phospholipid-cation complex extracted at low temperatures ($-45°$ to $-55°$C) had a Na:K ratio of 1.8, whereas the Na:K ratios in the whole brain tissue were 0.6. The extraction procedure used by these workers avoids the alteration of cations bound to lipids in situ and prevents the exchange of inorganic cations between the lipid bound and other compartments during extraction. If this extract is representative of the cation phospholipid complexes in the brain, it would imply the presence in cells of lipids which bind Na^+ better than K^+. As the total concentration of K^+ is higher in cells, there should be other substances present in the cells which have the ability to sequester potassium preferentially.

Proteins could well be involved in the retention of potassium in the cells. Several enzymes are known to be potassium-requiring entities (Dixon and Webb, 1958; Cantoni, 1960). There are at least 23 enzymes identified in animal, bacterial, or plant cells which are activated 10- to 20-fold by potassium (Lubin, 1964). Of great interest and relevance to this argument are the pronounced potassiumphilic properties of cyclic polypeptides such as valinomycin. Valinomycin, a 12 amino acid peptide, permits K^+ but not Na^+ to penetrate the membrane (Pressman, 1965). Another macrocyclic tetrolide, nonactin, forms a complex with K^+, K^+ being bound to the anionic groups of the molecule within its "doughnut-hole" configuration (Kilbourn et al., 1967). It is not inconceivable that the rise in the chick brain content of potassium may be related to, and be a consequence of, the overall increase in cellular proteins and the emergence of proteins such as valinomycin with high potassium-binding capacity.

Involvement of Multiple Biochemical and Structural Components

A gradual pattern of development is not limited to traits such as spontaneous and evoked electrical activity in the brain. In developing dog pups, it has been observed that differences between sleep and wakefulness as gauged by electroencephalograms (EEGs) were present from the second day but were clearly marked only at the end of the first week. Activated sleep was noted at 12 days. Response to auditory stimuli was present only during the third week (Diperri et al., 1964). In chick embryos, even though the spontaneous electrical activity matures around Days 11 to 13 of incubation, EEG patterns characteristic of behavioral sleep or attention appear after hatching (Peters et al., 1965). In rats, Crain (1952) described the general trend in the spontaneous electrocorticographic activity from birth to about 10 days towards increasing rhythmicity, regularity, and amplitude.

The gradual maturation of these traits indicates the following.

(1) These activities are likely to depend not on a single but on multiple components.

(2) Besides developments within cells, the establishment of interconnections between cells is crucial for manifestation of diverse properties. Maturation of most brain functions, including electrical activity patterns, would require the development and arborization of structures capable of establishing communication with other cells.

(3) Simultaneously enzymatic machinery for elaboration, storage, release, and degradation of neurotransmitters has also to evolve to render the intercellular connections functional.

Support to these views is available from a number of observations, some of which follow. The capacity to accumulate glutamic acid and glutamine parallels the growth of the brain from fetal life to maturity in rat, rabbit, guinea pig, cat, and dog (Himwich and Peterson, 1959). The time sequence of the development of the enzymes giving rise to and metabolizing GABA in the chick embryo cerebellum is correlated with the development and increase in the recognizable synaptic structures (Koriyama et al., 1968). Neurons of the cerebral cortex become selectively permeable to sodium at about Day 45 of gestation in guinea pig, when electrical activity can be recorded. There is a correlated series of sharply defined morphologic and biochemical changes in the brain of the growing animals at this stage (Flexner and Flexner, 1949). Histological changes in the mouse cerebral cortex parallel the developmental changes in the electrocorticogram. A fifth layer is formed at six days of age by the separation of small- and medium-sized pyramidal cells (Kobayashi et al., 1964). There is also considerable branching of dendrites of all types of pyramidal cells. In developing dog brains the most dramatic changes occur in the arborization of dendrites between two to four weeks of age (Fox et al., 1966).

Na-K-Mg-Activated ATPase

The generation of action potential is inherently linked with a momentary flux of sodium ions into the cell. In the recovery phase, the ionic equilibrium is reestablished. The intracellular concentration of Na^+ is lower than in the extracellular compartment. The sodium pump for active transport of Na^+ is believed to function through the membrane linked enzyme Na-K-Mg ATPase. This enzyme was first discovered by Skou (1965) in erythrocytes but it is also present in the membranes of nerves and mammalian brain (Skou, 1957, 1960; Jarnefelt, 1961; Schwartz et al., 1962). The activity of the enzyme is highest in the cerebral gray matter. It utilizes ATP for expulsion of Na^+ and uptake of K^+. Although the enzyme is active within 70 percent of maximal capacity in a wide range of Na^+ and K^+ concentrations, the system is most sensitive to Na^+ concentrations of the order encountered within the cell and to K^+ concentrations prevalent in the extracellular environment.

As the enzyme has an important role in the cation dynamics, its activity was followed in the chick embryo brain in the course of development. Almost the entire activity of the enzyme is recovered in the mitochondrial and 105,000-g pellet, the latter composed of nerve endings, synaptosomes, microsomal membranes, etc. The 105,000-g pellet fraction has, however, higher specific activity. It was observed that the activity of the enzyme was low on Day 6. It rose rapidly between Days 10 and 12 of incubation, at which time a plateau is attained and stays at this level till hatching (Fig. 1). This steep period of rise in the specific activity of the enzyme does not coincide with a generalized spurt in protein synthesis in the brain. Furthermore, the activity of the enzyme does not show a similar sharp increase in tissues other than brain—as, for example, heart.

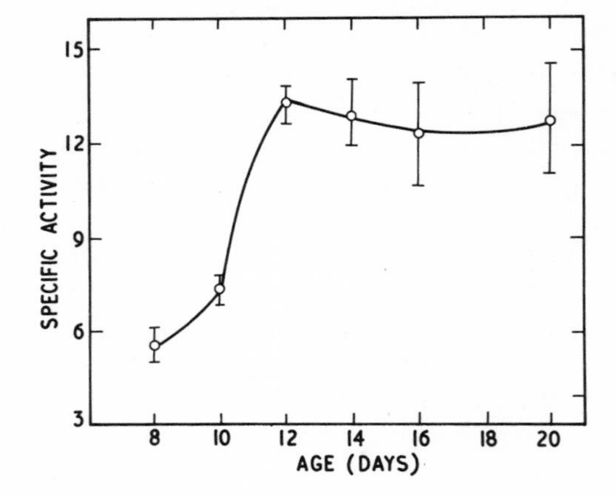

FIG. 1. Activity of microsomal Na-K-Mg ATPase at different ages in developing chick embryo brain. The values are mean of two sets, in each set eight to nine brains were pooled in age groups 6 to 10 days and two to three in age groups 12 to 20 days. (From Zaheer, Iqbal, and Talwar. 1968. *J. Neurochem.,* 15:1217.)

The enzyme preparation from chick embryo brain from Day 13 has the same properties and characteristics as the six-day-old embryo brain enzyme. It has similar pH optima and temperature for maximal activity. Both are optimally activated by 100 mM Na^+ and 20 mM K^+ concentrations. Both are inhibited in a parallel manner by Ca^{++} and ouabain. Experimental evidence rules out simple activation (removal of inhibitors) as the possible basis for increase of activity of the enzyme between Days 8 and 10, suggesting that these developmental changes in the activity of the enzyme may represent new synthesis of the enzyme (Zaheer et al., Talwar 1968).

The above studies show a temporal relationship between the optimal synthesis of this enzyme and the maturation of spontaneous and evoked electrical activity. In rats, the Na-K-activated, ouabain-sensitive ATPase appears on Day 21 of gestation and reaches the adult levels in its specific activity by Day 12 postpartum. The EEG activity is detectable on Day 21 of gestation and becomes more rhythmic and regular during the first week of postnatal life in these animals (Côte, 1964; Abdel-Latif et al., 1967).

Acetylcholinesterases

The factors discussed so far were those concerned with the development of resting potentials, excitability of the membrane, and cation dynamics. It is logical to expect in this context the ontogenesis of enzymes that are capable of synthesizing and metabolizing neurotransmitters, enabling the passage of the impulse across synaptic junctions. The activity of glutamic acid decarboxylase has been shown to emerge in synchrony with the development of synaptic structures (Koriyama et al., 1968).

The activities of choline acetylase and acetylcholinesterases are observed to increase 22- and 10-fold, respectively, in the chick spinal cord from Stage 22 (3.5 days) to Stage 30 (6 days) of development (Burt, 1968). There is a correlation between the Na-K-ATPase and acetylcholinesterase in the Electrophorus electric organ. The concentration gradient of the activities of both enzymes from the rostral to the caudal end is similar (Fahn, 1968). A good correlation has been found between the increase in 5-hydroxytryptophan decarboxylase, choline acetylase, acetylcholinesterase, and cell maturation as gauged by increase in dendrite volume, myelination, and appearance of mature electrophysiologic responses in the developing rabbit brain (McCaman and Aprison, 1964).

Acetylcholinesterase is observed to exist in several isozyme forms. When chick embryo brain enzyme preparations are run on polyacrylamide gels and the bands developed by the method of Maynard et al. (1964), in which acetylthiocholine is used as the substrate, at least four isozyme bands are detectable. Tissue preparations from embryos of Day 6 incubated eggs show only three bands. The fourth band develops around the ninth day (Fig. 2). This band has the lowest migration, and is inhibited by inhibitors such as eserine and PCMB. The function and differential regional distribution of these isoenzymes is not yet clear. Using alpha-naphthyl esters as substrate, Bernsohn et al. (1964) have found the presence of at least 14 isozymic bands of enzymes with esterase activity in the developing rat brain.

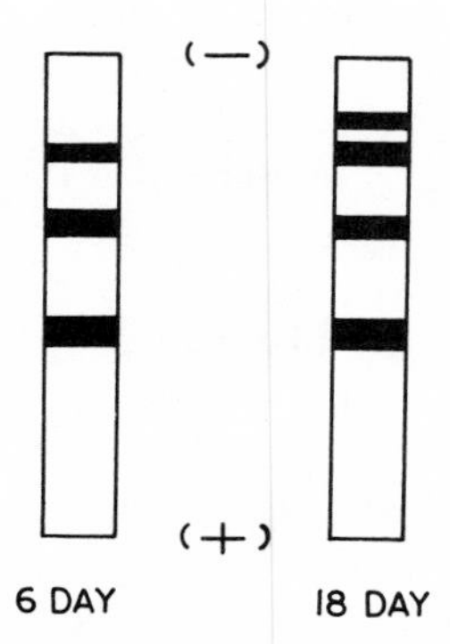

FIG. 2. **Zymogram pattern of acetylcholinesterase on polyacrylamide gels at different ages in developing chick embryo brain.**

Brain RNA: Facts and Fallacies

Content

Much has been said and speculated upon about the role and function of RNA in the brain (e.g., Hydén, 1958, 1960; Dingman and Sporn, 1961, 1964; Eigen, 1966; Gaito, 1966; Pevzner, 1966; and Talwar, 1969). In earlier years, attention was drawn to this constituent because of the apparently high content of RNA in the neurons, which was of the order of 1,500 $\mu\mu$g in a large neuron of the Deiters' nucleus (Hydén, 1960). RNA, however, varies with the cell type and within cells of the same type, the values range from 50 to 2,000 $\mu\mu$g for neurons. Edström and Pigon (1958) have suggested the existence of a proportionality between the area of the cell body surface and the content of RNA in the spinal ganglion cells. The glial cells have a much lower concentration of RNA, about 3.5 $\mu\mu$g per cell. It is because of this large disparity that the RNA/DNA ratios in the whole brain work out to be lower than those of liver or pancreas (Leslie, 1955), even though large neurons have a high RNA content. The high content of a component in a cell is not, per se, sufficient as an argument for the vital role of that component. It may also be pointed out that though the content of RNA in glia cells is low, the rate of synthesis of RNA in these cells is twice as rapid as in neurons (Daneholt and Brattgård, 1966).

Alterations Following Moderate and Intensive Stimulation

It was observed that the RNA content of the brain underwent an alteration in diverse type of situations. Moderate stimuli tended to cause a small increase, whereas intense and prolonged stimuli diminished the total amount of RNA. Hamberger et al. (1949) reported an increase in the RNA concentration of the vestibular ganglion cells on rotatory stimulation of the animals. Barr and Bertram (1949) found that on moderate electrical stimulation, the nucleolar satellite enlarges and moves away from the nucleolus coincident with the appearance of Nissl substance. Vraa-Jensen (1957) has found by histological techniques evidence for an increased content of RNA in the earlier stages of a functional stress. On topical application of dilute solutions of metrazol to isolated cerebral cortex preparations of dogs, three phases (Fig. 3) of electrical activity can be recorded (Virmani et al., 1963). In the prespike phase, marked by a slow-wave activity preceding the onset of spike activity, there is a tendency toward a rise in the RNA content. During the phase of spike activity, the

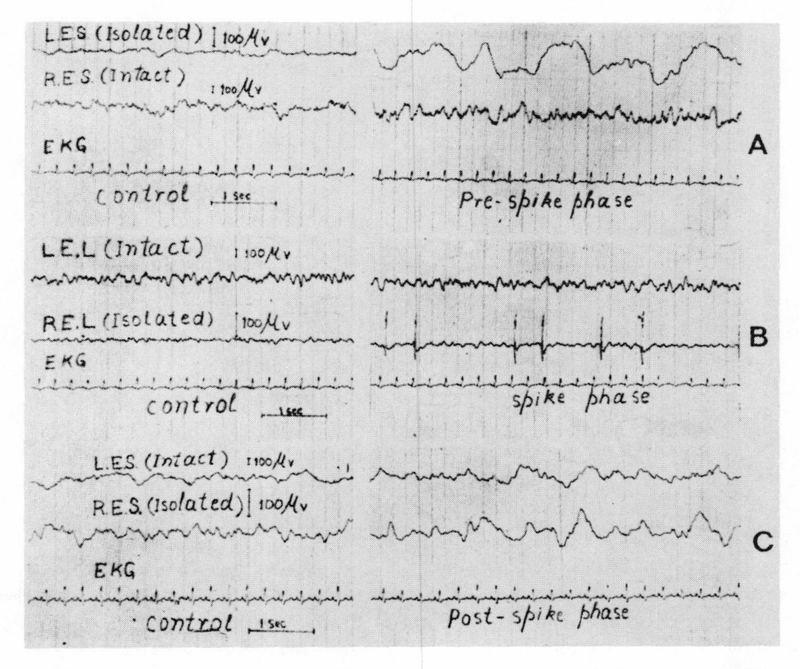

FIG. 3. Left half of the figure shows the EEG activity of intact and "neuronally isolated" cerebral cortex of dog. Sequential changes in electrical activity on topical application of dilute solutions of pentamethylene tetrazol (metrazol) are shown in the right half of the figure. L.E.L. (left ectolateral), R.E.L. (right ectolateral), L.E.S. (left ectosylvian), R.E.S. (right ectosylvian). (A)—Prespike phase characterized by high voltage slow waves, (B)—Intermittent bursts of spikes during spike phase, and (C)—Early post-spike phase with slow waves. (From Virmani et al. 1963. *Ind. J. Med. Res.,* 51:75.)

RNA decreases, whereas the values return to normal level (Fig. 4) in the postspike phase (Chitre and Talwar, 1963).

Intense convulsions lead to a fall of the brain RNA concentrations (Talwar et al., 1961; Noach et al., 1962). These changes were found to be organ-specific to some extent. No change occurred in the liver content of RNA, while the gastrocnemius muscle RNA increased during convulsions (Chitre et al., 1964). Hydén (1943) and Barbato and Barbato (1965) observed a significant fall in RNA after convulsions induced by insulin. Similarly, acoustic stimulation caused a decrease in the nucleoproteins of the nerve cells. Geiger and Yamasaki (1956) and Geiger et al. (1956) observed an accumulation of acid-soluble nitrogenous compounds along with a decline in nucleic acid content in the cat brain during stimulation through afferent nerves. These chemical changes were found to be propor-

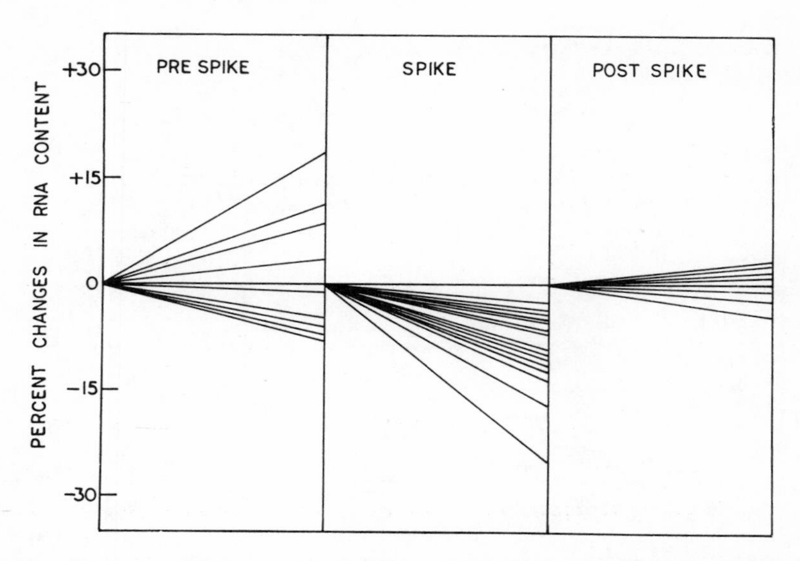

FIG. 4. Changes in RNA of neuronally isolated cerebral cortex of dogs in various phases of electrical activities induced by the topical application of metrazol. Each line represents one experiment. Contralateral areas were taken as controls in each experiment. (From Chitre and Talwar. 1963. *Ind. J. Med. Res.,* 51:80.)

tional to the number of effective stimuli. Vrba and Folbergrova (1959) reported that strenuous physical exercise—such as compulsive swimming in rats—caused a fall in the nucleic acids, protein, and orcinol-positive substances. Gomirato (1954) noted that motor exhaustion produced a decline in RNA in motor root cells. Einarson and Krogh (1955) found that the large pyramidal cells of the cortex were choromophobic in an epileptic patient who died a few hours after a series of grand mal seizures.

In ensemble, these observations suggest that moderate stimuli activate the synthesis of RNA, while convulsions or intense stimuli cause a fall of RNA. The stimulation is not limited to RNA, but is also observed for proteins and perhaps other constituents. These effects could be a consequence of the increased pool of precursors following permeability changes brought about by the input stimuli. The fall in RNA content is most likely due to a relative slowing down or arrest of synthesis of RNA owing to the limitation of ATP and other nucleoside triphosphates that are diverted for other essential energy needs of the cell (as, for example, the sodium pump). This viewpoint is supported by the observations of Orrego (1967), who stimulated brain slices electrically in vitro (1.5 V for 10 min) and observed a fall of 40 percent in the rate of synthesis of RNA. No change was noticed in the rate of breakdown of RNA, hence the net decline in RNA content was a result of relative inhibition of the synthesis rather than an increase in the rate of breakdown, though the latter can well be an additive factor in whole-brain experiments.

Types of RNA in the Brain: Physical and Metabolic Characteristics

Size

Analysis of RNA from brain on methylated albumin-Kieselguhr (MAK) columns or on sucrose density gradients (SDG) have shown patterns similar to those obtained for RNA from other tissues. There are three major classes of RNA: 4 S, 17 to 18 S, and 28 S, as seen by optical density profiles (Mahler et al., 1966). If a pulse of radioactivity is given, about five classes of RNA are detectable. These are 4 S, 7 to 14 S, 18 S, 28 S, and a component heavier than 28 S that has a distribution between 30 and 50 S (Toschi et al., 1966; Jacob et al., 1966). Vesco and Giuditta (1967) have reported two broad classes of RNA in the brain nuclei. One of them represents the precursors of ribosomal RNA, with sedimentation constants of 45 S and 32 to 35 S. The second category is a species of RNA with polydispersed character that has a size of 8 to 80 S. This may represent the messenger or template type of RNA.

Organ Specific Species

Even though the general size and sedimentation characteristics of brain RNA are similar to the RNA from other animal tissues, the molecular species are not completely identical. Brain RNA competes only partially with liver RNA for hybridization to homologous DNA (Talwar, 1969). There has been a recent report of an organ specific population of RNA in the Salmon brain (Mizuno et al., 1969).

Metabolism

The initial incorporation of radioactive precursors takes place in the fractions of RNA that sediment with the heavy pellet and represent (predominantly, although perhaps not exclusively) the nuclear RNA. RNA would be synthesized in the tissues that utilize DNA as a template and RNA polymerase as the transcribing enzyme. There is some evidence to suggest that the totality of RNA synthesized in the nuclei is not transferred to the cytoplasm. Bondy (1966) found that on Day 25, 16 percent of the radioactivity in RNA due to intracisternally administered (2-^{14}C) cytidine was still traceable in the nuclear fractions. Although the reutilization of the radioactive precursor cannot be completely excluded as a possibility in these experiments, there is a growing body of evidence indicating discrete control mechanisms for the transfer of RNA species from nuclear to the cytoplasmic compartments (see for discussion and other references, Talwar, 1969; Harris, 1970).

The turnover rate of brain ribosomal RNA is computed to be approximately 0.2 percent per hour, with a half-life of about 12.5 days (Bondy, 1966) or 13 to 15 days (Khan and Wilson, 1965). Dawson (1967) suggests an apparent half-life of about six days for brain ribosomal RNA (rRNA), a value that is not in agreement with the findings of other investigators cited above. This may, however, represent the minimum value. The half-life of liver rRNA is found to be about 5 to 7.2 days. The turnover rate of brain rRNA and tRNA would thus be at best equal to that of liver rRNA and tRNA, but it is more likely that it is only half as rapid.

The turnover rates of template RNA in cortex and subcortex show the presence of at least two populations of mRNA having a mean half life of either 2.6 hr (Orrego and Lipmann, 1967) or 10 to 20 hr (Appel, 1967).

RNA in Developing Brain

In developing rabbit pups, RNA per unit DNA rises sharply from Day 1 to about 10 to 11 days. This is the period at which the eyes open. In the period following the opening of the eyes, there is a fall in the RNA-to-DNA ratios of the brain. Prior blinding of the animals prevents the expected fall in the occipital cortex to a large extent (Talwar et al., 1966b; Talwar, 1969).

These observations indicate (1) a high rate of RNA synthesis in the brain in the early neonatal period; (2) a slowing down of the RNA synthesis at a defined period fairly early in neonatal development; and (3) a possible utilization of some species of early accumulated RNA for elaboration of developmental structures.

The fact that the rate of RNA synthesis is higher in brain during early neonatal period is supported by the findings of Orrego (1967), who showed that the incorporation of radioactive uridine into RNA decreased appreciably in developing rat brain from the third day of age to maturity. There was a fivefold decline in the apparent rate of RNA synthesis. Johnson (1967) made similar observations in mice. Cell suspensions prepared from newborn mouse brain synthesized RNA much faster than the preparations from older mouse brain. The decline started around the sixth day of age. The cause of the fall in the rate of RNA and protein synthesis in rats after birth is not clearly understood. It is possible that the template activity of the chromatin may get diminished in the course of development. Alternatively, in the course of development additional controls on

RNA polymerase may emerge and become superimposed. There may be a requirement of factors such as sigma (Burgess et al., 1969) for effective transcription of some operons.

RNA in Learning and Memory

Advances in molecular genetics inspired the idea that RNA may constitute the informational molecules for retention of experiences (Hydén, 1958; Dingman and Sporn, 1961, 1964; Eigen, 1966; Gaito, 1966). However, the situation appears to be substantially more complex than that envisaged previously (Richter, 1966; Talwar, 1969). We will summarize the salient grounds on which the premise was based. We give also the main evidence discounting the hypothesized key role of RNA in learning and storage of information. (Other chapters in this book, in particular those of Gaito, discuss this aspect more fully.)

ARGUMENTS FAVORING A ROLE FOR RNA

Induced Synthesis of RNA in Response to Learning and Stimuli

When rats were trained to obtain food by balancing on a wire perched at an angle, Hydén and Egyhazi (1962, 1963) found that there was a significant increase in the nuclear RNA of both neurons and glia of the Deiters' nucleus. There was apparently an induced synthesis of specific species of RNA, as the adenine-to-uracil ratios of the RNA were higher than in the controls. Plain vestibular stimulation increased the content of RNA, but did not cause a change in the base composition. Similar observations were made in cortical areas of rats when these animals were trained to use the opposite hand (Hydén and Egyhazi, 1964).

An increase in RNA, with alterations in base ratios, has also been reported in the catfish. Morpholine, camphor, and other odorants induced a rise in the nuclear RNA in the olfactory lobes of the fish cortex. The base ratios of RNA were found to be different for different odorants used in the perfusion water (Rappaport and Daginawala, 1968).

Interference by Base Analogues

The most elegant experiments in this context were reported by Dingman and Sporn (1961). Administration of 8-azaguanine interfered with the maze learning of rats, whereas previously learned experiences were not found to be impaired. This base analogue gets incorporated into RNA in place of guanine, resulting in the formation of abnormal RNA molecules.

Beneficial Effect of Compounds Stimulating RNA Synthesis

Chamberlain et al. (1963) found that the administration of tricyanoaminopropene to animals improved their acquisition rates of conditioned avoidance responses. This compound is believed to increase the brain RNA. Another compound, magnesium pemoline, has been claimed to enhance the brain RNA polymerase activity, and has also been observed to improve learning of a conditioned avoidance task (Plotnikoff, 1966). The effect of magnesium pemoline has been reported to be more on facilitation of the acquisition of training than on the ability to retain the experience (Frey and Polidora, 1967).

Observations on Planaria

The noxious effect of ribonuclease during regeneration of planaria on the retention of previously learned tasks and also the transfer of learned tasks to untrained planaria by feeding of trained organisms lent substance to the belief that RNA was perhaps the engram for retention of learned tasks (McConnell et al., 1959; Corning and John, 1961).

Transfer of Learning by RNA Extracts

Numerous reports have appeared in the literature attesting to the efficacy of RNA extracts as a vehicle for transfer of previously learned tasks. Babich et al. (1965) reported the transfer of a response to naive rats by injection of RNA extracted from trained rats. Similar results were obtained by Fjerdinstad et al. (1965) and Albert (1966). Brain extracts were employed successfully by Rosenblatt et al. (1966) and by Ungar and Cohen (1966) for transfer of conditioned responses or for transfer of morphine tolerance ability.

EVIDENCE AGAINST RNA AS AN ENGRAM. Though changes in the content and composition of RNA in response to stimuli have been observed in a large number of situations, there is no convincing demonstration of a direct relation between a stimulus and one or more species of the synthesized RNA. The rates of synthesis and breakdown of a constituent can change as a result of nonspecific effects. The relation, if any, between the transduction of electrical impulses and biosynthesis of RNA engrams is purely hypothetical and has not been elucidated so far.

The data on transfer of learning through RNA are open to controversy. In a joint letter, 24 investigators have disclaimed the validity of these contentions (Byrne et al., 1966). Similarly the work on planaria is also disputable. Bennett and Calvin (1964) have discussed the nonreliability of planarians for such experiments.

Work on tricyanoaminopropene has apparently not been followed in other laboratories, even though such prospects usually attract further investigations. Smith (1967) failed to obtain either a facilitation of learning or improvement of memory in human beings with the use of magnesium pemoline.

It may also be pointed out that massive doses of actinomycin-D sufficient to inhibit 96 percent of RNA synthesis in the mouse brain did not prevent the learning of mazes and retention of learned experience in mice (Cohen and Barondes, 1966).

Concluding Comments on the Function of RNA in the Brain

The general role of RNA in the brain is probably similar to that in other tissues—viz., its main function is in the synthesis of proteins. The neurons have prominent nucleolus, with an active machinery for synthesis of all types of RNA—ribosomal, transfer, and template (or messenger). The turnover of brain ribosomal RNA (rRNA) is apparently half as rapid as the liver rRNA. The rate of RNA synthesis is high in fetal and early neonatal stage and drops off (for unknown reasons) fairly early in the neonatal period. It is likely, though not proved, that during this period some species of RNA concerned with structural development are elaborated, after which the expression of cistrons corresponding to these species of RNA is shut off, or considerably slowed down.

There is a moderate turnover of RNA for the wear and tear of cellular proteins. RNA would also be continuously required for the synthesis of secretory polypeptides (such as release factors) that integrate the nervous and endocrine systems. Thus, apart from the development period, the predominant role of the major part of RNA in adult brain would be in the maintenance of structural and enzymatic proteins.

RNA should also have a part at some stage or the other in establishment of "boutons" or intercellular connections. Observations have been made on the presence of clusters of ribonucleoprotein (RNP) particles in proximal dendrites lying in close apposition to membranes and in a position subjacent to boutons of 3 to 6 μ in size (Bodian, 1965). These RNP particles may have a role in the formation of boutons and in creation of junctional contacts. RNA is also present in the synaptosomes (Austin and Morgan, 1967; Balazs and Cocks, 1967). D'Monte and Talwar (1967) obtained a nondialyzable extract from three areas of the monkey brain by gentle treatment of the tissues with urea solutions. The substance's chemical nature was that of ribonucleoprotein. It is likely that this material may correspond to the RNP granules associated with membrane structures. Ribonucleoprotein particles in synaptosomes and situated in proximity to membranous structures in dendrites may have a function in synthesis of appropriate holo or conjugated (sialo, glyco, lipo) proteins or other substances that enable the establishment of junctional and synaptic contacts.

Sensory Input and Synthesis of Proteins in Visual Cortex

Influence of Afferent Stimulation on Development

Studies on RNA and proteins in cortical areas of the developing rabbit pups revealed that the proteins-to-DNA ratio increases steadily in the occipital cortex of the animals till the period just preceding the opening of the eyes. Soon after opening of the eyes, there was a significant increase in the protein content of the occipital cortex. This rise was apparently related to the availability of afferent stimulation as blinding of the pups prior to the opening of eyes prevented to a large extent this normal increase in proteins. In these experiments the litters were divided into two sets. One set was blinded on Day 4 by severing of the optic nerve. Pups of both sets were allowed to develop till the periods set for analysis (Day 13 to 15), viz., till two days after the opening of the eyes of the control set of animals. The protein-to-DNA ratios in the occipital cortex of the blinded set of pups were found to be significantly lower than the control pups (Talwar et al., 1964; Talwar et al., 1966a; Talwar, 1969). These changes consequent on afferent stimulation were most marked in the occipital cortex, although other areas of cerebral cortex also showed similar changes. On the other hand, some areas (such as the cerebellar cortex) were not influenced in a parallel manner.

The dependance of the development of cerebral cortex on sensory stimulation has also been shown by other studies. The overall weight and protein content increased in rats (Bennett and Calvin, 1964). The diameters of the nuclei in cortical Layers II, III, and IV of the visual cortex were found to be smaller in mice with inherited retinal dystrophy, as compared to normals (Gyllensten and Lindberg, 1964). Light has also been observed to influence markedly the RNA and protein content of the retina in rabbits during early postnatal development. When normal light stimulation was available the proteins per unit volume of retina increased by 100 percent. On deprivation, the increase in RNA and proteins were arrested (Brattgård, 1951). Frogs when exposed to light after long periods of darkness, within seconds displayed signs of active protein synthesis in the nuclei of retinal ganglion cells (Chentsov et al., 1961). The volume of mitochondria and the number of RNP particles also increased. Succinic dehydrogenase activity of frog retina increased on illumination with bright intermittent light (Lukashevich, 1964). A significant increase in the activity of succinate oxidase in rods and cones of rats has been observed on transition from darkness to light (Epstein and O'Conner, 1966). Similarly, an increase in acetylcholinesterase activity in the eye exposed to light as compared to the control kept in darkness has also been reported (Glow and Rose, 1964).

Visual Stimulation and Synthesis of Proteins

A natural corollary of the above studies was to investigate the possible effect of visual stimulation on the synthesis of proteins in the occipital cortex of adult animals. These experiments have been done in the rabbit and the monkey (Talwar et al., 1966a, b). Adult animals of similar weight and sex were divided in two groups. The animals were injected with radioactive lysine (or other precursors) intracisternally. The control animal was kept in darkness while the experimental animal was exposed to a source of rhythmically flickering light. The optimal experimental conditions have been described elsewhere (Singh and Talwar, 1967). The intensity of light used was 1,614 lumens per square meter and the duration of exposure was from 45 to 120 min. The flicker frequency of the light is an important factor. In monkeys, the rate of incorporation of radioactive amino acids into proteins in Areas 17, 18, and 19 is not much changed from the darkness values if the animal is exposed to either plain light or light with a flicker frequency of 2 to 4 sec. Flicker frequency of 7 sec was found to give a clear stimulation. This is probably due to the fact that at this flicker frequency, the overall activity of the neurons in the visual cortex increases above the basal levels (Jung, 1958). Five types of neurons have been characterized in the mammalian occipital cortex. Type A neurons constitute about half of the total visual cortex neurons in cats and give a continuous background discharge independent of retinal stimuli. Types B and D are antagonistic neurons; B neurons are activated by light-on and D by light-off. C neurons are inhibited after light-on and light-off, and E neurons show a short preexcitatory inhibition by light-on followed by activation and stronger activation by light-off. This background information on electrophysiological properties of the visual cortex neurons was the rationale for our choice of rhythmically flickering light as a source of photic stimulation. Subsequent investigations justified the choice of these experimental conditions (Singh and Talwar, 1967).

It was observed that the incorporation of intracisternally given radioactive precursors into proteins in Areas 17, 18, and 19 was higher in animals exposed to rhythmically flickering light as compared to the animals kept in darkness (Talwar et al., 1964, 1966a; Singh and Talwar, 1967).

TABLE 1. Incorporation of L-(U-^{3}H) Lysine into Proteins of the Subcellular Fractions of the Occipital Cortex of Monkeys

Fraction	Specific radioactivity (counts/min/mg protein)		Change (%)	Relative specific radioactivity (counts/min/mg protein) counts/min/μg lysine) in acid-soluble pool		Change (%)
	Dark	Light		Dark	Light	
EXPERIMENT I						
(1) Total homogenate	91	170	+ 86.8	0.89	1.63	+ 83.0
(2) 105,000 g 60 min supernatant	163	296	+ 81.0	1.60	2.84	+ 77.5
(3) Triton extract of particulate fraction	124	297	+139.5	1.21	2.86	+135.0
EXPERIMENT II						
(1) 105,000 g 60 min supernatant	307.0	469.0	+ 52.7	1.74	2.97	+ 70.6
(2) Triton extract of particulate fraction	144.6	238.8	+ 65.1	0.82	1.51	+ 84.1
(3) Residual pellet proteins	117.0	213.6	+ 82.5	0.66	1.35	+104.5

500 and 800 μc/kg body weight of L-(U-^{3}H)lysine (specific activity 250 mc/mM) was injected intracisternally into monkeys in I and II, respectively. The control monkey was kept in darkness for 45 min, and the experimental exposed to a rhythmically flickering light (flicker frequency 7/sec) of 1, 614 lumens/m^2 intensity, at a distance of 25 cm for 45 min. (From Singh and Talwar, 1969.)

The increase was both in the soluble and particulate proteins (Table 1).

The soluble proteins of the visual cortex were fractionated on DEAE cellulose columns. Five peaks were obtained by batch elution (Singh and Talwar, 1969). Each of the DEAE-cellulose fractions was further resolved by electrophoresis on polyacrylamide gels. The gels were cut in slices of 2.5-mm thickness. The slices were pooled for counting of radioactivity. The protein bands were either eluted by maceration of the gel or the gel slices were directly dissolved in H_2O_2 and counted in a scintillation counter after addition of the scintillation mixture as described elsewhere (Singh and Talwar, 1969). It was found that, although there was a generalized increase in labeling of several proteins, the radioactivity in DEAE-cellulose Fraction II was mostly present in a group of proteins with fast migration characteristics (Fig. 5). The radioactivity in these peaks was substantially

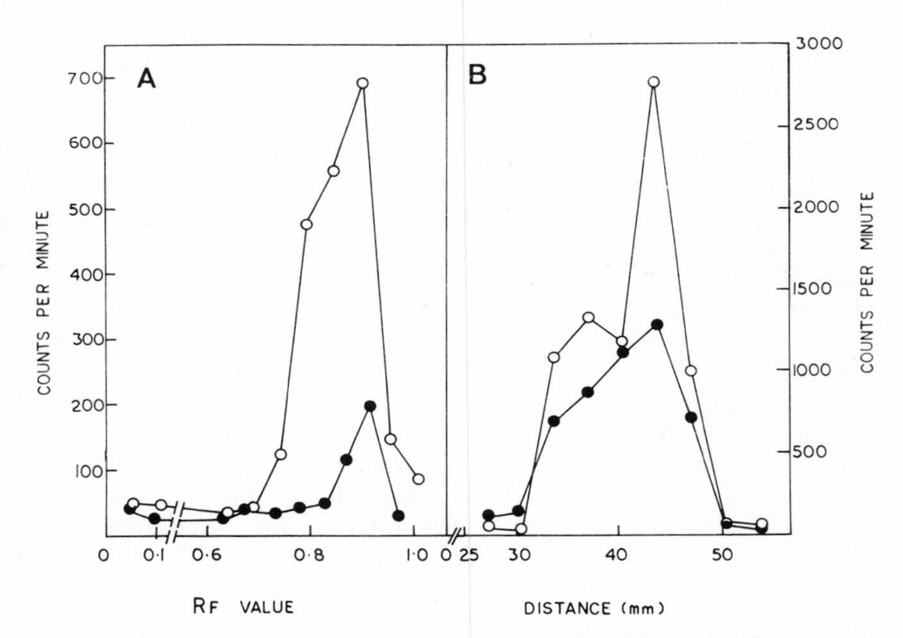

FIG. 5. Incorporation of L—(U—^{3}H) lysine into occipital cortex soluble proteins in monkeys kept in darkness (●— — —●) and exposed to rhythmically flickering light (O— — —O). The 105,000 g supernatant was fractionated on DEAE—cellulose columns. DEAE—cellulose fraction II was further resolved by electrophoresis on 6 percent polyacrylamide gel columns. The gels were sliced in 2.5 mm segments and radioactivity in bands determined by the methods described in the text. (A) The monkey received 800 μg/kg body weight of L—(U—^{3}H) lysine intracisternally. The time of exposure to light was 45 min, intensity 1614 lumens/m^2 and flicker frequency 7/sec. (B) The dose of radioactive isotope L—(U—^{3}H) lysine was 1 mc/kg body weight and the time of exposure was 75 min. (From Singh and Talwar. 1969. *J. Neurochem.,* 16:951.)

raised in preparations derived from light-exposed animals. The radioactivity peak does not have a symmetrical distribution. These samples may therefore represent a group of proteins with low molecular size and high negative charge. These bands give an immunological cross reaction with anti S-100 serum (obtained through the courtesy of Dr. L. Levine) that was prepared according to the criteria of Kessler et al. (1968).

Our observations on increased synthesis of proteins in visual cortex in response to sensory stimuli are supported by the work of Rose (1967), who found a significant increase in the relative specific radioactivity of visual cortex proteins in rats exposed to light for 3 hrs as compared to those kept in darkness. In these experiments tritiated lysine was used as the precursor. The increase in radioactivity of proteins was noticed only in visual cortex and not in motor cortex or liver on photic stimulation. On the other hand, placing of the rats in an activity wheel caused an increase in the motor cortex proteins. The changes were noticed only in proteins and nucleic acids but not in lipids of the cortical areas in these experiments. Appel et al. (1967) observed a decrease in the polysomes of brain on light-deprivation of animals. On stimulation with light, the polysomes increased together with an enhanced capacity for protein synthesis.

Rosenzweig et al. (1969) have reported a positive influence of environmental complexity and visual stimulation on total weight, cortical depth, and total acetylcholinesterase and cholineacetylase activities in rat brain. The differences were significantly greater in the occipital region of the cortex than in any other cortical area. Metzager et al. (1966, 1967), however, failed to get the effect of unilateral visual stimulation on protein synthesis in split-brain monkeys. They utilized tritiated water as the precursor instead of radioactive amino acids. Tritiated water would label all constituents by exchange reaction and would not measure uniquely the rate of de-novo protein synthesis. Moreover, they did not utilize the correct parameters (Singh and Talwar, 1967) for stimulation. The flicker frequency of light was variable; the animals were given intermittent shocks; and the experiments were either too short (10-min duration) or too long (3 hr) in view of our own experiences (Talwar and Singh, 1970).

The stimulation of amino acid incorporation into proteins during exposure of the animal to light is region-specific to a degree. All other regions of the brain do not show parallel changes (Talwar et al., 1966a). All metabolic activities are not activated in the occipital cortex under these conditions. Lack of stimulation of RNA has been reported earlier. No significant increase in the incorporation of ^{32}P-orthophosphate into phospholipids was obtained in optic lobes and cerebral hemispheres of pigeons when one eye was exposed to light and the other kept blindfolded (Jaffery and Talwar, unpublished data).

Brain-Specific Proteins

An acidic protein that is characteristic of the brain and absent in other tissues such as kidney, heart, and liver was first described by Moore (1965). This organ-specific protein is present in brains of all vertebrate species (Uyemura et al., 1967). The protein purified from beef brain has a low content of aromatic amino acids and is rich in acidic amino acids such as glutamic and aspartic acid (Moore, 1965). It is not precipitated by full saturation with ammonium sulphate at pH 7.4, hence represented as S-100. S-100 protein has high electrophoretic mobility and migrates faster than albumin.

It now appears that S-100 may not represent a single protein, but a group of proteins. The single band obtained on 7 percent polyacrylamide gels resolves into two bands in 10 to 15 percent gels and into five bands in 20 percent gels (Sharma and Talwar, unpublished data). McEwen and Hydén (1966) have suggested that S-100 protein may exist in a number of isomeric forms. The heterogeneity of the fraction has been shown on acrylamide-agarose gel electrophoresis and by ultracentrifugation (Gombos et al., 1966). The relative proportions of the two major bands constituting the S-100 fraction have been estimated in brains of a number of species (Tardy et al., 1968). Pork has the highest proportion (86.3 percent) of the slower migrating band, whereas in sheep, rabbit, rat, and guinea pig brain, the fastest-moving band constitutes more than 95 percent of the total S-100 fraction.

S-100 proteins are present in almost all parts of the nervous system, both peripheral and central (Moore et al., 1968). It is, however, predominantly a neurological protein. By fluorescent antibody techniques, Hydén and McEwen (1966) showed its localization: around the nuclei of oligodendrocytes and scattered throughout the glial membrane system. There was also some fluorescence observed in the neuronal nuclei, which led to the speculation that it may have a regulatory role in the neurons (Campbell, 1966). The localization of S-100 proteins in the neurons requires further confirmation. From localization studies, argument can also be advanced that the protein is initially synthesized in the neurons and transferred later to neuroglia instead of the transfers occurring from glial cells to neurons. The former hypothesis can, however, be discarded, as there is good evidence demonstrating the ability of glial cells to synthesize and secrete these proteins. Benda et al. (1968) have successfully propagated cell lines from rat glial tumors that synthesize S-100 proteins. In humans, astrocytomas were found to be particularly rich in S-100 proteins (Benda, 1968a, b).

The rate of synthesis of S-100 protein in vivo has been reported to be very rapid (McEwen and Hydén, 1966). It is also synthesized in appreciable amounts in-vitro systems (Rubin and Stenzel, 1965).

The biological function of S-100 group of proteins is not clearly known. The fact that they are glial proteins, rapidly synthesized and present along the membranes, suggest that they may have a role in glia-neuronal intercommunication. They may also have a vital role in events taking place at nerve endings and at synaptic junctions. Antiserum to S-100 protein was observed by De Robertis (1967) to produce disruption of nerve endings in vitro. When it was given in vivo to mollusc neurons that were recorded intracellularly, the electrophysiological properties of the neurons were disturbed. It also caused cytolysis. Perez and Moore (1968) measured the quantitative changes in S-100 protein during wallerian degeneration of rabbit tibial nerve. S-100 protein decreased progressively in the degenerated segment of the transected nerve. After 28 days it was about 2 percent of the amount present in the corresponding portion of the nerve taken from control rabbit. These authors also suggest the primary localization of the S-100 protein in axons.

Proteins have been implicated in the consolidation of a learning task (Agranoff, 1967; Agranoff et al., 1965; Flexner and Flexner, 1949; Flexner et al., 1965, 1967). All learning processes inherently have a sensory input. The impact of sensory stimuli on the biosynthesis and metabolism of macromolecules such as proteins would be a fruitful area for further investigation. Present studies have shown increased synthesis of proteins in the visual cortex of monkey on exposure of the animal to a rhythmically flickering light stimulus. Among the proteins whose synthesis is stimulated under these conditions are also the brain-specific group with low isoelectric pH and fast migration characteristics.

Acknowledgments

This work has been supported by research grants of the Indian Council of Medical Research, The Population Council (New York), and the World Health Organization (Geneva). We acknowledge with thanks the excellent help in bibliography given by Miss Lorraine Schulte of the UCLA brain information service.

References

Abdel-Latif, A. A., Brody, J., and Ramahi, H. (1967). J. Neurochem., 14: 1133.

Agranoff, B. W. (1967). Sci. Amer., 216:115. ——— Davis, R. E., and Brink, J. J. (1965). Proc. Nat. Acad. Sci. U.S.A., 54:788.

Albert, D. J. (1966). Neuroscychologica, 4:79.

Appel, S. H. (1967). Nature, 213:1253. ——— Davis, W., and Scotts (1967). Science, 157:836.
Austin, L., and Morgan, I. G. (1967). J. Neurochem., 14:377.
Babich, F. R., Jacobson, A. L., Bubash, S., and Jacobson, A. (1965). Science, 149:656.
Balazs, R., and Cocks, W. A. (1967). J. Neurochem., 14:1035.
Barbato, I. W. M., and Barbato, L. (1965). J. Neurochem., 12:60.
Barr, M. L., and Bertram, E. G. (1949). Nature (London), 163:676.
Benda, P. (1968a). Rev. Neurol (Paris), 118:364. ——— (1968b). Rev. Neurol (Paris), 118:368. ——— Lightbody, J., Sato, G., Levine, L., and Sweet, W. (1968). Science, 161:370.
Bennett, E. L., and Calvin, M. (1964). Neurosciences Res. Progr. Bull., 2 (No. 4):3.
Bernsohn, J., Barron, K. D., and Hess, A. R. (1964). Progr. Brain Res., 9:161.
Bodian, D. (1965). Proc. Nat. Acad. Sci. U.S.A., 53:418.
Bondy, S. C. (1966). J. Neurochem., 13:955.
Brattgård, S. O. (1951). Exp. Cell. Res., 2:693.
Breyer, U. (1965). J. Neurochem., 12:131.
Brinley, F. J. (1963). In: Pfeiffer, C. C., and Smythies, J. R., eds., International Review of Neurobiology, Vol. 5. New York: Academic Press.
Burgess, R. R. Travers, A. A., Dun, J. J., and Bautz, E. K. F. (1969). Nature (London), 221:43.
Burt, A. M. (1968). J. Exp. Zool., 169:107.
Byrne, W. L., Samuel, D., Bennet, E. L., Rosenzweig, M. R., Wasserman, E., Wagner, A. R., Gardner, F., Galambos, R., Berger, B. D., Margules, D. L., Fenichel, R. L., Stein, L., Carson, J. A., Enesco, H. E., Chorover, S. L., Holt, C E., Schiller, P. H., Chiappetta, L., Jarvik, M. E., Leaf, R. C., Dutcher, J. D., Horovitz, Z. P., and Carlson, P. L. (1966). Science, 153: 658.
Campbell, T. L. (1966). Science, 152:232.
Cantoni, G. L. (1960). In: Florkin, M., and Mason, H. S., eds., Comparative Biochemistry, Vol. 1. New York: Academic Press.
Chamberlain, T. J., Rothschild, G. H., and Gerard, R. W. (1963). Proc. Nat. Acad. Sci. U.S.A., 49:918.
Chentsov, Iu. S., Boroviagni, V. L., Brodskii, V. Ia. (1961). Biofizika, 6:590.
Chitre, V. S., and Talwar, G. P. (1963). Ind. J. Med. Res., 51:60. ——— Chopra, S. P., and Talwar, G. P. (1964). J. Neurochem., 11:439.
Cohen, H. D., and Barondes, S. H. (1966). J. Neurochem., 13:207.
Corning, W. C., and John, E. R. (1961). Science, 134:1363.
Côte, L. (1964). Life Sci., 3:899.
Crain, S. M. (1952). Proc. Soc. Exp. Biol. Med., 81:49.
Daneholt, B., and Brattgård, S. O. (1966). J. Neurochem., 13:913.
Dawson, D. M. (1967). J. Neurochem., 14:939.
De Robertis, E. (1967). Science, 150:907.
Dingman, W., and Sporn, M. B. (1961). J. Psychiat. Res., 1:1. ——— and Sporn, M. B. (1964). Science, 144:26.
Diperri, R., Himwich, W. A., and Peterson, J. (1964). Progr. Brain Res., 9:89.
Dixon, M., and Webb, E. J. (1958). In: Enzymes. New York: Academic Press.
Edström, J. E., and Pigon, A. (1958). J. Neurochem., 3:95.
Eigen, M. (1966). In: Schmitt, F. O., and Melnechuk, T., eds., Neuroscience Research Symposium Summaries. Cambridge, Mass.: M.I.T. Press.

Einarson, L., and Krogh, E. (1955). J. Neurol. Neurosurg. Psychiat., 18.1.
Epstein, M. H., and O'Conner, J. S. (1966). J. Neurochem., 13:907.
Fahn, S. (1968). Experientia, 24:544.
Fjerdingstad, E. J., Nissen, T., and Roigaard-Petersen, H. H. (1965). Scand. J. Psychol., 6:1.
Flexner, L. B., and Flexner, J. B. (1949). J. Cell. Comp. Physiol., 34:115.
——— Dela Haba, G., and Roberts, R. B. (1965). J. Neurochem., 12:535.
——— Roberts, R. B. (1967). Science, 155:1377.
Folch, J., Lees, M., and Sloane-Stanley, G. H. (1957). In: Richter, D., ed., Metabolism of the Nervous System. London: Pergamon Press.
Fox, M. W., Inman, O. R., and Himwich, W. A. (1966). J. Comp. Neurol., 127:199.
Frey, P. W., and Polidora, V. J. (1967). Science, 155:1281.
Gaito, J. (1966). In: Gaito, J., ed., Macromolecules and Behavior, 1st ed. New York: Appleton-Century-Crofts.
Garcia-Anstt, Jr., E. (1954). Proc. Soc. Exp. Biol. Med., 86:348.
Geiger, A., and Yamaski, S. (1956). J. Neurochem., 1:93.
———, Yamaski, S., and Lyons, R. (1956). Amer. J. Physiol., 184:239.
Glasky, A. J., and Simon, L. N. (1966). Science, 151:702.
Glow, P. H., and Rose, S. (1964). Nature (London), 202:422.
Gombos, G., Vincendon, G., Tardy, J., and Mandel, P. (1966). CR Acad. Sci. (D) (Paris), 263:1533.
Gomirato, G. (1954). J. Neuropath. Exp. Neurol., 13:359.
Gyllensten, L., and Lindberg, J. (1964). J. Comp. Neurol., 122:79.
Hamberger, C. A., Hydén, H., and Nilsson, G. (1949). Acta Otolaryngol. Suppl., 75:124.
Harris, H. (1970). In: Wolstenholme, G. E. W., and Knight, J., eds., Control Processes in Multicellular Organisms–A Ciba Foundation Symposium. London: J. and A. Churchill.
Hayden, R. O., Garoutte, B., Wagner, J., and Aird, R. B. (1961). Proc. Soc. Exp. Biol. Med., 107:754.
Himwich, W. A., and Peterson, J. C. (1959). Biol. Psychiat., 1:2.
Hydén, H. (1943). Acta Physiol. Scand., 6: Suppl. 17. ——— (1958). In: Proceedings of International Congress of Biochem. 4th Congress Symposium III, Brucke, F., ed. London: Pergamon Press. ——— (1960). In: Brachet, J., and Mirsky, H. E., eds., The Cell, Vol. IV. New York: Academic Press.
——— and McEwen, B. S. (1966). Proc. Nat. Acad. Sci. U.S.A., 55:354.
——— and Egyhazi, E. (1962). Proc. Nat. Acad. Sci. U.S.A., 48:1366.
——— and Egyhazi, E. (1963), Proc. Nat. Acad. Sci. U.S.A., 49:618.
——— Egyhazi, E. (1964). Proc. Nat. Acad. Sci. U.S.A., 52:1030.
Iqbal, Z., Sharma, S. K., and Talwar, G. P. Unpublished data.
Jacob, M., Stevenin, J., Jund, R., Judes, C., and Mandel, P. (1966). J. Neurochem., 13:619.
Jaffery, N. F., and Talwar, G. P. Unpublished data.
Jarnefelt, J. (1961). Biochem. Biophys. Acta., 48:104.
Johnson, T. C. (1967). J. Neurochem., 14:1075.
Jung, R. (1958). Exp. Cell Res. Suppl., 5:262.
Katzman, R., and Wilson, C. E. (1961). J. Neurochem., 7:113.
Kessler, D., Levine, L., and Fasman, G. (1968). Biochemistry (Wash.), 7:758.
Kahn, A. A., and Wilson, J. E. (1965). J. Neurochem., 12:81.

Kilbourn, B. T., Dunitz, J. D., Pioda, L. A. R., and Simon, W. (1967). J. Molec. Biol., 30:559.
Kobayashi, T., Inman, O. R., Buno, W., and Himwich, H. E. (1964). Progr. Brain Res., 9:87.
Koriyama, K., Sisken, B., Ito, J., Simonsen, D. G., Haber, B., and Eugene, R. (1968). Brain Res., 11:412.
Leslie, I. (1955). In: Chargaff, E., and Davidson, J. N., eds., The Nucleic Acids., Vol. II. New York: Academic Press.
Lubin, M. (1964). In: Hoffman, H. F., ed., The Cellular Functions of Membrane Transport. Englewood Cliffs, N.J.: Prentice-Hall.
Lukasherich, T. P. (1964). Akad. Nauk u SSSR, 156:1436.
Mahler, H. R., Moore, W. J., and Thomson, R. J. (1966). J. Biol. Chem. 241:1283.
Maynard, E. A. (1964). J. Exp. Zool., 157:251.
McCaman, R. E., and Aprison, M. H. (1964). Progr. Brain Res., 9:220.
McConnell, J. V., Jacobson, A. L., and Kimble, D. P. (1959). J. Comp. Physiol. Psychol., 52:1.
McEwen, B. S., and Hydén, H. (1966). J. Neurochem., 13:823.
Metzager, H. P., Cuenod, M., Grynbaum, A., and Waelch, H. (1966). Life Sci., 5:1115. ——— Cuenod, M., Grynbaum, A., and Waelch, H. (1967). J. Neurochem., 14:183.
Mizuno, S., Tano, S., and Shirahata, S. (1969). J. Biochem. (Tokyo), 66:119.
D'Monte, B., and Talwar, G. P. (1967). J. Neurochem., 14:743.
Moore, B. W. (1965). Biochem. Biophys. Res. Commun., 19:739. ——— Perez, V. J., and Gehring, M. (1968). J. Neurochem., 15:265.
Noach, E. L., Bunk, J. J., and Wijling, A. (1962). Acta Physiol. Pharmacol. Neerl., 11:54.
Orrego, F. (1967). J. Neurochem., 14:851. ——— and Lipmann, F. (1967). J. Biol. Chem., 242:665.
Perez, V. J., and Moore, B. W. (1968). J. Neurochem., 15:971.
Peters, J., Vonderahe, A., and Schmid, D. (1965). J. Exp. Zool., 160:255.
Pevzner, L. Z. (1966). In: Gaito, J., ed., Macromolecules and Behavior, 1st ed. New York: Appleton-Century-Crofts.
Plotnikoff, N. (1966). Science, 151:703.
Pressman, B. C. (1965). Proc. Nat. Acad. Sci. U.S.A., 53:1076.
Rappaport, D. A., and Daginawala, H. F. (1968). J. Neurochem., 15:991.
Richter, D. (1966). In: Aspects of Learning and Memory. London: Heinemann.
Rose, S. P. R. (1967). Nature (London), 215:253.
Rosenblatt, F., Farrow, J. T., and Herblin, W. F. (1966). Nature (London), 209:46.
Rosenzweig, M. R., Bennett, E. L., Diamond, M. C., Wu, S-Y, Slagle, R. W., and Saffrane, E. (1969). Brain Res., 14:427.
Rubin, A. L., and Stenzel, K. H. (1965). Proc. Nat. Acad. Sci. U.S.A., 53:963.
Schwartz, A., Bachelard, H. S., and McIlwain, H. (1962). Biochem. J., 84: 627.
Sharma, K. N., Dua, B., Singh, B., and Anand, B. K. (1964). Electroenceph. Clin. Neurophysiol., 16:503.
Sharma, N. C., and Talwar, G. P. Unpublished data.
Singh, U. B., and Talwar, G. P. (1967). J. Neurochem., 14.675. ——— and Talwar, G. P. (1969). J. Neurochem., 16:951.

Skou, J. C. (1957). Biochem. Biophys. Acta, 23:394. ——— (1960). Biochem. Biophys. Acta, 42:6. ——— (1965). Physiol. Rev., 45:596.
Smith, R. G. (1967). Science, 155:603.
Talwar, G. P. (1969). In: Bittar, E. E., ed., The Biological Basis of Medicine. New York: Academic Press. ——— Chopra, S. P., and Goel, B. K. (1964). VI Int. Congr. Biochem. New York, 5:419. ——— Chopra, S. P., Goel, B. K., and D'Monte, B. (1966a). J. Neurochem., 13:109. ——— Goel, B. K., Chopra, S. P., and D'Monte, B. (1966b). In: Gaito, J., ed., Macromolecules and Behavior, 1st ed. New York: Appleton-Century-Crofts. ——— Sadasivudu, B., and Chitre, V. S. (1961). Nature (London), 191:1007. ——— and Singh, U. B. (1970). In: Lajtha, A., ed., Handbook of Neurochemistry. New York: Plenum Press.
Tardy, J., Gombos, G., Vincedon, G., and Mandel, P. (1968). C.R. Acad. Sci. Paris, 267:669.
Thudichum (1884). The Chemical Constitution of the Brain. London: Baillière, Tindall and Cox.
Toschi, G., Dore, E., Angeletti, P. U., Levi-Montalcini, R., and De Haen, Ch. (1966). J. Neurochem., 13:539.
Ungar, G., and Cohen, M. (1966). Int. J. Neuropharmacol., 5:183.
Uyemura, K., Tardy, J., Vincendon, G., Mandel, P., and Gombos, G. (1967). C.R. Soc. Biol. Paris, 161:1396.
Vesco, C., and Giuditta, A. (1967). Biochem. Biophys. Acta, 142:385.
Virmani, V., Sherma, K. N., Talwar, G. P., Anand, B. K., and Singh, B. (1963). Indian J. Med. Res., 51:75.
Vraa-Jensen, G. (1957). In: Richter, D., ed., Metabolism of the Nervous System. London: Pergamon Press.
Vrba, R., and Folbergrova, J. (1959). J. Neurochem., 4:338.
Wooley, U., and Campbell, N. K. (1962). Biochem. Biophys. Acta, 57:384.
Zaheer, N., Iqbal, Z., and Talwar, G. P. (1968). J. Neurochem., 15:1217.

Chapter 3
BRAIN FUNCTION AND RNA

EDWARD GLASSMAN and JOHN E. WILSON

University of North Carolina, Chapel Hill, North Carolina

There are many pitfalls in trying to correlate chemical changes in the nervous system with a particular behavioral experience, such as learning. First, there is a tremendous uncertainty concerning the amount of information that is stored in the nervous system. This can be estimated by the performance of the animal, but unfortunately performance factors such as the physical condition and the emotional state of the animal at the time of testing can interfere greatly with such an assay. It is thus almost impossible a priori to relate the amount of performance with the amount of information stored, and then to expect a quantitative relationship to exist between the amount stored and the chemical changes that might be observed.

Second, there is some uncertainty concerning the states of memory. Initially, memory is presumed to be retrievable from a short-lived form (short-term memory), but as this declines, memory is then retrieved from a more permanent form (long-term memory). The evidence for this is chronicled elsewhere (John, 1967). It is now becoming accepted that it is the early stages of the formation of long-term memory that are sensitive to agents that interfere with memory consolidation. Thus the steps in the learning process and memory storage can be represented operationally as follows:

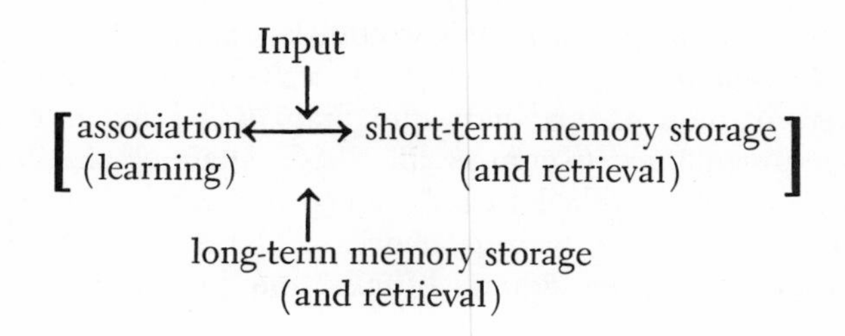

The role of chemicals in this process is not known, but certain ideas seems to be generally accepted.

First, it is extremely likely that chemicals do play a role in encoding information in the nervous system, but probably not by encoding the experiential information within their own structure. More likely, the chemicals participate in intraneuronal events that enable the neurons to become involved in the interneuronal connections that form the pathways that encode information.

Second, because short-term memory formation is not sensitive to inhibitors of RNA or protein synthesis it probably is not dependent on the synthesis of these macromolecules. There is no knowledge of whether preexisting chemicals are utilized, but it is difficult to imagine the formation of new pathways without the involvement of chemical changes of some type, presumably conformational, at the synapse or in the cell.

Third, because the formation of long-term memory is prevented by inhibitors of protein and RNA synthesis given before or immediately after training, this process may depend on the synthesis of these macromolecules. The sensitivity to these inhibitors is short-lived, however, and thus the perpetuation of long-term memory is not specifically dependent on the continuous synthesis of RNA and protein.

There are many reports of a correlation between change in RNA or protein and behavior (see review by Glassman, 1969). This work has generated two specific questions concerning the significance of such findings. First, the sequence of physiological and chemical events that lead to the reported chemical changes is not known. Second, it is not clear what role(s) these macromolecules play in the function of the nervous system. Before these questions can be approached, however, a number of related problems must be solved. For example, what is the total inventory of cells participating in a particular chemical response to experiential events? What component of the behavior is responsible for triggering the chemical response? Does the response generate the formation of new or unique species of molecules? Clearly, an extensive and coordinated research project involving many neurobiological disciplines is necessary to answer these questions.

Such a project has been initiated within the Neurobiology Program of the University of North Carolina. We have found that mice injected with radioactive uridine in a double-isotope-labeling technique and then trained for 15 min to avoid a shock by jumping to a shelf incorporated about 50 percent more radioactivity into brain RNA and into polysomes than did untrained mice (Zemp et al., 1966; Adair et al., 1968a). The untrained animals were yoked to the trained ones and were subjected to the same injections, lights, buzzers, shocks, and handling. No differences were found in liver or in kidney. Localization by autoradiography has shown that trained mice incorporated more ^{3}H-uridine into many clusters of neurons throughout the limbic system (Kahan et al., 1970; see also Zemp et al., 1967).

Sucrose density gradient centrifugation showed that the increased radioactivity associated with the RNA of the trained mouse was quite heterogenous with respect to sedimentation rate and a unique species of RNA was not detectable (Zemp et al., 1966). The patterns of radioactivity were of similar shape for both trained and untrained mice and resembled those found after RNA synthesis has been stimulated in liver by hydrocortisone or in uterus by estrogen. These findings suggest that a general increase in the synthesis of rapidly labeled RNA had occurred and that the increased radioactivity may be in messenger RNA or in preribosomal RNA. As expected, there was increased incorporation into polysomes of the brain during the training experience (Adair et al., 1968a). The base ratio of this radioactive RNA ($G + C/A + U = 0.78$ to 0.86) is DNAlike (Wu, 1968). It is not known whether the radioactive RNA is involved in the synthesis of new proteins or in the replenishment or increase in the amount of proteins already present.

These studies have been reported in detail elsewhere and will not be discussed further except to emphasize two important conclusions. First, autoradiography has shown that the bulk of the increased incorporation into RNA in structures occurs throughout the limbic system (Kahan et al., 1970). Lesser changes occur elsewhere. We shall reserve judgment as to the significance of the response of this region until we can pinpoint the functional and anatomical relationships between the responding cells and the rest of the brain. The second conclusion is that sucrose gradient centrifugation patterns suggest that the increased incorporation into RNA is part of a general increase in metabolism similar to that induced into liver cells by hydrocortisone or in the uterus by estrogen (Zemp et al., 1966). Thus the chemical response induced in these brain cells is similar to that in other organs and no unique biochemical reactions have been observed. It appears that cells in all tissues respond to stimulation by a widespread, though possibly selective increase in the synthesis of RNA, but each organ responds appropriately to its own specific stimuli. Thus liver responds to hydrocortisone, uterus to estrogen, and brain to this experience, although the involvement of a hormonal intermediate in this response in the brain has not been excluded.

In this regard, knowledge of the intermediate chemical and physiological steps that lead to increased incorporation into RNA and polyribosomes are extremely crucial to many theories. In mammals, this process usually involves an hormonal intermediate. Uridine incorporation into brain polysomes during training in mice with various endocrine glands removed has been studied by Adair et al. (1968b) and Coleman et al.

(1971a). The goal was to determine whether the biochemical response that occurs during 15 minutes of avoidance training is dependent on the release of hormone from a functioning gland during this time. Long-term debilitating effects of gland removal were avoided. The data indicate that the adrenal, the pituitary, the testes, and the ovary are not necessary for a mouse to learn the avoidance task, or for the increased incorporation of uridine into RNA that accompanies it. The possible effects of other glands are being investigated. In addition, changes in histones, norepinephrine, 3′, 5′-cyclic AMP, adenyl cyclase, RNA polymerase, and protein kinases are also being studied. These are substances believed to be related to increased incorporation of uridine into RNA in other mammalian tissues, and they may be important to this brain response as well.

The main purpose of the rest of this chapter is to describe briefly work being performed in our laboratory by Mr. Barry Machlus on the effect of a short training experience on the phosphorylation of the so-called acidic proteins in the nuclei of rat brain cells. Mr. Machlus used a modification of the training procedures described by Coleman et al. (1971a). This consists of a one-way step-up avoidance procedure. Briefly, the training apparatus consisted of a runway, the beginning of which was 40 cm long and had black walls and a shock grid floor. The remainder of the runway was a safe area that was painted white and had a wooden floor raised 7 cm above the level of the grid floor. Rats were held by the tail and lowered onto the grid floor of the start area facing away from the safe area. Five seconds later the foot shock was activated and remained on until the rat escaped by jumping onto the white platform. If the rat stepped back onto the grid it received foot shocks until it remounted the platform. The trial lasted 30 seconds. At this time the rat was picked up by the tail and the next trial started. Training continued for 5 min. Learning curves and further details are shown in Coleman et al. (1971a). They have shown that 15 to 20 minutes of this training procedures produces significantly increased incorporation of radioactive uridine into polysomes of rat brain.

For yoked rats the apparatus was modified by placing a cardboard barrier between the start area and the safe area. Each yoked animal was paired with the training record of a previously trained rat. On each trial the yoked animal was treated exactly as its trained counterpart except that at the point at which the trained animal had escaped or avoided the foot shock by mounting the white platform, the yoked animal was lifted by the tail and placed on the grid. Thus yoked rats differed from trained subjects in that the yoked animal could not learn to escape or avoid the foot shock. Additional behavioral experiences are needed for comparison since no single behavior is adequate as a control for the training experience.

The double isotope labeling method was used throughout. One rat was injected intracranially with 0.1 mC of $H_3{}^{33}PO_4$. After a 30-min wait, one of the animals was trained for 5 min, while the other was kept quiet. The brains of both rats were homogenized together. To monitor the differences in the amounts of radioactive phosphate injected into each brain, the ratio of ^{32}P to ^{33}P in 5′-AMP isolated from the original homogenate was used as a correction factor. Fractions containing nuclei were isolated and treated with 0.2 N HCl to extract histones and the associated acidic proteins.

These proteins were precipitated with acetone, dried with ethanol and ether, and the dried proteins were dissolved in 1 ml of 7 percent guanidium chloride. About 250 γ of protein were applied to a 0.9 × 15 cm Amberlite IRC-50 column, and eluted with 100 ml of a linear gradient of guanidium chloride from 7 percent to 14 percent, and then with 25 ml of 40 percent guanidium chloride in 0.1 M potassium phosphate, pH 6.8. One-ml samples were collected. The flow rate was one drop every 30 seconds, and protein recovery was over 95 percent. Optical density was determined at 400 mμ and radioactivity was determined in a Packard Scintillation counter, Model 3375.

The eluates showed 4 main peaks when the optical density was determined. These corresponded, in order of elution, to the acidic proteins (which are not retarded by the column under the conditions used), to the lysine rich histones, the slightly lysine rich histones, and to the arginine rich histones. In many experiments, the acidic proteins from the trained rat brain showed approximately 100 percent more radioactive phosphate than those from the untrained animal. However, preliminary data indicate that the reverse is true for histones, that is, the histones from the brain of the trained rat show less incorporation of radioactive phosphate. The significance of this must await further work, especially in view of the fact that whole brains were used.

Additional fractionation of the acidic protein on Amberlite IRC-50 columns and by electrophoresis on polyacrylamide gels show that this is a heterogenous mixture of proteins, not all of which are phosphorylated. Treatment of these proteins with RNAase, pronase, trypsin, chymotrypsin, phosphodiesterase, and lipase indicated that the increased incorporation of radioactive phosphate is in protein, not RNA; indeed amino acid analysis of acid hydrolysates reveals that the increased incorporation of phosphate in trained rats is in phosphoserine. This is also shown by relative increased radioactivity.

Mr. Machlus has attempted to localize the area of the brain involved in this chemical response by grossly dissecting the brain into four parts. This was accomplished by first removing the cerebellum. The brain stem was then removed by a cut at the colliculus posterior. The cerebrum was then dissected into two parts by a horizontal cut through the fissure rhinalis so that the bulbus olfactoris remained with the upper part. Only the lower half of the cerebrum, the part containing many but not all limbic system structures (amygdala, entorhinal cortex, hypothalamus, mid-brain tegmentum, and posterior-ventral hippocampus) showed increased incorporation of radioactive phosphate into the acidic nuclear proteins. More work is necessary for more exact localization, possibly by autoradiography. It is of interest that this is also the region involved in the increased incorporation of radioactive uridine into RNA of mice undergoing jump box training as shown in previous studies (Zemp et al., 1967; Kahan et al., 1970).

The behavioral trigger is elusive as ever. The increased incorporation of radioactive phosphate into these proteins does not take place in "yoke" shocked rats. The increased incorporation does occur when the animal is being trained or if the behavior is extinguished (see also Coleman et al., 1971b). The significance of this is not clear, since rats trained for three or for seven days and then made to *perform* the task with shock reinforcement also show the higher levels of incorporation of radioactive phosphate into these acidic nuclear proteins. This was unexpected since Adair et al. (1968b) reported that mice performing the jump box after three days of prior training did not show increased incorporation of radioactive uridine into RNA. This prompted further examination of the methods used by Adair. Although the evidence is still preliminary, it does appear that if the methods used by Adair et al. (1968a, b) to isolate polysomes are changed slightly, some increased incorporation of radioactive uridine into brain polysomes of performing mice can be detected. The fact that Adair et al. did not detect them is probably due entirely to technical reasons. These data will be published in full detail elsewhere (in preparation).

In any case, the discussion of the behavioral trigger by Glassman and Wilson (1970), in which it was concluded that the insight development phase exclusively contained the behavioral trigger, will have to be modified. It appears that there is some aspect of the performance of the avoidance conditioning task after it has been learned that can trigger these chemical phenomena, although it is clear that yoke and shocked animals that have never learned the task do not show this chemical response. It is tempting to speculate that a specific emotional response is actually the trigger, an emotional response generated by the training experience, but which can also be evoked to a lesser degree when prior trained animals perform the task at a later time. This is consistent with the fact that the chemical response appears to take place in the limbic system, an area of the brain thought to be involved in emotional responses and arousal.

The functional significance of the chemical response is not clear. An important finding is its location in components of the limbic system. One is tempted to postulate the synthesis of chemicals in this region that are then distributed throughout the brain, where they render permanent

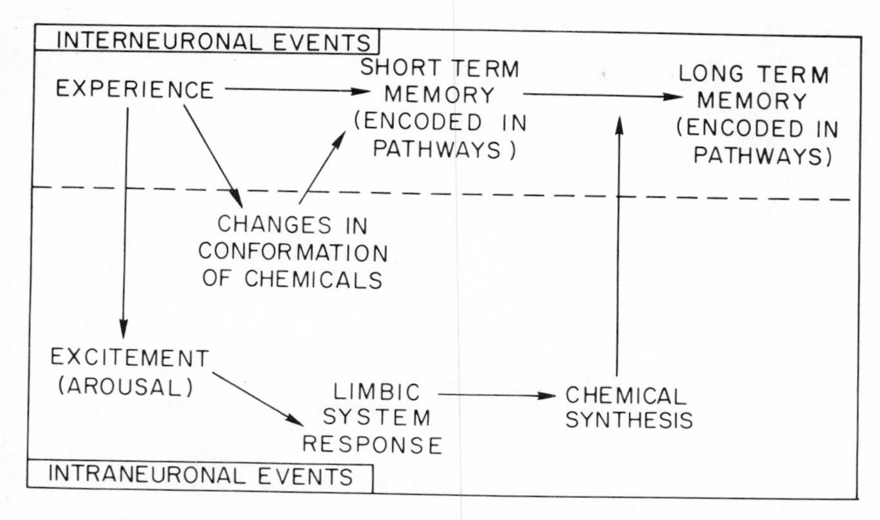

FIG. 1. **The effect of changes in macromolecules on memory storage.**

the new synaptic relationships that were generated to form the neuronal pathways and networks that encode short-term memory. In other words, as a working hypothesis it is suggested that during training, conformational changes of short duration (hours) take place in neurons so that new pathways and networks are formed. Only if the animal is sufficiently aroused or specifically emotionally affected is the limbic system stimulated to synthesize molecules that lead to the preservation of the new pathways and networks throughout the brain, either by preventing the reversal of the conformational changes, or by instituting new changes that are permanent (see Fig. 1). This idea has as its basis the concept that the pathways and networks in which short-term and long-term memory are encoded are identical. The main difference between these two phases of memory storage is the type of chemical mechanism involved to insure the proper level of connectivity between the neurons in the pathway.

The confirmation, modification or rejection of these ideas will depend on future data. For the present, they remain for us interesting ideas on which to plan future experiments.

Acknowledgments

This investigation was partially supported by research grants from the U.S. Public Health Service (GM-08202; GM-07699; NS-07457; MH-18136), from the National Science Foundation (GB-18551), and from the Geigy Pharmaceutical Corporation.

Dr. Glassman was supported by a Research Career Development Award (GM-K3-14,911) from the Division of General Medical Sciences, National Institutes of Health.

References

Adair, L. B., Wilson, J. E., Zemp, J. W., and Glassman, E. (1968a). Proc. Nat. Acad. Sci. U.S.A., 61:606. ——— Wilson, J. E., and Glassman, E. (1968b). Proc. Nat. Acad. Sci. U.S.A., 61:917.

Coleman, M. S., Pfingst, B., Wilson, J. E., and Glassman, E. (1971a). Brain Res., in press. ——— Wilson, J. E., and Glassman, E. (1971b). Nature (London), in press.

Glassman, E. (1969). Ann. Rev. Biochem., 38:605. ——— and Wilson, J. E. (1970). In: Biochemistry of Brain and Behavior, Bowman, R. E., and Datta, S. P., eds. New York: Plenum Press.

John, E. R. (1967). Mechanisms of Memory. New York: Academic Press.

Kahan, B. E., Krigman, M. R., Wilson, J. E., and Glassman, E. (1970). Proc. Nat. Acad. Sci. U.S.A., 65:300.

Wu, B. E. (1968). Ph.D. thesis. Department of Biochemistry, University of North Carolina at Chapel Hill.

Zemp, J. W., Wilson, J. E., Schlesinger, K., Boggan, W. O., and Glassman, E. (1966). Proc. Nat. Acad. Sci. U.S.A., 55:1423. ——— Wilson, J. E., and Glassman, E. (1967). Proc. Nat. Acad. Sci. U.S.A., 58:1120.

Chapter 4

MACROMOLECULES AND BRAIN FUNCTION

JOHN GAITO

York University, Toronto, Ontario, Canada

Because the macromolecules, DNA and RNA, are ultimately involved in basic cellular functions, it is obvious that they play some role in all brain functioning. The real question, however, is whether the role of these macromolecules is a unique one for behavior. Early investigators suggested a unique function for the nucleic acids in learning and memory with base changes in RNA being the most popular mechanism in representing memory for experiential events (Hydén, 1959; Gaito, 1961; Dingman and Sporn, 1961). Other molecular changes were offered also (Gaito, 1961; Dingman and Sporn, 1961). Other individuals who suggested the uniqueness of RNA for learning events included Cameron (1963), McConnell (1962), Landauer (1964), and Pribram (1966). At present those individuals who think of RNA as providing a unique contribution during learning have discarded the base change ideas and emphasize instead selective activation of DNA sites to produce specific types of RNA (e.g., Hydén and Lange, 1965; Bonner, 1966; Gaito, 1966). (For a critical review of these approaches, see Dingman and Sporn, 1964 and Gaito, 1969.)

In the molecular approach to learning, there has been much speculation but little conclusive research data. Unfortunately, the available research data are subject to multiple interpretations. Thus there is no *conclusive* evidence to support the notion that one of these macromolecules has a unique, primary role in learning.

However, Hydén's work, and that of other individuals (e.g., Palladin and Vladimirov, 1956; Pevzner, 1966), does show interesting results which can be used to suggest some of the molecular events underlying learning and other behavioral events. RNA amounts per cell in humans increase over age (up to approximately 40 years) and then decrease in later life (Hydén, 1961). Hydén (1967) reported similar results in brain cells of rats. Apparently the template activity of DNA for RNA synthesis decreases with increasing age (Harbers, et al., 1968). Animal work shows transient changes involving increase during functional activity, such as sensory stimulation, motor activity, and learning, or decreases during functional inactivity such as sleep. Within a short period of time, the amount of RNA is restored to its previous level. The normal level (equilibrium or steady state) would presumably have to increase slowly over long periods of time to be consistent with the long-term RNA increment reported by Hydén. Associated with the transient changes in RNA levels (and possibly with the long-term trend also) are increases in proteins and phospholipids (Hydén, 1961; Palladin and Vladimirov, 1956). Thus increases in RNA, proteins, and lipids go together and suggest that these substances all play an important role in behavior.

These results, along with basic molecular biology data, suggest a DNA-activation approach to behavior, i.e., in the complex sequence of events during learning and other behavior there are three important points about which we have some information. These are: (1) the activation of DNA such that (2) specific gene sites can function in RNA synthesis which (3) then triggers protein synthesis. This paradigm of DNA becoming active, DNA making RNA, and protein being synthesized would be important events in the sequence for learning and other behavior. The events that precede and follow these three are obscure at this time.

Research Concerning Macromolecular Involvement in Brain Function

In our approach to the question of the role of macromolecules in brain function during behavior, we have concentrated our research efforts in two broad areas: (1) an attempt to determine brain loci contributing to learning in rats (Sprague-Dawley and Wistar albino strains) and (2) determination of precise molecular events that occur during behavior.

Brain Loci

Numerous studies in the literature suggested that during functional activity of cells, synthesis of RNA and proteins occurred. Thus if a specific area of the brain contributed directly to a certain behavior, then one would expect that the RNA-DNA and protein-DNA ratios, and possibly the protein-RNA ratio as well, would be modified in that brain area of an animal involved in the behavior. Because in rats the brain areas contributing to auditory stimulation and to motor activity are known, two experiments were conducted with auditory and motor stimulation. An orderly change in the ratios occurred over a 30-min period (an inverted U curve with the theoretical maximum occurring at 15 min); these results suggested that the ratios might be useful indices for evaluating more complex behavior (Gaito, et al., 1968a).[1]

Another set of experiments was instituted in which these ratios were utilized in one way shock avoidance conditioning. Complementing the ratio indices was the use of radioisotope incorporation, viz., the specific activity of RNA, protein, and tissue-pool fractions and the relative specific activity of RNA and protein fractions. Three sets of experiments were conducted with these two types of indices:

(1) Under normal conditions (Gaito, et al., 1968a;[2] Gaito, et al., 1968b[3]).

(2) During presumed enhancement of learning with magnesium pemoline (Gaito, et al., 1968).[4]

(3) During presumed impaired learning with hypophysectomy (Gaito, et al., 1966).

Eleven experiments were completed and the results pointed consistently to the involvement of the medial ventral cortex (MV) in this type of avoidance conditioning (Fig. 1). The posterior ventral cortex (PV) may be implicated also; significant results did not appear as consistently as with MV, however. Research in other laboratories in which different procedures were used has also suggested that these cortices or adjacent tissue may contribute uniquely to learning events (Adey, et al., 1960; Barondes and Cohen, 1966; Flexner, et al., 1967).

Because the shock-avoidance conditioning involved the use of an extreme stressor (the electric shock), the results did not conclusively implicate the ventral cortex uniquely in learning behavior. Therefore, it appeared that other tasks not involving extreme stress might be useful. Four experiments were conducted with a six-unit water T maze learning task (Gaito, et al., 1969).[5] The main brain tissues showing differences between learning and nonlearning animals were the anterior dorsal cortex (AD) and MV. The former probably reflected the motor activity—the swimming—required in the maze; the latter, either the learning process or the effects of the stressful situation. The results of these experiments when compared to those obtained earlier in shock avoidance experiments indicated that MV consistently showed differences between learning and nonlearning rats.

[1] Issued originally in more detail as York University Technical Report MPL1, 1965.
[2] Issued originally in more detail as York University Technical Report MPL4, 1966.
[3] York University Technical Report MPL8, 1967.
[4] York University Technical Report MPL5, 1966.
[5] York University Technical Report MPL9, 1967.

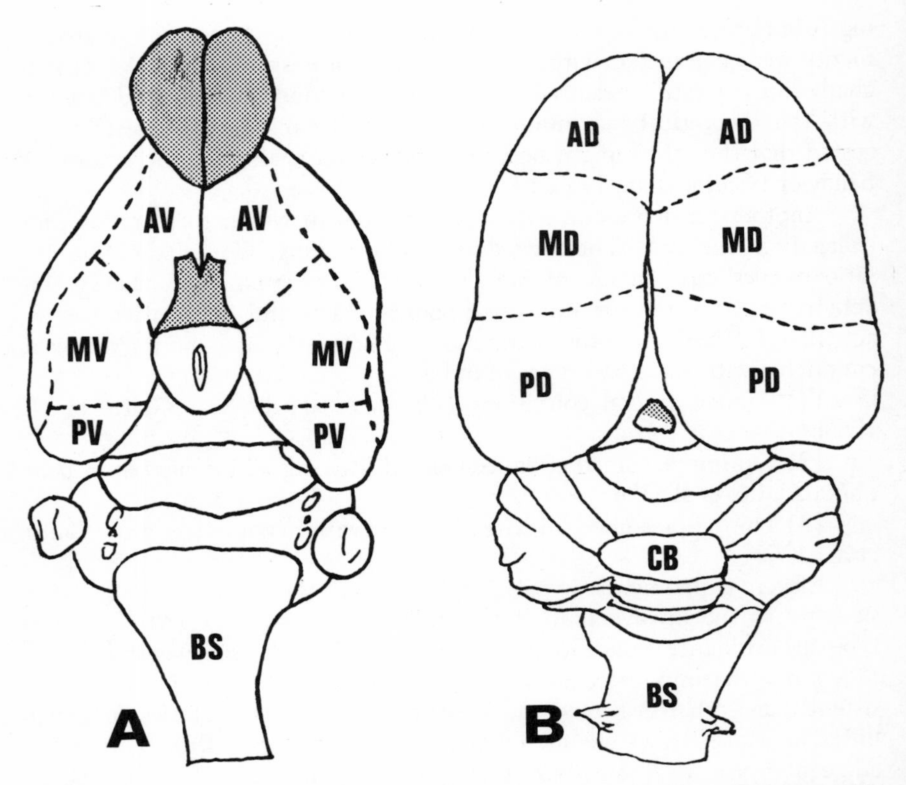

FIG. 1. Rat brain: (a) ventral view; (b) dorsal view. AV, anterior ventral cortex; MV, medial ventral cortex; PV, posterior ventral cortex; AD, anterior dorsal cortex; MD, medial dorsal cortex; PD, posterior dorsal cortex; CB, cerebellum; BS, brain stem.

The shock avoidance conditioning and water maze experiments suggested that portions of the ventral cortex of the rat brain were contributing significantly in these learning tasks. However, the results were not adequate to indicate conclusively that these tissues contribute *uniquely* to learning. The difference in learning and nonlearning rats may have been due to brain processes induced by stress agents (shock in the shock avoidance experiments and water in the water maze experiments). Although a group of animals was used as shock controls in some experiments (Gaito, et al., 1968a), the results were not definite. Furthermore, the use of shock controls or nonlearning controls does not eliminate completely the possibility that these animals may be experiencing some learning. Even though these animals are not given an opportunity to learn an active avoidance response, receiving shock without being able to run or just being in the conditioning apparatus could present a situation conducive to the initiation of some type of learning. Similar comments are applicable to the water maze task. Therefore, to provide further information relative to the importance of the ventral cortex in these two learning tasks, studies concerned with electrolytic lesions were begun. Three lesions experiments were completed (Mottin and Gaito, 1969). In Experiment 1 rats lesioned in MV or medial dorsal (MD) brain areas were not significantly different from nonlesioned animals in shock avoidance conditioning behavior. The extent of the lesion was small and when the lesion in MV was increased in Experiment 2, the lesioned animals were inferior in performance on the shock avoidance task. However, because the lesion in this latter experiment permeated the dorsal cortex as well as MV, the difference could be attributed to the dorsal lesion. Thus, to evaluate this possibility, in Experiment 3 a large lesion was inflicted in this portion of the dorsal cortex. No significant differences were observed in shock avoidance conditioning nor in learning a watermaze pattern. The results of the three experiments, when considered with the previous neurochemical work, suggested that MV tissue (which includes cortical and subcortical tissue) is involved in some manner in shock avoidance conditioning, and possibly in watermaze learning. Considering that the lesions impinged upon the amygdala and this structure has been implicated as having an important role in fear regulation (Goddard, 1964), it is reasonable to speculate that these and the previous results may reflect amygdala activity since some stress is involved in both shock avoidance conditioning and in watermaze learning.

Molecular Neurochemical Events: Whole Brain and Critical Loci

As the experiments concerning brain loci were progressing, attention was directed to the products of DNA activity. The brain loci experiments indicated that RNA changes accompanied learning aspects. An important question concerning these changes is: Are the RNAs produced during behavior, and specifically during the learning, *qualitatively* different than those produced during control conditions? Although many researchers would answer in the affirmative, there is no *conclusive* evidence to document this possibility. Unfortunately, the methods used have been too gross to detect qualitative changes.

One means of attempting to answer this question is by DNA-RNA hybridization procedures (Bonner, 1966; Gaito, 1966). If one heats a solution of rat brain DNA at 95°C for 10 min, the double-stranded DNA will split into single strands. If this DNA is then poured into nitrocellulose membranes, these membranes will "trap" single strands but will allow any double strands to pass through. If a membrane with attached DNA is placed in a solution of RNA, those RNA molecules which are complementary in base sequence to DNA sites will become firmly attached and be resistant to ribonuclease (RNAase) treatment. If this DNA-RNA hybrid is put in another solution of the same RNA, no further hybridization will occur because all DNA sites complementary to the RNA will be occupied. On the other hand, if this hybrid is added to a different solution of RNA which is complementary to other DNA sites, further hybridization will occur.

Putting this procedure within a behavioral framework, the rationale is as follows. If there exist unique species of brain RNA which are synthesized during behavior, e.g., learning, and RNA from the brain of a nonlearning animal is hybridized with single-strand DNA, then when RNA from the brain of a learning animal is added to this hybrid, the unique RNA species should adhere to the DNA. An important aspect of this *successive competition hybridization procedure* is that only the RNA from learning animals is labeled. Thus the presence of label in the twice-hybridized DNA will suggest that RNA species not present in the brain of nonlearning animals have been synthesized in learning animals during the task.

At the present time three sets of behavioral studies are underway in the Molecular Psychobiology Laboratory at York University in which double hybridization procedures, supplemented by appropriate single hybrids, are involved. These are: (1) learning in a shock avoidance task; (2) discrimination learning with split-brain preparation involving permanent separation of cerebral hemispheres by surgery; and (3) visual stimulation.

In all experiments, DNA is extracted by repeated cold phenol treatments and purified by phenol, RNAase, and pronase steps. RNA is extracted by cold and hot phenol, then purified by cold phenol and DNAase treatments. The hybridization procedures are those of Gillespie and Spiegelman (1965). DNA trapped on membranes is incubated with an RNA sample at 66° C for 12 hr, washed, and treated with RNAase. For double hybrids, this procedure is repeated.

Prior to the initiation of behavioral experiments, a preliminary experiment with gastrointestinal DNA indicated that all, or almost all, of the DNA adhered to nitrocellulose membranes (Schleicher and Schuell, B-6, 25 mm). Other preliminary experiments indicated that the amount of RNA hybridizing with 50 μg of DNA seemed to reach a peak with an input of 50 μg of RNA and that hybridization for 24 hr gave the same results as were obtained at 12 hr. Thus most experiments had 50 or 100 μg of RNA hybridizing with 50 μg of DNA for 12 hr.

Rapid progress has occurred in the one-way active shock avoidance study. In Experiment 1 (Machlus and Gaito, 1968a),[6] labeled RNA from

[6] Issued originally in more detail as York University Technical Report MPL12, 1967.

avoidance conditioned rats competed with unlabeled RNA from non-behaving control animals. Label appeared consistently in the double hybrid, suggesting that the brains of learning animals contained RNA species qualitatively different than those in the brains of nonlearning rats. Results with single hybrids for the two groups of rats, and with double hybrids in which the RNA from nonlearning rats was labeled and the RNA from learning animals was unlabeled, provided support for this conclusion. In a second experiment (Machlus and Gaito, 1968b), similar results occurred when brain RNA from shock avoidance trained rats competed with brain RNA from rats in a motor activity task, suggesting that the qualitatively different RNA species were not due to the motor aspects of the avoidance task.

Experiments 3 and 4 utilized more adequate controls than did the previous experiments (Machlus and Gaito, 1969). A shock avoidance rat was injected intracranially with 1 mC of uridine-5^3H in 100 ml of physiological saline. There were two shock yoke control animals in each litter; one was injected with uridine-5-^3H and the other, with unlabeled uridine. Then, 90 min later, the three littermates were placed in the shock chamber of a one-way active shock avoidance apparatus. The shock chamber was divided into three parts so that each of the rats was isolated. The shock avoidance rat was able to respond to the conditioned stimulus by running from the chamber; the shock rats could not leave this chamber. After 15 min of adaption in the shock chamber of the apparatus, the shock avoidance animal was given 15 trials in 15 min and sacrificed by immersion in liquid nitrogen. All shock avoidance animals showed 10 or more avoidance responses in 15 trials. The shock rats received shock when the avoidance rat was shocked; these controls did not receive training and were sacrificed at the end of 30 min in the shock chamber, at the same time as the avoidance rat.

Labeled RNA was extracted from the whole brain of the shock avoidance rat and from one shock rat. Unlabeled RNA was extracted from one-half of the brain of the rat that was injected with unlabeled uridine; DNA was extracted from the other half of the brain. This DNA was used for all hybrids because preliminary work indicated that hybridization results with DNA from littermates were similar to those in which DNA and RNA were from the same animal.

With the three rats in each litter, four hybrids were obtained, as shown in Table 1. Hybrids 3 and 4 were the crucial ones involving successive competition hybrids; Hybrids 1 and 2 were single hybrids which were utilized as a check on the results with Hybrids 3 and 4. Experiments 3 and 4 differed only in one respect; in Experiment 3, 100 μg of RNA was hybridized with 50 μg of DNA; 500 μg of RNA was annealed to 20 μg of DNA in Experiment 4.

TABLE 1. Estimated Amounts of RNase Resistant RNA in Single and Double Hybrids

Hybrids	Experiment 3		Experiment 4		Experiment 5		Experiment 6	
	Mean	S.D.	Mean	S.D.	Mean	S.D.	Mean	S.D.
1. DNA-RNA_{SA}*	1.40	.173	0.62	.100	2.68	.374		
2. DNA-RNA_S*	1.08	.114	0.44	.084	1.80	.281		
3. DNA-RNA_S-RNA_{SA}*	0.64	.153	0.27	.075	0.76	.200	0.00	0.00
4. DNA-RNA_S-RNA_S*	0.02	.019	0.00	.000	0.00	0.00	0.00	0.00
5. DNA-RNA_{SA}-RNA_{SA}*					0.00	0.00		
6. DNA-RNA_{SA}-RNA_S*					0.00	0.00		

RNA_{SA}, RNA from shock avoidance animal; RNA_S, RNA from nonlearning shocked animal; asterisk indicates the presence of labeled precursor in RNA; S.D. is standard deviation. DNA and RNA were from brain in Experiments 3, 4, and 5; in Experiment 6, from liver. There were 12 observations for each hybrid in Experiments 3 and 5, 8 in Experiment 4, and 16 in Experiment 6. In Experiments 3, 5, and 6, 50 μg of DNA was hybridized with 100 μg of RNA; in Experiment 4, 20 μg of DNA, with 500 μg of RNA.

In all experiments, the cpm for the hybrids were obtained by two runs in a Beckman Liquid Scintillation Spectrometer (LS-100) at 100 min or 2 percent error. The first one was used for the analysis; the second run provided for a consistency check. The cpm were converted to dpm by using a quench correction curve. The specific activity of RNA for each littermate was determined from a 50-μg sample in which the cpm were converted to dpm. The specific activity in the various experiments varied between approximately 1,000 and 2,500 cpm/μg RNA. The counting efficiency was approximately 30 percent. There appeared to be no difference in specific activities between shock avoidance and shock rats. The amount of labeled RNA hybridizing in each hybrid was obtained by comparison with the specific activity of the RNA for the rat from which the hybridized RNA was obtained.

The results of all experiments are presented in Table 1 as amounts of RNAase-resistant RNA present in each hybrid. Two samples were obtained for each of six litters in Experiment 3 (12 in all), but only one sample was possible for each of 8 litters in Experiment 4. The results in both experiment were clear. In all samples Hybrid 3 showed values much greater than zero, Hybrid 3 was greater than Hybrid 4, and Hybrid 4 was zero or near zero. Likewise, in all samples, the amount of RNA hybridized for the shock avoidance rat (Hybrid 1) was greater than that for the shocked rat (Hybrid 2). The amounts hybridized with 20 μg of DNA in Experiment 4 were proportionately the same as those achieved in Experiment 3 with 50 μg of DNA. The single and double hybrids data suggested that qualitatively different species of RNA were present in the brains of shock avoidance learning rats than were present in the littermate shocked animals. Thus the results of the four experiments consistently suggested the presence of "unique" brain RNA, using RNA-DNA ratios of 1, 2, and 25.

Another study, Experiment 5, was conducted with further controls; six litters of four rats each were used. A second shock avoidance rat (injected with unlabeled uridine) was added. With the four rats in each litter, six hybrids were obtained (Table 1). Hybrids 3 to 6 were the crucial ones involving successive competition hybridization. Two samples were obtained for each litter in Experiment 5 (twelve in all). In all twelve samples Hybrid 3 showed values of μg RNA much greater than zero whereas the other double hybrids were at background levels. Likewise, in single hybrids, for all twelve cases, the amount of RNA hybridized for the shock avoidance rat was greater than that for the shocked rat. With cpm or dpm as the dependent variable, the results were the same. These results are consistent with those of the previous experiments and suggest the presence of "unique" RNA species during shock avoidance conditioning.

Another experiment (Experiment 6) was undertaken to determine if the "unique" species were present in the liver during conditioning. Four litters of three rats each were used. Two rats were injected with labeled uridine (shock avoidance trained and shock control) and the other with unlabeled uridine (shock control). Only Hybrids 3 and 4 of Table 1 were used (DNA-RNA_s-RNA_s*, and DNA-RNA_s-RNA_s*). Samples (50 μg) of RNA from shock avoidance and shock animals indicated the synthesis of labeled RNA in the liver; however, no label was detected in either hybrid for 16 samples (four observations for each of the four litters).

Experiment 7 was conducted to investigate the possibility that the "unique" RNA species were localized in specific brain areas. Three pairs of littermate rats were used. One of each pair received labeled uridine; the other, unlabeled uridine. Both rats were trained to avoid shock. The brain of each rat was dissected into three parts: cerebral hemispheres (CH), brain stem (BS), and cerebellum (CB). Three single hybrids were obtained, one for each of the three parts. Six double hybrids were used, one each for the six possible sequences of the three parts in successive competition: RNA_{CH}-RNA_{CB}*, RNA_{CH}-RNA_{BS}*, RNA_{CB}-RNA_{CH}*, RNA_{CB}-RNA_{BS}*, RNA_{BS}—RNA_{CH}*, RNA_{BS}-RNA_{CH}*. Three observations were obtained for each. The amount of RNAase-resistant RNA hybridizing for the three brain parts were: CH, 2.43 μg; BC, 2.37 μg; CB, 2.46 μg. No double hybrid showed greater than background label.

Because in-vivo labeling of RNA is not of a uniform nature, the estimation of the amounts of RNA hybridized by liquid scintillation spectrometry must be considered as an approximation. However, the results have been consistent in this series of studies. The differences have been substantial in indicating the presence of "unique" RNA, and this conclusion occurs whether one uses cpm, dpm, or amount of RNA as the dependent variable of concern.

The results of these experiments suggest that during this behavioral task, RNA species were produced throughout the brain which were qualitatively different from those present in the brains of nonlearning rats and were not present in the liver. During the first 105 min of the incorporation period, the learning and nonlearning animals were treated the same. Thus the differences probably reflect the synthesis of RNA during the last 15 min during which learning was occurring. Other individuals have reported what appears to be an increase in the rate of synthesis of RNA in the brains of mice after 15 min of training in a shock avoidance task.

Thus the results of this series of studies suggest the synthesis of "unique," qualitatively different species of RNA during this learning task, a conclusion which is consistent with those of Hydén and others who hypothesize that RNA has a unique role in learning events. Other experiments, however, are being conducted to evaluate the possibility that these results are artifactual in nature, because the hybridization techniques are very difficult ones and have serious pitfalls.

Experiments are underway with other behavioral tasks (discrimination learning, visual stimulation [7]) and further experiments are anticipated in which the shock avoidance task will be used. These experiments will employ the double saturation curve and double saturation competition curve procedures of Miyagi et al. (1967) so as to provide comparisons between control and experimental rats over varying—including large—amounts of RNA. Other experiments are attempting to determine the physical and chemical characteristics of this RNA.

The study using the split-brain rat is an interesting one; this preparation will provide behavioral, humoral, and biochemical controls within the same animal. Rats will be trained on a visual discrimination with one eye occluded. Nucleic acids of the learning hemisphere will be compared to those nonlearning hemispheres of the same animal by successive competition hybridization.

The results already obtained lead naturally to studies concerned with events at the DNA level in an attempt to determine what regulatory events underlie the synthesis of the "unique" RNA. Much research has suggested the possibility that quantitative and/or qualitative changes in histones (nuclear proteins) affect the template activity of DNA. For example, a reduction in the amounts, or acetylation or phosphorylation, of

[7] As this volume goes to press, part of the visual stimulation study is complete. Using improved hybridization procedures, unique species were not detectible during visual stimulation, and RNA/DNA ratios of 50 and 75 did not saturate the DNA sites. This lack of saturation is inconsistent with the results of the shock avoidance experiments in which RNA/DNA ratios of 1, 2, and 25 appeared to saturate DNA sites. Further research is underway to reconcile these discrepancies.

histones in a chromosomal fraction increases the amount of RNA synthesized. (See review by Bonner et al., 1968.)

Preliminary research has begun on this problem. The first behavioral experiments, however, were inconclusive because of the low specific activity and low yield of histones. At present we are improving our procedures so as to overcome these problems.

Another research direction suggested by the results of the hybridization studies is one concerned with the ultimate fate of the "unique" RNAs. Possibly these RNAs code for specific proteins which proceed to certain parts of the cell—e.g., the synaptic area—to perform some function. We plan soon to study certain chemical events occurring in the synaptic area during learning. Ultimately, it is hoped that we can determine the relationship between learning and nonlearning states and the three sets of chemical events: histones and other possible regulators of gene activity, the RNAs produced, and certain synaptic chemicals (e.g., the choline acetylase-acetylcholine-acetylcholinesterase system).

Discussion

The research concerned with brain loci suggests that the ventral cortex contributes in some fashion to the learning tasks of shock avoidance and watermaze conditioning. Results of other research efforts also suggest the involvement of this tissue. However, the effects which have resulted in both types of learning tasks may be due to the stress involved in each; thus these research efforts may not have located the brain area unique for these two tasks but rather the brain area that shows contributions to stressful situations.

The research reported in the literature so far has indicated quite clearly that quantitative changes in RNA and proteins do occur during learning and other behaviors (Gaito and Bonnet, 1971). In general, increases occur with mild stimulation and decreases occur during drastic or extended stimulation. There has been no research, however, which definitely indicates the occurrence of qualitative changes in RNA and/or proteins during behavior. Unfortunately, the procedures used have not been ones that could permit detection of qualitative changes. For example, the base ratio analysis of Hydén does not exclude the possibility that *only* quantitative changes in RNA are occurring.

The DNA-RNA hybridization procedures provide a means whereby qualitatively different species of RNA can be trapped and differentiated; essentially a base sequence analysis is provided by the annealing of RNAs to specific DNA sites, without indicating to the researcher, however, the exact sequence involved. Thus the results reported in this chapter with this technique represent the first systematic set of experiments which have attempted to determine directly whether qualitative RNA changes occur in learning.

An interesting aspect of the hybridization approach is the concern as to what these "unique" RNAs are. It is too early to suggest the exact role that these RNAs play in learning (assuming, of course, that they are not artifactual). Recently the idea has been advanced that much of the DNA of higher organisms is made up of sequences which recur anywhere from a thousand to a million times per cell (Britten and Kohne, 1968). During DNA-RNA hybridization, these repeated sequences in DNA anneal rapidly with RNA, even when the RNA concentration is low. With the short period of incubation (12 hr) and the low concentration of RNA that have been used in this laboratory, it is probable that the "unique" RNA is annealing to the repeated DNA sequences (J. Bonner, personal communication). The "unique" RNA species may be messenger RNA or nuclear RNA (activator RNA) which functions to derepress other DNA sites (J. Bonner, personal communication; Britten and Davidson, 1969). If they are nuclear RNA, e.g., a type of activator RNA, they may function to increase the synthesis of RNA from DNA sites which have been continuously active, but at a low level; or, they may activate DNA sites which were repressed. In the former case, only the activator RNAs would be "unique" (i.e., qualitatively different); in the latter event, both the activator RNA and the messenger RNA from the derepressed site would be "unique."

It should be obvious that the above treatment concerning the DNA complex is concerned with a narrow sequent of the overall problem of behavior and brain function. For example, in a specific learning situation a sequence of biological events is involved, say *a, b, c . . . x*. Event *a* could be a stimulation of receptors, *b* the transmission of nerve impulses into the central nervous system, etc. Another event, say *g*, would involve DNA; *h*, RNA synthesis; and *i*, protein synthesis. These three segments represent a small portion of the overall sequence. However, an understanding of certain aspects of *g, h* and *i* may help to suggest the antecedent and postcedent conditions prevailing.

Acknowledgments

The experiments reported herein were supported by grants from the National Research Council of Canada and from the Office of Naval Research of the U.S.A. (Contract Nonr—4935[00]). The author wishes to thank the members of the Molecular Psychobiology Laboratory of York University (undergraduate and graduate students, faculty members) for their contributions toward the objectives of the Laboratory.

References

Adey, W. R., Dunlop, C. W., and Hendrix, C. E. (1960). Amer. Med. Assoc. Arch. Neurol., 3:74.

Barondes, S. H., and Cohen, H. D. (1966). Science, 151:594.

Bonner, J. (1966). In: Gaito, J., ed., Macromolecules and Behavior, 1st ed. New York: Appleton-Century-Crofts. ——— Dahmus, M. E., Fambrough, D., Huang, R. C., Marushige, K., and Yuan, D. Y. H. (1968). Science, 159:47.

Britten, R. J., and Davidson, E. H. (1969). Science, 165:349. ——— and Kohne, D. E. (1968). Science, 161:529.

Cameron, D. E. (1963). Brit. J. Psychiat., 109:325.

Dingman, W., and Sporn, M. B. (1961). J. Psychiat. Res., 1:1. ——— and Sporn, M. B. (1964). Science, 144:26.

Flexner, L. B., Flexner, J. B., and Roberts, R. B. (1967). Science, 155:1377.

Gaito, J. (1961). Psychol. Rev., 68:288. ——— (1966). Molecular Psychobiology. Springfield, Ill.: Thomas. ——— (1969). In: Bourne, G. H., ed., The Structure and Function of Nervous Tissues, Vol. 2. New York: Academic Press. ——— and Bonnet, K. (1971). Psychol. Bull., 75:109. ——— Davison, J. H., and Mottin, J. (1968). Psychon. Sci., 13:257. ——— Davison, J. H., and Mottin, J. (1969). Psychon. Sci., 14:46. ——— Mottin, J., and Davison, J. H. (1968). Psychon. Sci., 13:41 (a). ——— Mottin, J., and Davison, J. H. (1968). Psychon. Sci., 13:259 (b). ——— Mottin, J., Davison, J. H., and Rigler, J. (1966). *York University Technical Report MPL 6.*

Gillespie, D., and Spiegelman, S. (1965). J. Molec. Biol., 12:829.

Goddard, G. V. (1964). Psychol. Bull., 62:89.

Harbers, E., Domagk, G. F., and Muller, W. (1968). Introduction to Nucleic Acids. New York: Reinhold.

Hydén, H. (1959). Proc. 4th Int. Congr. Biochem., Vienna, 1958. Biochemistry of the Central Nervous System, Vol. III. New York: Pergamon. ——— (1961). Sci. Amer., 205:62. ——— (1967). In: Quarion G. C., Melnechuk, T., and Schmitt, F. O., eds., The Neurosciences. New York: Rockefeller Univ. Press. ——— and Lange, P. W. (1965). Proc. Nat. Acad. Sci. U.S.A., 53:946.

Landauer, T. K. (1964). Psychol. Rev., 71:167.

Machlus, B., and Gaito, J. (1968a). Psychon. Sci., 10:253. ——— and Gaito, J. (1968b). Psychon. Sci., 12:111. ——— and Gaito, J. (1969). Nature, 222:573.

McConnell, J. V. (1962). J. Neuropsychiat., 3:42.

Miyagi, M., Kohl, D., and Flickinger, R. A. (1967). J. Exp. Zool., 165:147.

Mottin, J., and Gaito, J. (1969). Psychon. Sci., 15:12.

Palladin, A. V., and Vladimirov, G. E. (1956). In: Proc. Internat. Conf. on the Peaceful Uses of Atomic Energy, Vol. 12. New York: United Nations.

Pevzner, L. Z. (1966). In: Macromolecules and Behavior, 1st ed., Gaito, J., ed. New York: Appleton-Century-Crofts.

Pribram, K. (1966). In: Macromolecules and Behavior, 1st ed., Gaito, J., ed. New York: Appleton-Century-Crofts.

Chapter 5

Inhibitors of Cerebral Protein or RNA Synthesis and Memory

LARRY R. SQUIRE and SAMUEL H. BARONDES

University of California (San Diego), La Jolla, California

An accumulating body of evidence suggests that cerebral RNA and protein synthesis may play an important role in memory storage (see reviews by Glassman, 1969; Barondes, 1970). The purpose of this report is to present and critically review evidence derived from studies of the effects of inhibition of cerebral protein or RNA synthesis on memory.

Since RNA and protein synthesis are known to be important in many biological regulatory mechanisms, it would be surprising if they were not involved in memory in some way (Dingman and Sporn, 1964). One possibility is that specific and unique macromolecules represent specific bits of information (Gaito, 1961). A more parsimonious alternative is that memory is not stored by the synthesis of new and unique macromolecules, but by increased synthesis of some common macromolecules (Barondes, 1965) or by alterations in the metabolic efficiency of neurons, which might involve a whole class of biochemical events (Pfaff, 1969). It has further been proposed that new interneuronal connections could be the basis for information storage (Hebb, 1949; Eccles, 1953) and, specifically, that synthesis of one or a few protein molecules might be involved in the establishment of new functional synaptic connections (Barondes, 1965). Results of behavioral experiments with agents which specifically inhibit cerebral protein synthesis strongly support this hypothesis, but definitive confirmation cannot be obtained from such studies because of the complex effects which may be produced by these inhibitors.

Some Pharmacological Properties of the Inhibitors

To determine, by pharmacological means, whether the synthesis of a protein or group of proteins is specifically required for memory storage, one would optimally select drugs which specifically inhibit the synthesis of these molecules and which do not have other actions on the brain. Unfortunately, the agents thus far used do not satisfy this requirement. First, they inhibit the synthesis of all protein molecules—both "constitutive" protein synthesis required to replace proteins which are normally being degraded and also "inducible" protein synthesis, which might subserve adaptive processes such as memory storage. Prolonged inhibition of cerebral protein synthesis might lead to depletion of critical "constitutive" proteins by preventing their normal replacement. In this case, a behavioral deficit resulting from protein synthesis inhibition may be due to a depletion of "constitutive" protein rather than to inhibition of protein synthesis specifically required for memory formation. This problem may not be great, since inhibition in some experiments (e.g., with subcutaneous cycloheximide; Barondes and Cohen, 1968b) is fairly brief. Moreover, no detectable behavioral or electrophysiological abnormalities are observed following cycloheximide treatment (Cohen et al., 1966; Cohen and Barondes, 1967). However, the possibility that effects on memory are due to interference with "constitutive" protein synthesis cannot be entirely ruled out.

A second problem is one which complicates all pharmacological research: the drugs studied may have another action besides the one for which they are being used. The protein synthesis inhibitor, puromycin, for example, has actions on cerebral electrical activity, cerebral mitochondria, and the respiration of cerebral cortical slices (Table 1), which effects are not observed when equivalent or greater inhibition of cerebral protein synthesis is produced by cycloheximide or acetoxycycloheximide. The results of a number of studies of effects of inhibitors of cerebral protein or RNA synthesis on brain physiology or biochemistry are summarized in Table 1. Some of the parameters studied were affected and others were not. Of these affected, some may be secondary to inhibition of protein synthesis, e.g., inhibition of ganglioside synthesis by puromycin (Kanfer and Richards, 1967) or acetoxycycloheximide (Barondes and Dutton, 1969). Others, e.g., in-vitro inhibition of cyclic AMP phosphodiesterase by puromycin (Appleman and Kemp, 1966) are unrelated to inhibition of protein synthesis. The mechanism of still others is not known: e.g., depletion of brain acetylcholine by actinomycin D and cycloheximide (Weiner, personal communication); or abnormalities in electrical activity in an isolated ganglion following puromycin, acetoxycycloheximide, or puromycin ami-

TABLE 1. Some Effects of Inhibitors of Protein or RNA Synthesis on the Nervous System

	Puro.	PAN	AXM, CXM	Actino.
Abnormal cerebral electrical activity	+(1-3)	+(3)	0(1,2)	+(4)
Mitochondrial abnormalities	+(5,6)		0(6)	
Inhibits respiration in cerebral cortex slices	+(7)			0(7)
3′5′ AMP phosphodiesterase inhibition	+(8)			
Ganglioside synthesis inhibition	+(9)		+(10)	
Impulse activity in stretch receptor neuron	+(11)			0(11,12)
Abnormal electrical activity in isolated ganglion	+(13)	+(13)	+(13)	
Lowers brain acetylcholine			+(14)	+(14)
Neuronal necrosis			0(6,15)	+(4,16,17)

Puro., puromycin; PAN, puromycin aminonucleoside; AXM, acetoxycycloheximide; CXM, cycloheximide; Actino, actinomycin-D.
[1] *Cohen et al. (1966).*
[2] *Cohen and Barondes (1967).*
[3] *Agranoff (1969a).*
[4] *Nakajima (1969).*
[5] *Gambetti et al. (1968a).*
[6] *Gambetti et al. (1968b).*
[7] *Jones and Banks (1969).*
[8] *Appleman and Kemp (1966).*
[9] *Kanfer and Richards (1967).*
[10] *Barondes and Dutton (1969).*
[11] *Toschi and Giacobini (1965).*
[12] *Edström and Grampp (1965).*
[13] *Paggi and Toschi (1970).*
[14] *Weiner (personal communication, 1970).*
[15] *Barondes, unpublished data.*
[16] *Appel (1965).*
[17] *Koenig and Lu (1967).*

nonucleoside (Paggi and Toschi, 1970). The relevance of these findings to the effects of these drugs on memory will be considered in the ensuing discussion.

Cycloheximide and Acetoxycycloheximide (the Glutarimide Derivatives)

When inhibition of 95 percent of cerebral protein synthesis was established by intracerebral injection of 20 μg of acetoxycycloheximide 5 hr before learning a shock-escape position habit, mice learned normally but exhibited impaired retention one or seven days later (Barondes and Cohen, 1967b). This effect depended both on the establishment of an extensive degree of cerebral protein synthesis inhibition and on the use of a low learning criterion. When less than 80 percent of cerebral protein synthesis was inhibited during training, retention was not affected. Even 80 to 85 percent inhibition had only a small amnesic effect (Barondes and Cohen, 1967b). When training proceeded to a criterion of 9 out of 10 correct responses in the position discrimination, retention was normal in spite of inhibition of 90 to 95 percent cerebral protein synthesis by either cycloheximide (Barondes and Cohen, 1967a) or acetoxycycloheximide (Barondes and Cohen, 1967b). Retention was impaired only when 90 to 95 percent of cerebral protein synthesis was inhibited and training in the position discrimination was conducted to a criterion of three out of four correct responses (Barondes and Cohen 1967b).

Following intracerebral injection of 20 μg of acetoxycycloheximide, mice were trained to escape shock by choosing the lighted limb of a T-maze to a criterion of five out of six responses; they learned normally but exhibited amnesia one day or seven days later (Cohen and Barondes, 1968a). This effect was also antagonized by more extensive training. When training was continued to a criterion of 9 out of 10 correct responses, amnesia was still observed, but with a criterion of 15 out of 16 correct responses, retention was substantial (Cohen and Barondes, 1968a). Perhaps additional training enables the slight residual capacity for cerebral protein synthesis to mediate long-term memory.

Following establishment of 95 percent cerebral protein synthesis inhibition by subcutaneous injection of cycloheximide, mice given brief training in a water reward, position discrimination learned normally, but were impaired one day or seven days later (Cohen and Barondes, 1968b). Thus the amnesic effects of protein synthesis inhibition are not limited to learning under a specific type of motivation. They have been demonstrated in both aversive and appetitive tasks.

Interpretation of all these amnesic effects depends on the fact that neither acetoxycycloheximide nor cycloheximide impairs original learning. In the light-dark discrimination, for example, acquisition curves for acetoxycycloheximide and saline-treated groups are superimposable (Cohen and Barondes, 1968a). In cases in which original learning is impaired by these agents, it is difficult to determine whether memory or other aspects of performance has been disrupted. Thus reports that cycloheximide impaired the learning of leg position in the headless cockroach (Brown and Nobel, 1967, 1968) cannot yet be clearly interpreted. In the studies with mice, however, original learning was not affected by protein synthesis inhibition. Moreover, evidence has been presented that the results of these studies cannot be easily explained by gradually developing systematic toxicity, gradually developing nonspecific cerebral abnormalities resulting from prolonged cerebral protein synthesis inhibition, or by state-dependent learning (Barondes and Cohen, 1967b). It therefore appears likely that both acetoxycycloheximide and cycloheximide inhibit the synthesis of protein required for adequate long-term retention.

In the course of continued studies on the effects of cycloheximide on learning, it was found that cycloheximide injected either subcutaneously or intracerebrally exerted marked effects on the activity of mice (Segal et al., 1970; Squire et al., 1970). Following subcutaneous injection of cycloheximide at a dose which has amnesic effects, mice were hyperactive for 45 min and then hypoactive for several hours. Following intracerebral injections, which take much longer to produce a high level of inhibition, only depression was observed. These findings raised the possibility that the amnesic effects of cycloheximide might be due to its effect on activity rather than to the inhibition of cerebral protein synthesis specifically required for memory storage. However, the amnesic effects and the activity effects can be dissociated. First, isocycloheximide, a stereoisomer of cycloheximide which does not inhibit protein synthesis and does not produce amnesia, produced similar activity changes (Segal et al., 1970). Second, amphetamine, which antagonizes the amnesic effect of cycloheximide, did not antagonize the effect of cycloheximide on activity (Segal et al., 1970). Thus, the effects of cycloheximide on activity do not appear to be responsible for its amnesic action. The possibility cannot be excluded, of course, that cycloheximide has some other effect, unrelated to protein synthesis inhibition, which interferes with long-term retention.

To determine the time after training when the protein synthesis apparently required for long-term memory is sufficiently complete to support long-term retention, inhibition has been established at different times after learning. The delay in onset of generalized inhibition following intracerebral injection, which was attributed to the time required for diffusion (Barondes and Cohen, 1967b), precluded the use of this route of injection for such studies. However, following subcutaneous injection of 240 μg acetoxycycloheximide, 90 percent of overall cerebral protein synthesis was inhibited within 10 min (Barondes and Cohen, 1968a). With use of this technique for producing rapid onset of inhibition, it was found that, in contrast with the marked amnesic effect of injections 5 min before training, injections given immediately or 5 min after the completion of training in a light-dark discrimination to a criterion of five out of six correct responses had only a slight effect on retention when the animals were tested seven days later. When the drug was injected 30 min after learning, retention was not affected (Barondes and Cohen, 1968a). The results of these experiments suggest that the protein synthesis required for formation of long-term memory has progressed substantially during training or within minutes after training.

Other studies have indicated that the time after training when memory remains sensitive to protein synthesis inhibition may vary depending on the task and speices used. Goldfish trained in a shuttle-box task motivated by shock exhibited impaired retention three days after training if 0.1 μg of acetoxycycloheximide was injected intracranially immediately after training (Agranoff, et al., 1966). This dose of acetoxycycloheximide was sufficient to inhibit 80 to 90 percent of cerebral protein synthesis for several hours after injection (Brink et al., 1966). Memory was also affected somewhat by injections of 0.2 μg acetoxycycloheximide 1 hr after training, but injections given 3 hr after training had no effect (Agranoff et al., 1967; Agranoff and Davis, 1968).

Mice trained in a one-trial, step-through, passive avoidance procedure (Jarvik and Kopp, 1967) and tested seven days later were impaired by subcutaneous injections of cycloheximide 10 or 30 min after the acquisition trial, but not when injections were delayed for 2 hr (Geller et al., 1969). The degree of impairment of long-term retention produced by giving cycloheximide 20 sec after the acquisition trial could be markedly increased when the learned response was weakened by increasing the interval between the step-through response and the punishing shock (Geller et al., 1970). Perhaps strengthening the conditioned passive avoidance response, like overtraining of discrimination problems, may enable the slight residual capacity for protein synthesis to subserve long-term memory. The retrograde amnesia gradient produced by acetoxycycloheximide in mice in these studies of passive avoidance training is similar to that in goldfish and differs markedly from that observed in mice given light-dark discrimination training (Barondes and Cohen, 1968a). It appears that even an hour or more after the completion of training, in some situations, the cerebral protein synthesis involved in long-term memory storage has not progressed sufficiently to store long-term memory adequately.

Since original training is not disrupted by these agents, but retention 24 hr after training is severely impaired, it has been of considerable interest to determine how long after original learning these amnesic effects first appear. This question has been investigated by training animals during the establishment of protein synthesis inhibition by acetoxycycloheximide or cycloheximide and retesting them at various intervals thereafter. For both light-dark discriminations and position habits, mice learned normally, and had normal memory for 3 hr; but from 3 to 6 hr after training, they exhibited memory impairment. This impairment reached its maximum at 6 hr and was permanent thereafter (Barondes and Cohen, 1967b, 1968a; Cohen and Barondes, 1968a, b). These results have led to the suggestion that for a few hours following learning, memory may depend on a short-term, temporary storage system which is not affected by cerebral protein synthesis inhibition.

The results with goldfish differ from the results with mice given discrimination training. Goldfish given 0.2 μg acetoxycycloheximide immediately after training and tested at various intervals thereafter had normal retention 4 or 6 hrs after training, but exhibited progressively impaired retention from 24 to 72 hrs after training (Agranoff, 1970). Cycloheximide (10 μg), given immediately after training, produced impaired retention which was first observed between 4 and 8 days after training. Thus the decay of memory following the establishment of protein synthesis inhibition requires several days in goldfish. In mice, the decay of memory following acetoxycycloheximide or cycloheximide is complete within 6 hrs (Barondes and Cohen, 1967b, 1968a; Cohen and Barondes, 1968a). Thus, with goldfish, there is a relatively long retrograde amnesic gradient (1 hr) and a relatively long decay time for memory (one to three or four to eight days) following initiation of protein synthesis inhibition immediately after training. With mice given discrimination training, the retrograde amnesia gradient is brief (5 min) and the decay of memory following training is rapid (3–6 hr). After passive avoidance training in mice, the retrograde amnesic gradient is more prolonged. These differences between mice and goldfish are not yet clearly understood, but they may provide clues for our understanding of short-term and long-term memory processes.

Currently, conclusions concerning the role of protein synthesis inhibition in the establishment of long-term memory rest primarily on tasks involving discrimination learning and passive avoidance learning in mice and shuttlebox avoidance in goldfish. Which other forms of long-term behavior modification may also require protein synthesis are not yet clear. It has recently been reported that habituation of exploratory activity in a small compartment, which endures for at least two weeks, is retained normally when protein synthesis inhibition was initiated immediately after the first exploratory session (Squire et al., 1970). This finding contrasts strikingly with the marked amnesic effect observed when cycloheximide injections were given immediately after passive avoidance training (Geller et al., 1969). It is possible that, in the case of exploratory habituation, but not in the case of passive avoidance tasks or discrimination learning the small amount of protein synthesis remaining after treatment with cycloheximide (5—10 percent) may be sufficient to subserve normal long-term retention. Alternatively, it is an intriguing possibility that some forms of behavior modification may be independent of cerebral protein synthesis. These problems deserve further investigation.

PUROMYCIN

Amnesic effects have also been produced with puromycin, another inhibitor of protein synthesis, but these results have been even more difficult to interpret than those with the glutarimide derivatives, since it is now known that puromycin has multiple effects on cerebral function. In the earliest studies with this agent, inhibition of protein synthesis in mice established by subcutaneous injections of puromycin was not sufficient to produce behavioral impairment (Flexner, et al., 1962, 1964). Intracerebral injections of 180 μg of puromycin, sufficient to inhibit more than 80 percent of cerebral protein synthesis for 8–10 hr after injection (Flexner et al., 1964, 1965), produced amnesia when given one day after training (Flexner et al., 1963). Intracranial administration of puromycin to goldfish before training also impaired retention (Agranoff, et al., 1965, 1966; Shashoua, 1968). Furthermore, it was found with mice trained in a position discrimination to a criterion of 9 out of 10 correct responses 5 hr following intracerebral injection of 180 μg of puromycin, they learned normally and performed well 15 min after training, but their performance rapidly deteriorated so that by 3 hr after training it was markedly impaired (Barondes and Cohen, 1966).

Amnesic effects of puromycin given after training have also been demonstrated. In mice, performance was disrupted by bitemporal injections of puromycin given as long as several days after training, but not one week after training (Flexner, et al., 1963, 1965). Performance also could be disrupted by multiple, widespread injections given as long as 60 days after training (Flexner, et al., 1963, 1965). In goldfish, puromycin was effective if given within an hour after training, but ineffective if given longer than an hour after training (Agranoff and Klinger, 1964: Agranoff et al., 1965, 1966; Davis et al., 1965). Repeated injections of puromycin after daily training sessions also impaired subsequent performance (Potts and Bitterman, 1967). In the quail, puromycin given immediately after training impaired retention three days later (Mayor, 1969).

These results with puromycin are different from those obtained with cycloheximide and acetoxycycloheximide in three respects. First, with puromycin retention could be disrupted by treatment many days after learning (Flexner et al., 1963, 1965). With the glutarimide derivatives, retention was impaired only when protein synthesis inhibition was initiated within less than an hour after learning (Agranoff et al., 1966, 1967; Barondes and Cohen, 1968a; Geller et al., 1969). Second, with puromycin, retention in mice was disrupted after training in a spatial task to a criterion of 9 out of 10 correct responses, but no impairment was observed in this task at this criterion with other inhibitors (Barondes and Cohen, 1967a, b). Third, the decay of memory following learning in mice pretreated with puromycin was complete within 3 hr (Barondes and Cohen, 1966), but required about 6 hr with the glutarimide derivatives (Barondes and Cohen, 1967b, 1968a; Cohen and Barondes, 1968a, b).

Earlier explanations of the behavioral differences between acetoxycycloheximide and puromycin in terms of differential tendencies of these drugs to degrade species of messenger RNA (mRNA) essential for expression of memory (Flexner and Flexner, 1966; Flexner et al., 1966, 1967) have not proved to be correct. According to the hypothesis, mRNA required for expression of memory is degraded by puromycin, but conserved in the presence of acetoxycycloheximide. Thus puromycin given after training destroys memory (Flexner and Flexner, 1966; Flexner, et al., 1966), but acetoxycycloheximide causes only a transient memory deficit corresponding to the transient sickness which it produces (Flexner, et al., 1966). It is now clear, however, that after brief rather than prolonged training acetoxycycloheximide does produce a sustained loss of memory (Barondes and Cohen, 1967b). Furthermore, it now appears that puromycin given after training does not destroy memory, as had been suspected. Memory can be restored by further treatment (see below).

Further investigations of differences between puromycin and the glutarimide derivatives have led to other explanations of the discrepancies in the behavioral effects of these drugs. Puromycin produces abnormal electrical activity in the hippocampus, but cycloheximide does not (Cohen et al., 1966). Furthermore, after administration of pentylenetetrazol, the threshold for overt seizures in mice was reduced by puromycin, but not by cycloheximide (Cohen and Barondes, 1967). Electrical abnormalities have also been observed in goldfish following intracranial injections of puromycin (Agranoff, 1969a).

The electrical abnormalities produced by puromycin may explain some of the differences between the behavioral effects produced by this drug and other protein synthesis inhibitors. When puromycin was given 5 hr before training, the anticonvulsant diphenylhydantoin reduced the amnesic effect usually observed 3 hr after training. The amnesic effect of acetoxycycloheximide, first observable about 6 hr after training, was not affected by this treatment (Cohen and Barondes, 1967). Indications of the importance of the changes in electrical activity produced by puromycin also come from studies of the amnesic effects of KCl. This compound, which only moderately inhibits protein synthesis (Bennett and Edelman, 1969), produced gross abnormalities in cerebral electrical activity and impaired performance when given 24 hr after training (Avis and Carlton, 1968; Hughes, 1969).

Puromycin also produces mitochondrial swelling in mice (Gambetti, et al., 1968a, b). It is suggested that the release of abnormal peptidyl-puromycin from ribisomes following puromycin treatment (Nathans, 1964; Williamson and Schweet, 1965) might be responsible for this effect. It has also been demonstrated that both cycloheximide and acetoxycylo-heximide, which do not prematurely release growing peptide chains (Ennis and Lubin, 1964), actually reduce the production of peptidyl-puromycin (Colombo et al., 1965), and in fact, can antagonize the amnesic effects of puromycin in mice (Flexner and Flexner, 1966; Barondes and Cohen,

1967a). Accordingly, the appearance and long survival in brain of peptidylpuromycin, rather than inhibition of protein synthesis per se, may be responsible for some of the amnesic effects of puromycin (Flexner and Flexner, 1967; 1968b).

Further support for this possibility comes from experiments indicating that the amnesic effects of puromycin are reversible. Memory lost by bitemporal injection of puromycin one day after training could be restored by bitemporal injections of saline given 30 to 60 days later (Flexner and Flexner, 1967). The effects of saline are observed most clearly when puromycin is injected one day or longer after training. When puromycin was injected before or immediately after training, saline improved performance only marginally (Flexner and Flexner, 1968a). Five or ten days, but not two days, following bitemporal puromycin injection, bifrontal saline injections were also effective in improving retention (Flexner and Flexner, 1969a). Retention could also be improved by intraperitoneal injection of certain adrenergic drugs (Roberts, et al., 1970). The effect of saline injections on mice injected with puromycin in the temporal region of the brain one day after training has been confirmed in another laboratory (Rosenbaum, et al., 1968).

It has been suggested that saline and adrenergic compounds might somehow act on existing peptidyl-puromycin complexes, which are now believed to be the basis of some of the behavioral effects of puromycin on memory (Flexner and Flexner, 1968b; Roberts, et al., 1970), but this suggestion remains tentative in the absence of direct evidence that these substances affect the concentration of peptidyl-puromycin in brain.

These results indicate that puromycin may have two different modes of producing amnesic effects. When it is administered before or immediately after training in mice or in goldfish, deficits in retention may result primarily from inhibition of protein synthesis and from abnormal cerebral electrical activity. When puromycin is given to mice later after training, the formation of memory is apparently not affected. Rather, performance is reversibly impaired. In contrast, the amnesic effects of cycloheximide and acetoxycycloheximide are observed only when the drug is given before or shortly after training and are believed to depend specifically on the inhibition of protein synthesis required for the formation of memory. In keeping with this view, saline injections did not influence the amnesia produced by intracerebral injections of acetoxycycloheximide before training (Rosenbaum et al., 1968).

It has recently been reported that when puromycin was neutralized with bases of certain cations (K^+, Li^+, Ca^{++}, Mg^{++}) rather than with Na^+, it no longer had an amnesic effect when given one day after training (Flexner and Flexner, 1969b). It is suggested that these cations may bind to anionic sites of neuronal membranes and thereby protect mice from the disruptive effects of peptidyl-puromycin. Further experiments are required to test this hypothesis adequately. Since the amnesia produced when puromycin is injected before or immediately after training is thought to be due to interference with the formation of memory rather than to the presence of peptidyl-puromycin complexes (Flexner and Flexner, 1968a), it is important to determine whether neutralization with cations at these injection times can antagonize amnesic effects.

Because puromycin has multiple effects on the brain, behavioral studies with this drug appear to shed little light on the role of protein synthesis in memory. These experiments are of interest, however, in another regard. The demonstration that bitemporal injection of puromycin one to three but not seven or more days after training impaired retention (Flexner et al., 1963) indicates that some aspect of stored memory changes between Days 1 and 7 after training. The nature of this change remains obscure. The fact that widespread, multiple injections of puromycin seven days after learning can block expression of memory, whereas bitemporal injections are ineffective at this time, was initially interpreted to indicate that memory changes its locus in the brain after training. One to three days after training, the memory trace was postulated as residing in the temporal region, but by seven days after training it was thought to have spread to other regions and therefore to have become invulnerable to bitemporal injections (Flexner et al., 1963). An alternative that does not require the concept of movement of memory from one brain region to another is that memory is distributed over a large number of sites. For a few days after training, but not longer, the temporal regions of the brain are specifically involved in processing this stored information. Results with anticholinesterases (Deutsch et al., 1966; Deutsch and Leibowitz, 1966; Hamburg, 1967; Wiener and Deutsch, 1968; Squire, 1970; Squire et al., 1971) and anticholinergic drugs (Deutsch and Rocklin, 1967) and with actinomycin D (Squire and Barondes, 1970) provide further support for a change in memory between one and seven days after learning.

The experiments involving delayed injection of puromycin were initially conceived as a way of determining the role of protein synthesis in the maintenance of long-term memory storage. It now appears that the release of peptidyl-puromycin into cells, and not protein synthesis inhibition itself, may be responsible for the behavioral effects of puromycin given long after training. Since these amnesic effects are reversible, these experiments do not address the question of whether continuing protein synthesis is required for the maintenance of long-term memory storage. Unfortunately, little is currently known about this aspect of memory. Acetoxycycloheximide inhibits protein synthesis for many hours (Barondes and Cohen, 1967b), but it does not affect memory when it is given long after training (Barondes and Cohen, 1968a). Since it has not been possible to inhibit cerebral protein synthesis markedly for days without causing death, it has not yet been possible to determine if there is a maintenance process in long-term memory storage which requires replacement of proteins with half-lives as short as several days.

Puromycin and Detention

Goldfish are ordinarily not affected by puromycin given more than 1 hr after training (Agranoff et al., 1965, 1966; Davis et al., 1965). It has been reported, however, that if goldfish are left in the training tank for 1, 24, or 48 hr after training, puromycin injections given after this detention period markedly impair memory (Davis and Agranoff, 1966; Davis, 1968). Puromycin also impairs memory when given 24 hours after training if fish are simply replaced in the training environment for 5 min just prior to injection (Davis and Klinger, 1969). The prolongation of the interval after training during which puromycin can impair memory was originally interpreted to indicate that the formation of permanent memory was somehow delayed by long exposures to the training environment (Davis and Agranoff, 1966). Since retention is also impaired when puromycin follows a short (5 min) exposure to the training environment at a time when permanent memory has presumably already formed, this formulation has been revised. It has now been suggested that arousal produced by brief exposure to the training environment somehow interacts with the effects of the amnesic agent to produce behavior at retest which is incompatible with active avoidance responding (Davis and Klinger, 1969). A similar interpretation has been offered in the case of passive avoidance to account for long gradients of "amnesia" produced by ECS (Chorover and Schiller, 1966; Schneider and Sherman, 1968). The generality of these interactive effects is not yet clear. It would be useful to determine whether they could be demonstrated in a discrimination task in a T-maze where some plausible interactive effects (e.g., on activity) would not be expected to alter performance.

Interference with RNA Metabolism

Since protein synthesis is regulated either by synthesis of messenger RNA or by increased translation of already existent RNA, it is important to investigate the possibility that RNA synthesis might play a role in the establishment of long-term memory. Two drugs have been used to investigate this question: 8-azaguanine, which is incorporated into the RNA molecule thereby producing an abnormally functioning structural analogue of RNA (Lasnitzki et al., 1954), and actinomycin D, which inhibits transcription by binding to guanine nucleotides on the DNA molecule (Reich et al., 1962). Both these drugs, however, have properties that limit their usefulness for behavioral studies.

Dingman and Sporn (1962) gave 8-azaguanine intracisternally to rats either 15 min before or 30 min after the acquisition of a water maze habit. Animals injected before training were impaired in original learning, but injection after training had no effect on retention 15 min later. Because 8-azaguanine impaired acquisition, it is difficult to evaluate the role of RNA synthesis in memory processes. These results may be due to impairment of some other aspect of performance. The same difficulties in interpretation apply to the effects of 8-azaguanine on spinal fixation time, on avoidance conditioning (Chamberlain et al., 1963), and on the learning of a fixed-interval schedule (Jewett et al., 1965).

Actinomycin D differs from 8-azaguanine in that it has no effect on acquisition in many tasks (Barondes and Jarvik, 1964; Cohen and Barondes, 1966; Squire and Barondes, 1970). However, the toxicity produced by doses of actinomycin D large enough to cause marked inhibition of RNA synthesis has also made interpretation of studies with this agent difficult. Mice given 60 μg of actinomycin D intracerebrally—sufficient to inhibit about 95 percent of cerebral RNA synthesis—learned a spatial task normally to a criterion of 9 out of 10 correct responses and remembered normally 1 hr and 4 hr after training (Cohen and Barondes, 1966). Since these doses caused severe illness noticeable 10 hr after injection and produced death within 24 hr, retention could not be measured at longer intervals after training. It was thus not possible to determine if RNA synthesis might be required for long-term memory. Mice trained in a passive avoidance, step-through procedure (Barondes and Jarvik, 1964) after intracerebral injection of 50 μg actinomycin also showed normal retention 1 and 3 hr after training while inhibition was at 80 to 90 percent but, again, the toxicity of the drug prevented tests for retention at longer intervals after training. In another passive avoidance test, rats given 25 μg of actinomycin D intraventricularly exhibited normal retention 24 hr later (Goldsmith, 1967), but the level of inhibition of RNA synthesis achieved during training in this study may have been too low to permit the conclusion that RNA synthesis is not required for long-term memory. With the use of brief training, a procedure that makes learning more susceptible to the effects of acetoxycycloheximide (Barondes and Cohen, 1967b) and a dose of actinomycin D (20 μg) sufficient to inhibit about 75 percent of RNA synthesis during training, it was found that animals learned normally and exhibited normal retention 24 hr later (Barondes and Cohen, 1967b). Again, it is possible that the degree of inhibition achieved was insufficient to impair memory.

Recently, it has been demonstrated that actinomycin D (100–200 μg) disrupted the discrimination of home water from unfamiliar water in salmon, as measured by EEG recorded from the olfactory bulb (Oshima et al., 1969). Electrophysiological records taken during infusion of water from one familiar and two unfamiliar sources were distinct in normal fish, but were disrupted 4–7 hr after administration of actinomycin D. At 9 to 28 hr after injection, the records were again differentiable. Interpretation of these results is complicated by the possibility that actinomycin D in these doses might have produced reversible toxicity which disrupted electrical activity for a few hours. In the absence of behavioral tests, it is not possible to conclude that memory for home water has been lost during this interval. The electrical response of the olfactory bulb may still carry sufficient information to permit behavioral discrimination. Thus these results neither confirm nor deny that RNA synthesis participates in formation of long-term memory.

Nakajima (1969) has reported that five days after very small intracerebral doses (1 μg) of actinomycin D, both acquisition and retention of a spatial task were impaired. Within five days after injection of this small dose, there was evidence of abnormal cerebral electrical activity and extensive chromatolysis in the hippocampus. These effects of actinomycin D, and other evidence of cerebral abnormalities produced by actinomycin D (Appel, 1965; Koenig and Lu, 1967), support the suggestion that behavioral effects produced by this drug may be due to actions other than simple suppression of RNA synthesis. Perhaps both hippocampal cellular damage and the establishment of epileptogenic activity within the hippocampus contribute to behavioral impairment. Studies reporting that 20 μg actinomycin D given 4 hr before training impaired later retention (Meerson et al., 1966) are also not clearly interpretable because of these other effects of actinomycin D.

One might minimize the possibility that toxicity produced by actinomycin D is a major factor in behavioral deficits, if it could be demonstrated that actinomycin D given immediately after training impairs retention a few days later, but that actinomycin D given a few hours after training does not impair retention a few days later. This procedure might clarify the interpretation because, a few days after training, the toxicity resulting from injections a few hours apart should have developed to about the same point. For this reason, the demonstration of a short gradient of retrograde amnesia produced by intracranial injection of 2 μg of actinomycin D in goldfish (Agranoff et al., 1967) is of particular interest. Fish injected immediately after training or within an hour after training were impaired in retention tests three days later. Fish injected 3 hr after training were not affected. It does not seem possible to attribute this gradient to toxicity at the time of retesting produced by the drug, because the toxicity produced by injection 1 or 3 hr after training should be nearly the same 3 days later. However, interpretation of these results is still difficult. Retrograde amnesia gradients of this same time scale can be produced by many kinds of manipulations—e.g., by ECS (Kopp et al., 1966), by brain puncture (Bohdanecka et al., 1967; Dorfman et al., 1969), and by puromycin (Agranoff et al., 1965, 1966; Davis et al., 1965). Thus the fact that such a gradient can be produced by actinomycin D, which is known to have multiple effects, need not necessarily lead to the conclusion that these effects are a result of a depression of cerebral RNA synthesis. They may instead be due to acute effects of intracranial injection of actinomycin D on some other critical brain function.

Recent experiments in which brief training in mice was used (Squire and Barondes, 1970) indicate that actinomycin D given at different times after training can produce effects similar to those produced by puromycin (Flexner et al., 1963). Mice injected bitemporally with 1 to 30 μg of actinomycin D 3 hr before or one day after training in a position discrimination, in which training was conducted to a criterion of only two correct responses, exhibited no retention when tested 27 hr after injection. However, mice injected with 1 μg bifrontally one day after training or with 1 or 30 μg bitemporally seven days after training exhibited significant savings. Abnormal cerebral electrical activity and other neuronal abnormalities are known to be produced by both puromycin (Cohen et al., 1966; Gambetti et al., 1968a, b) and by actinomycin D (Appel, 1965; Koenig and Lu, 1967; Nakajima, 1969), so that these results also neither confirm nor deny that RNA synthesis is necessary for long-term memory storage. These findings are primarily of interest because they support the suggestion that aspects of the memory storage process continue to change for several days after training. Between one and seven days after training, stored information somehow becomes insusceptible to the cellular damage, abnormal electrical activity, and depression of RNA synthesis produced by actinomycin D.

Available evidence has failed to demonstrate conclusively that RNA synthesis is necessary for memory storage. The evidence now indicates that the agents used in these experiments produce multiple effects. This fact has been a major difficulty in obtaining an adequate test of the hypothesis.

Short-Term and Long-Term Memory

Results with inhibitors of protein synthesis suggest that, for a few hours after training, retention is supported by a short-term memory system which does not depend on protein synthesis. Since the protein synthesis apparently required for long-term memory has already progressed substantially within a few minutes after training (Barondes and Cohen, 1968a), it has been suggested that separate long-term and short-term memory may coexist. This arrangement of short-term and long-term memory has been proposed previously (Albert, 1966; McGaugh, 1966). Coexistent short-term and long-term memory systems may either be completely separate or they might depend on each other in a variety of ways.

The amnesia that develops within 6 hr after training acetoxycycloheximide-injected mice has been interpreted to be due to the natural decay of a short-term memory process in mice that were prevented from developing a long-term memory process. The results are also consistent with the possibility that decay of memory during the hours after training represents both the decay of a short-term memory process and also gradually diminished access to a "weakly established" long-term memory process. Some changes of the type which subserve long-term memory might have been established at the time of training by the slight capacity (5 to 10 percent) for protein synthesis which remains after cycloheximide or acetoxycycloheximide injection. Such a weakly established long-term memory process might be particularly vulnerable to interference and could rapidly become inaccessible. Rapid forgetting thus need not depend entirely on decay of a temporary short-term process. It could also involve interference with a weakly established long-term system.

Evidence favoring this interpretation lies in the variability with which memory decays after treatment (Table 2). After the animals are treated with amnesic agents, memory may decay as rapidly as 45 to 180 min after training (Barondes and Cohen, 1966) or as slowly as four to eight days (Agranoff, 1969b). Some of this variability might be attributable to species differences or to task differences, but it might also be due to some extent to differences in the strengths of long-term processes that were formed at the time of training. It thus may prove difficult to attribute the decay of memory after training entirely to the decay of a separate short-term process, because the lifetimes of such processes should not be differentially affected by agents supposedly interfering only with the development of long-term memory. Decay of memory after training may thus reflect both the decay of a separate short-term system and, superimposed on this process, interference with the retrievability of a weakly established long-term system, formed at the time of training. The duration of the hypothetical short-term process is not known. Clearly it might be as long as 3–6 hr, since this time interval has been obtained in several studies in which different procedures and different amnesic agents were used, but it could conceivably be shorter. In any case, some kind of short-term process seems plausible by the present evidence, and is also likely because a period of time would appear to be required before newly synthesized macromolecules could affect interneuronal connections.

TABLE 2. Time of Appearance of Amnesia Following Treatment with Various Amnesic Agents

Onset of Amnesia After Training	Time Agent Given Relative To Training	Species	Amnesic Agent	Reference
45-180 min	5 hr before	Mice	Puromycin	Barondes and Cohen, 1966
3-6 hr	30 min before	Mice	CXM	Barondes and Cohen, 1968b; Cohen and Barondes, 1968a
3-6 hr	5 min, 30 min, or 5 hr before	Mice	AXM	Barondes and Cohen, 1967b, 1968a; Cohen and Barondes, 1968a
3-6 hr	5 min after	Rats	Cathodal polarization	Albert, 1966
6-12 hr	Immediately after	Goldfish	KCl	Davis and Klinger, 1969
10-20 hr*	1 day after	Mice	Puromycin	Flexner et al., 1967
1-3 days†	Immediately after	Goldfish	AXM	Agranoff, 1970
1-3 days	Immediately after	Goldfish	Puromycin	Davis and Agranoff, 1966
4-8 days†	Immediately after	Goldfish	CXM	Agranoff, 1969b

CXM=cycloheximide; AXM=acetoxycyloheximide
Time of appearance of amnesia after treatment.
†*The results with AXM were obtained in one task, and the results with CXM were obtained in a modified version of the same task.*

Recent experiments have indicated that "long-term" amnesia in mice produced by subcutaneous injection of cycloheximide before discrimination training could be antagonized by additional manipulations initiated at a time (3 hr after training) when "short-term" retention could still be exhibited (Barondes and Cohen, 1968b) but not at later times when "short-term" retention was no longer demonstrable. Mice given foot shock, amphetamine, or corticosteroids 3 hr after training exhibited good retention 24 hr later in spite of 90 to 95 percent inhibition of protein synthesis during training established by administration of cycloheximide 30 min before training. When these manipulations were introduced 6 hr after training, they had no effect. The induction of long-term memory by these manipulations depended on the relatively short duration of action of cycloheximide and on the recovery of cerebral protein synthesis within 3 hr after training. It was not obtained when cerebral protein synthesis inhibition was reestablished at this time by injections of acetoxycycloheximide.

These experiments show that establishment of long-term memory may occur long after training of cerebral protein synthesizing capacity has recovered and "short-term" memory is still present. It was suggested that the "arousal-inducing" manipulations might somehow induce conversion of information from a still-intact short-term system to a long-term storage system which could now be mediated by recovered cerebral protein synthesizing capacity (Barondes and Cohen, 1968b). Alternatively, if one prefers a formulation in which "short-term" and "long-term" memory are separate, one may argue that these manipulations may act by somehow strengthening the already existent but "weak" long-term memory process.

Conclusion

The experiments described here indicate clearly that inhibitors of cerebral RNA or protein synthesis have profound effects on memory storage. The major difficulty in interpreting these results is that of distinguishing between effects primarily due to inhibition of cerebral RNA or protein synthesis and effects which might be due to other actions of these drugs. Since learning and memory for several hours after learning are intact despite very marked inhibition of cerebral RNA or protein synthesis during training, it seems safe to conclude that the acquisition process and some form of short-term memory are not dependent on cerebral RNA or protein synthesis. Greater difficulties arise in interpreting the profound effects that these inhibitors have on "long-term" memory storage. In the case of actinomycin D (which produces neuronal necrosis and abnormalities in cerebral electrical activity) and in the case of puromycin (which also alters cerebral electrical activity), conclusions must be made with considerable caution. These drugs are not well suited for use in investigating the role of specific macromolecules underlying memory. Of the agents available, the glutarimide derivatives, cycloheximide and acetoxycycloheximide, appear to be the drugs of choice for these studies.

However, in the light of the recent findings that cycloheximide affects activity and that cycloheximide and acetoxycycloheximide alter the electrical responsiveness of isolated rat ganglion and diminish rat brain acetylcholine; firm conclusions concerning the mechanism of these agents' action on long-term memory cannot be drawn. At the present time, the effects of cycloheximide on activity appear to be unrelated to its amnesic effects, because isocycloheximide, which does not produce amnesia, affects activity similarly (Segal, et al., 1970). The significance of the other possible side effects must await further investigation. Thus the significance of altered neural responsiveness in isolated rat ganglion following incubation with acetoxycycloheximide (Paggi and Toschi, 1970) is not clear, since these effects (in contrast with the immediate effects of puromycin or puromycin aminonucleoside) do not appear until 3 hr of continued incubation with acetoxycycloheximide. This drug is quite effective, however, in impairing long-term memory of a discrimination task if given 5 min before training. It is marginally effective if given 5 min after training (which requires 6–8 min), and completely ineffective if given 30 min after training (Barondes and Cohen, 1968a). Since these injection times are only a few minutes apart, one would expect that by a few hours after training, altered responsiveness of neural tissue should have developed to about the same extent in each case. Since the behavioral effects are strikingly dependent on the time relative to training when the drug is given, it is difficult to understand how these behavioral effects could be attributed to a late-developing electrophysiological abnormality.

We conclude that the hypothesis that the amnesic effect of the glutarimide derivatives is related to inhibition of cerebral protein synthesis specifically required for long-term memory storage has much to support it, but is not rigorously proven. Many arguments have been presented in support of this hypothesis (Barondes, 1970). However, the evidence that these drugs may have other relevant actions on the brain makes conclusions all the more difficult. We must be equally cautious about either ignoring or overemphasizing the role of "side effects," in the absence of further study.

Acknowledgments

This work has been supported by Grants MH 12773 and MH 06418 from the U.S. Public Health Service.

References

Agranoff, B. W. (1969a). In: Protein Metabolism in the Nervous System, Lajtha, A., ed. New York: Plenum Press. ——— (1969b). Presented at Dalhousie Symposium, Dalhousie, Nova Scotia. ——— (1970). In: Molecular Approaches to Memory and Learning, Byrnes, W. L., ed. New York: Academic Press. ——— and Davis, R. E. (1968). In: The Central Nervous System and Fish Behavior, Ingle, D., ed. Chicago: Univ. of Chi. Press. ——— Davis, R. E., and Brink, J. J. (1965). Proc. Nat. Acad. Sci. U.S.A., 54:788. ——— Davis, R. E., and Brink, J. J. (1966). Brain Res., 1:303. ——— Davis, R. E., Casola, L., and Lim, R. (1967). Science, 158: 1600. ——— and Klinger, P. D. (1964). Science, 146:952.

Albert, D. J. (1966). Neuropsychologia, 4:65.

Appel, S. H. (1965). Nature (London), 207:1163.

Appleman, H. M., and Kemp, R. G. (1966). Biochim. Biophys. Res. Commun., 24:564.

Avis, H., and Carlton, P. L. (1968). Science, 161:73.

Barondes, S. H. (1965). Nature (London), 205:18. ——— (1970). Int. Rev. Neurobiol., 12:177. ——— and Cohen, H. D. (1966). Science, 151:594. ——— and Cohen, H. D. (1967a). Brain Res., 4:44. ——— and Cohen, H. D. (1967b). Proc. Nat. Acad. Sci. U.S.A., 58:157. ——— and Cohen H. D. (1968a). Science, 160:556. ——— and Cohen, H. D. (1968b). Proc. Nat. Acad. Sci. U.S.A. 61:923. ——— and Dutton, G. R. (1969). J. Neurobiol., 1:99. ——— and Jarvik, M. E. (1964). J. Neurochem., 11:187.

Bennett, G. S., and Edelman, G. M. (1969). Science, 163:393.

Bohdanecka, M., Bohdanecky, Z., and Jarvik, M. E. (1967). Science, 157:334.

Brink, J. J., Davis, R. E., and Agranoff, B. W. (1966). J. Neurochem., 13: 889.

Brown, B. M., and Noble, E. P. (1967). Brain Res., 6:363. ——— and Noble, E. P. (1968). Biochem. Pharm., 17:2371.

Chamberlain, T. J., Rothschild, G. H., and Gerard, R. W. (1963). Proc. Nat. Acad. Sci. U.S.A., 49:918.

Chorover, S. L., and Schiller, P. H. (1966). J. Comp. Physiol. Psychol., 61: 34.

Cohen, H. D., and Barondes, S. H. (1966). J. Neurochem., 13:207. ——— and Barondes, S. H. (1967). Science, 157:333. ——— and Barondes, S. H. (1968a). Nature (London), 218:271. ——— and Barondes, S. H. (1968b). Commun. Behav. Biol., A1:337. ——— Ervin, F., and Barondes, S. H. (1966). Science, 154:1557.

Colombo, B., Felicetti, L., and Baglioni, C. (1965). Biochim. Biophys. Res. Commun., 18:389.

Davis, R. E. (1968). J. Comp. Physiol. Psychol., 65:72. ——— and Agranoff, B. W. (1966). Proc. Nat. Acad. Sci. U.S.A., 55:555. ——— Bright, P. J., and Agranoff, B. W. (1965). J. Comp. Physiol. Psychol., 60:162 ——— and Klinger, P. D. (1969). Physiology and Behavior, 4:269.

Deutsch, J. A., Hamburg, M. D., and Dahl, H. (1966). Science, 151:221. ——— and Leibowitz, S. F. (1966). Science, 153:1017. ——— and Rocklin, K. (1967). Nature (London), 216:89.

Dingman, W., and Sporn, M. B. (1962). J. Psychiat. Res., 1:1 ——— and Sporn, M. B. (1964). Science, 144:26.

Dorfman, L. J., Bohdanecka, M., Bohdanecky, Z., and Jarvik, M. E. (1969). J. Comp. Physiol. Psychol., 69:329.
Eccles, J. C. (1953). The Neurophysiological Basis of Mind. Oxford: Oxford University Press.
Edström, J. E., and Grampp, W. (1965). J. Neurochem., 12:735.
Ennis, H. L., and Lubin, M. (1964). Science, 146:1474.
Flexner, J. B., and Flexner, L. B. (1967). Proc. Nat. Acad. Sci. U.S.A., 57: 1651. ——— and Flexner, L. B. (1969a). Proc. Nat. Acad. Sci. U.S.A., 62:729. ——— and Flexner, L. B. (1969b). Science, 165:1143. ——— Flexner, L. B., and Stellar, E. (1963). Science, 141:57. ——— Flexner, L. B., Stellar, E., de la Haba, G., and Roberts, R. B. (1962). J. Neurochem., 9:595.
Flexner, L. B., and Flexner, J. B. (1966). Proc. Nat. Acad. Sci. U.S.A., 55: 369. ——— and Flexner, J. B. (1968a). Science, 159:330. ——— and Flexner, J. B. (1968b). Proc. Nat. Acad. Sci. U.S.A., 60:923. ——— Flexner, J. B., de la Haba, G., and Roberts, R. B. (1965). J. Neurochem., 12: 535. ——— Flexner, J. B., and Roberts, R. B. (1966). Proc. Nat. Acad. Sci. U.S.A., 56:730 ——— Flexner, J. B., and Roberts, R. B. (1967). Science, 155:1377. ——— Flexner, J. B., Roberts, R. B., and de la Haba, G. (1964). Proc. Nat. Acad. Sci. U.S.A., 52:1165. ——— Flexner, J. B., and Stellar, E. (1965). Exp. Neurol., 13:264.
Gaito, J. (1961). Psych. Rev., 68:288.
Gambetti, P., Gonatas, N. K., and Flexner, L. B. (1968a). J. Cell Biol., 36: 379. ——— Gonatas, N. K., and Flexner, L. B. (1968a). Science, 161:900.
Geller, A., Robustelli, F., Barondes, S. H., Cohen, H. D., and Jarvik, M. E. (1969). Psychopharmacologia, 14:371. ——— Robustelli, F., and Jarvik, M. E. (1970). Psychopharmacologia, 16:281.
Glassman, E. (1969). Ann. Rev. Biochem., 38:605.
Goldsmith, L. J. (1967). J. Comp. Physiol. Psychol., 63:126.
Hamburg, M. D. (1967). Science, 156:973.
Hebb, D. O. (1949). The Organization of Behavior. New York: Wiley.
Hughes, R, A. (1969). J. Comp. Physiol. Psychol., 68:637.
Jarvik, M. E., and Kopp, R. (1967). Psychol. Rep., 21:221.
Jewett, R. E., Pirch, J. H., and Norton, S. (1965). Nature (London), 207: 277.
Jones, C. T., and Banks, P. (1969). J. Neurochem., 16:825.
Kanfer, J., and Richards, R. L. (1967). J. Neurochem., 14:513.
Koenig, H., and Lu, C. (1967). Trans. Am. Neurol. Ass., 92:250.
Kopp, R., Bohdanecky, Z., and Jarvik, M. E. (1966). Science, 153:1547.
Lasnitzki, I., Matthews, R. E. F., and Smith, J. D. (1954). Nature (London), 173:346.
Mayor, S. J. (1969). Science, 166:1165.
McCaugh, J. (1966). Science, 153:1351.
Meerson, F. Z., Kruglikov, R. I., and Goryacheva, I. A. (1966). Dokl. Akad. Nauk. SSSR, 170:741.
Nakajima, S. (1969). J. Comp. Physiol. Psychol., 67:457.
Nathans, D. (1964). Proc. Nat. Acad. Sci. U.S.A., 51:585.
Oshima, K., Gorbman, A., and Shimada, H. (1969). Science, 165:86.
Paggi, P., and Toschi, G. (1970). J. Neurobiol. (in press).
Pfaff, D. W. (1969). Psychol. Rev., 76:70.
Potts, A., and Bitterman, M. E. (1967). Science, 158:1594.

Reich, E., Goldberg, I. H., and Rabinowitz, M. (1962). Nature (London), 196:743.
Roberts, R., Flexner J. B., and Flexner, L. B. (1970). Proc. Nat. Acad. Sci. U.S.A., 66:310.
Rosenbaum, M., Cohen, H. D., and Barondes, S. H. (1968). Commun. Behav. Biol., A2:47.
Schneider, A. M., and Sherman, W. (1968). Science, 159:219.
Segal, D. S., Barondes, S. H., and Squire, L. R. (1970). Science (in press).
Shashoua, V. E. (1968) Nature (London), 217:238.
Squire, L. R. (1970). Psychon. Sci., 19:49. ——— and Barondes, S. H. (1970). Nature (London), 225:649. ——— Geller, A., and Jarvik, M. E. (1970). Commun. Behav. Biol., A5:249. ——— Glick, S. D. and Goldfarb, J. (1971). J. Comp. Physiol. Psych., 74:41.
Toschi, G., and Giacobini, E. (1965). Life Sci., 4:1831.
Wiener, N., and Deutsch, J. A. (1968). J. Comp. Physiol. Psych., 66:613.
Williamson, A. R., and Schweet, R. (1965). J. Molec. Biol., 11:358.

Chapter 6

Biological Assays for the Molecular Coding of Acquired Information

GEORGES UNGAR

Baylor College of Medicine, Houston, Texas

There have been three main experimental approaches to the problem of molecular mechanisms in the processing of acquired information. The first approach, initiated by Hydén (1959), is the attempt to demonstrate, by means of chemical and physical methods, that acquisition and fixation of information are accompanied by chemical changes in the brain (see reviews by Booth, 1967, 1970). The second approach has shown that interference with the synthesis of RNA and protein can impair the fixation of acquired information (as reviewed by Cohen, 1970). The results of the two methods converge toward the conclusion that information processing is associated with an increased turnover of RNA and proteins in the brain and that this increase is an essential condition of the fixation of information. Some of the results suggest also the possibility of qualitative RNA and protein changes but the precise nature of these changes would be practically impossible to specify with the techniques available at present. However, recent attempts by immunological methods (Janković et al., 1968) and by DNA-RNA hybridization techniques (Machlus and Gaito, 1969) may offer promising leads. Up to now, the only evidence for actual molecular coding of information in the nervous system has been produced by the third approach, which is the object of this chapter. This method consists of detecting the chemical correlates of information processing by biological assay rather than by chemical or physical means. Bioassays, although primarily pharmacological methods, have been widely used in all cases in which a chemical process had to be demonstrated, but the nature of the material was either unknown or, if known, was inaccessible by existing analytical techniques. The approach has always been regarded as a temporary expedient to be superseded as soon as more accurate and more reliable methods become available. In spite of this consideration bioassays played an indispensable and irreplaceable role at the initial and most decisive phases of the study of hormones, vitamins, and neurotransmitters.

It is interesting to note that the use of bioassays and the results obtained by them have often been controversial. The early experiments on hormones at the end of the last century met with considerable skepticism and were the object of ridicule. The idea of chemical transmission at synaptic junction was received with scorn, and the concept took a quarter of a century to gain recognition. New ideas can be delayed by controversy, but they can never be killed by it if they contain a parcel of truth. All really new departures are heresies and the history of science has taught us that today's heresy often becomes tomorrow's dogma.

The Biological Assay Method

Bioassays have been used to test the hypothesis that acquisition of information is accompanied by chemical changes and that these chemical changes represent a record of the information. Such a hypothesis would not have been possible without the knowledge of a chemical coding of genetic information. In its crudest forms, the molecular concept of memory is a somewhat slavish imitation of the molecular theory of heredity. Both are based on the information content of molecular structures but, while the stability of the genome requires the invariant structure of DNA, the plasticity of memory must utilize the more diverse and flexible derivative codes of RNA and protein.

The reasoning behind the bioassay approach is the following: if acquired information is recorded in terms of molecular structure, we should be able to communicate the information to other individuals by treating them with substances containing the corresponding molecule. The reasoning is analogous to that which prompted the use of bioassays in the study of hormonal actions or neurohumoral transmitters.

In practice, the experiments include the following steps:

1. Input of information into donor animals by appropriate training procedures.
2. Extraction of the information-containing material from the donors.
3. Administration of the extract to recipient animals.
4. Testing of the recipients for acquisition of the information originally imparted to the donors.

In the following pages I shall summarize the representative experiments and discuss the principal problems still debated, such as the reliability and specificity of the method.

Experiments in Planarians

The first attempt to prove the chemical coding of learned information by biological assay was made by McConnell in 1962. He trained planarian worms (*Dugesia dorotocephala*) in a Pavlovian conditioning paradigm. The conditioned stimulus (CS) was light; the unconditioned stimulus (US), electric shock. The trained subjects responded with a characteristic contraction to the presentation of the CS. When they were fully trained, their bodies were cut into pieces and fed to naive recipients while a control group was fed fragments of untrained donors. It was then observed that recipients of trained worms had a significantly higher rate of responses to CS than the controls. These results have never been contested and were, in fact, confirmed by several groups of investigators, particularly John (1964) and Kabat (1964), as were extended to other forms of learning by Westerman (1963). In later experiments (Zelman et al., 1963; Jacobson et al., 1966b) the recipients were injected with RNA preparations made from trained worms.

The controversy over these experiments centered around the significance of the phenomenon. One contention was that planarians do not learn and are, therefore, unsuitable for studying any aspect of learning (Bennett and Calvin, 1964; Jensen, 1965). These objections have been answered by the experiments of Jacobson et al. (1967) and Block and McConnell (1967). The problem has been reviewed by McConnell and Shelby (1970) and, on the whole, the results seem to have stood the test of time.

It may have been hoped that simple organisms like planarians would play in neurobiology the role played by *E. coli* in the study of the genetic code. It is probably true that the mechanism of neural information processing could be better understood in simple systems, but the Aplysia, the leech, the cockroach, or the crayfish may represent a more appropriate material than the planarians.

Experiments in Vertebrates

In 1965, within an interval of a few weeks, four publications came out describing experiments in which the bioassay approach was used in mammals (Reinis; Fjerdingstad et al.; Ungar and Oceguera-Navarro; Babich et al.). In spite of strong opposition (Gross and Carey, 1965; Gordon et al., 1966; Luttges et al., 1966; Kimble and Kimble, 1966; Byrne et al., 1966), mainly based on the inability to replicate the results of Babich et al., more and more workers were attracted to the field. By December 1969, there was in the literature more than 80 publications (72 reporting positive and 10 reporting negative results) and about a dozen more were scheduled to appear within the next few months. Table 1 lists the most representative papers and the reader is referred to reviews by Gurowitz (1969), Rosenblatt (1970), and Ungar (1970b, c, d, e).

The experiments vary widely in technical competence, appropriateness of design, adequacy of controls, the number of animals used, and sobriety of interpretation. I should like to summarize here only those experiments that have been proved consistently successful and have been repeated in an adequate number of animals.

FOOD-SEEKING BEHAVIOR ON SIGNAL. In one of the first experiments, Reinis (1965) trained donors to gain access to food on a combined light-sound signal. The correct response was to find the food within 20 sec after appearance of the signal. When brain extracts from the trained donors were injected into naive recipients, the latter gave 63.3 percent correct responses, as compared with 17.1 percent in controls injected with brain extracts from untrained donors. Related experiments were done by Dyal et al. (1967), Dyal and Golub (1968), Golub and McConnell (1968), and Krylov et al. (1969).

HABITUATION. The first design used in this laboratory was habituation to sound, i.e., the suppression of the startle response to a sound stimulus when repeated a large number of times. Recipients of brain extracts

TABLE 1. Principal publications reporting results of biological assays

	Positive results	Negative results
1965	Reinis; Fjerdingstad et al.; Ungar and Oceguera-Navarro; Babich et al.; Nissen et al.	Gross and Carey
1966	Rosenblatt et al.; Jacobson et al.; Albert; Byrne and Samuel	Gordon et al.; Luttges et al.; Kimble and Kimble; Byrne et al.
1967	Essman and Lehrer; Byrne and Hughes; Dyal et al.; Revusky and DeVenuto; Gay and Raphelson; Ungar; Gibby and Crough; Adam and Faiszt	Lambert and Saurat; Hoffman et al.
1968	Røigaard-Petersen et al.; Ungar et al.; Dyal and Golub; Golub and McConnell; Reinis; Faiszt and Adam; Rosenthal and Sparber; Gibby et al.; Chapouthier and Ungerer; Kleban et al.; McConnell et al.; Daliers and Rigaux-Motquin	Bozhko; Corson and Enesco; Lagerspetz et al.
1969	Fjerdingstad; Wolthuis et al.; Krylov et al.; Giurgea et al.; Golub et al.; Moos et al.; Zippel and Domagk	

from habituated animals showed a decrease of startle responses to less than 50 percent at the first trial 24 hr after injection. Control recipients treated with untrained donor brain showed between 90 and 100 percent (Ungar and Oceguera-Navarro, 1965). It was shown later (Ungar, 1967a, b) that the transfer is stimulus-specific because recipients injected with extracts from sound-habituated donors kept their startle responses to airpuff and those treated with airpuff-habituated donor brain showed suppression of startle responses to airpuff but not to sound.

DISCRIMINATION IN TWO-WAY MAZES. In these experiments, initiated by Fjerdingstad et al. (1965), the animal can reach food or water or escape electric shock by running into the arm of the maze signaled by a light or some other distinctive feature. Besides the original authors, Rosenblatt et al. (1966), Ungar (1967a, b), Ungar and Irwin (1967), and Wolthius et al. (1969) also obtained good results with this type of design. Rosenblatt (1970) is now using a more complex system that gives the animal multiple clues for its choice.

CONDITIONED AVOIDANCE. Donors were trained to avoid electric shock by jumping into another compartment on a conditioned stimulus (light, sound) presented for 5 to 10 sec. Good results were obtained with this design by Ungar (1967a, b), Adam and Faiszt (1967), Krylov et al. (1969), and Chapouthier and Ungerer (1969). In all these experiments, however, the recipients were reinforced with the unconditioned stimulus and the difference between experimental and control recipients was evaluated in terms of savings. It was not clear whether the savings were caused by a transfer of information or by facilitation of learning. This could be decided only by experiments in which the recipients were tested without reinforcement. Such experiments were done in this laboratory by Fjerdingstad (1970), who used goldfish in a shuttle box.

PASSIVE AVOIDANCE. In the first experiments of this type, Gay and Raphelson (1967) reversed the dark preference of rats by giving them electric shocks in the dark compartment of a three-chamber system. We have been using this design now for over two years in this laboratory with excellent results (Ungar et al., 1968) which have been confirmed by Golub et al. (1969) and Wolthius (1969). One of the advantages of this method for bioassay purposes is the good dose-response relationship, which has allowed a semiquantitative evaluation of the active material present in the brain of trained animals and helped us to obtain a highly purified preparation.

Another similar paradigm was the avoidance of step-down from a platform (Ungar, 1970b). Good results were obtained but the dose-response relationship was less satisfactory and the evaluation of the results less objective than in dark avoidance.

MISCELLANEOUS DESIGNS. I have deliberately omitted the experiments of Babich et al. (1965), which proved difficult to replicate (see below) and the left-right discrimination paradigms which have been found to raise very complex problems (Ungar, 1967b; Ungar and Irwin, 1967; Rosenblatt, 1970).

There are some promising designs which have, however, not been suffciently explored yet: training for alteration (Wolthuis, 1970; Fjerdingstad, 1969), saccharine avoidance in irradiated animals (Revusky and DeVenuto, 1967; Moos et al., 1969) and the fixation of cerebellar asymmetry (Giurgea et al., 1969). The most important of recent contributions is the work of Zippel and Domagk (1969) on the possibility of transferring color and taste discrimination in the fish. The bioassay approach has also been extended to birds by using simple maze learning in newly hatched chicks (Rosenthal and Sparber, 1968).

Reliability of the Method

Biological assays are notorious for their inaccuracy. They are particularly unreliable at the early stage of research, i.e., before the optimal conditions become clearly defined. Thus, in 1965, as soon as the paper of Babich et al. appeared (the only one of the four original papers to be published in the United States), a series of articles came out reporting failure to replicate the experiments. Most of them were limited to a more or less faithful reproduction of the Babich experiments (Gross and Carey, 1965; Gordon et al., 1966), but others (Kimble and Kimble, 1966; Luttges et al., 1966) introduced other procedures.

Critical Conditions. An analysis of the negative experiments (Ungar, 1970a), compared with the successful assays, led me to consider a number of critical factors.

Training of Donors. The training of the donors should not only aim at the retention of information as judged by a certain performance criterion but it should also leave sufficient time for the synthesis of the information-carrying molecules. The exact duration of optimal training varies with the paradigm but in general it is between six and 12 days. In several of the unsuccessful experiments the training was shorter and in some the criterion of performance was too low. Not all animals are good donors and those that are conspicuously slow to learn should be eliminated (Reinis, 1966). It should be noted also that in some cases unduly prolonged training seemed to decrease the quality of the donors (Ungar et al., 1968). This observation is difficult to explain but it is possible that the burst of synthesis that take place at the early stages of training subsides once the information is consolidated.

Preparation of Extracts. There are certainly many critical steps in the preparation of the extracts, especially when the material to be extracted is unknown. In some of the early studies and in most of the negative experiments, the extraction was aimed at isolating RNA, which may not be the active substance. It is preferable to start out with crude extracts and purify them stepwise, guided by bioassay. It is also advisable to assume that the active substance is unstable and should be protected against enzymic or spontaneous degradation that may take place at temperatures above 5°C.

DOSE. In all the positive experiments published (with the exception of Rosenblatt's (1970) study), only large doses of brain gave good results. In this laboratory, we use rats as donors and mice as recipients so that we can give these doses equivalent to three or four times the weight of their own brain. We seldom had good results with less than two brain equivalents. In my survey (Ungar, 1970a), out of the 236 experimental recipients listed in the negative experiments, I found that only 40 (17 percent) received more than one brain equivalent.

TEST SCHEDULE. The schedule of testing, particularly the interval between administration of the extract and the first trial, is an extremely critical factor. In almost all the successful experiments, the effect was not detectable in less than 24 hr and often only 48 to 72 hr afterwards. In most of the negative experiments, testing was done early—between 4 and 24 hr and usually only once.

RECIPIENT SELECTION. Selection of the recipients is an important factor in success. One should ascertain that the organisms are able to learn by training the task to be transferred to them. They should not be too young or too old and should exhibit certain traits essential to the demonstration of the behavior to be transferred. For example, transfer of dark-avoidance can be demonstrated only in animals that, prior to injection of the extract, had a definite preference for the dark.

BLIND CONDITIONS. All testing should be done under blind conditions. The results, especially if they are not automatically recorded, may be influenced by the bias of the experimenter. This may, of course, work both for negative and for positive results.

TRIAL AND ERROR. As a final and most important rule, the optimal conditions must be found out by trial and error, varying the conditions until the best results are found. One has to keep in mind that these are difficult experiments that require careful planning and reasonable competence in both the behavioral and chemical techniques involved. The question such work is trying to answer is too important to be dealt with in a few hasty trials with leftover animals, and with the work being done by unsupervised beginners.

Evaluation of the Results

Like all bioassay methods, estimation of the information-carrying substances requires statistical analysis for validation of the results. The reliability of a bioassay is inversely related to the complexity of its test object: assays done with isolated tissues can reach a reasonable accuracy but those done in whole animals depend on too many variables to give more than semiquantitative results. The complexity becomes even more troublesome when the assay is based not on physiological or morphological changes but on behavioral patterns. Furthermore, in the present case, the reliability depends not only on the test object—the recipient animal—but also on the variability of the donor that supplies the material to be assayed, the preparation of the extract, and the various other factors just mentioned.

Under these conditions, it is not surprising that not all individual experiments are successful and that the validity of the method has to be assessed by statistical treatment of pooled results. Table 2 summarizes the evaluation of the results obtained in my laboratory during the first period of our work (1964 to 1967). For each experimental design, the table lists the recipients that behaved as if they had received from the donors information relevant to the test to which they were submitted (+) and those that showed no change (0) or exhibited a change opposite to the one expected (−). The list includes recipients of trained extracts (TRB) and untrained extracts (NRB) from all the experiments whether they were unsuccessful or failed. In each case, I computed by the chi χ^2 method the

TABLE 2. Behavioral changes in animals injected (1964-1967) with extract of brain taken from trained (TRB) or untrained (NRB) donor rats

	No. of Recipients				*P*†
	TRB		NRB*		
	+	0 or −	+	0 or −	
Habituation	41	1	0	22	< .001
Conditioned avoidance					
Reinforced	10	5	2	11	< .01
Escape to light					
Reinforced	15	9	5	15	< 02
Unreinforced	80	12	2	38	< .001
Left or right escape	110	67	2	28	< .001
Audiovisual discrimination	30	9	2	14	< .001
TOTALS	286	103	13	128	< .001
%	73.5	26.5	9.2	90.8	
N	389		141		

+ = *Change in the direction expected from the training of donors.*
0 = No change; − = change in opposite direction.
**In control animals, the direction of the change was counted as in the corresponding experimental animals.*
†χ^2 *method.*

probability of the two groups belonging to the same population. It is seen that the difference is significant for each group of experiments and very highly significant for the pooled results of all groups.

During the last two years, 1967 to 1969, we have concentrated on the dark-avoidance paradigm (see above and Ungar et al., 1968). To the end of 1969, 149 brain extracts were prepared, 115 from trained donors and 34 from untrained animals. Each extract was made from a pool of 20 to 100 rat brains and was tested in 6 to 18 recipient mice. Figure 1 shows the mean time (in percent) spent by the recipients in the dark box before injection, in the course of screening (S), and after injection in the experimental (E) and control groups (C). The difference between the groups

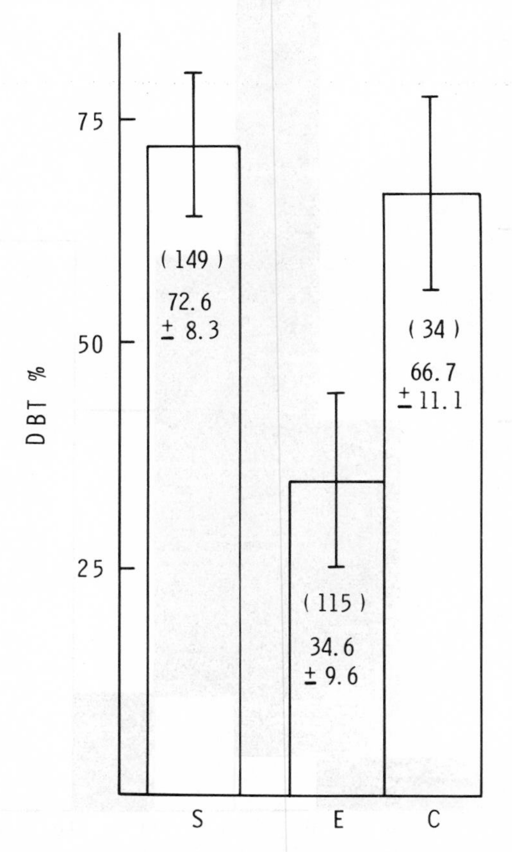

FIG. 1. **Effect of brain extracts from dark avoidance trained rats (E) and untrained control rats (C) on the item spent in the dark (DBT) by recipient mice. S = screening values, before injection of the extracts. Ordinate = percent time in dark box out of a total of 180 sec ± S.D. Number of extracts between parentheses; each extract was tested on 6 to 12 mice.**

by the t test is highly significant and the same reliability is indicated by the χ^2 test after dividing the results among groups that spent less than 50 percent of the time in the dark and those that spent 50 percent or more.

Figure 2 shows the same data in frequency distribution. It indicates two distinct populations: one, the recipients of trained extracts, with a peak about 60 sec (33 percent) and the other, representing the animals treated with untrained extracts, about 140 sec (78 percent). There can, therefore, be hardly any doubt that the treatment of the recipients with

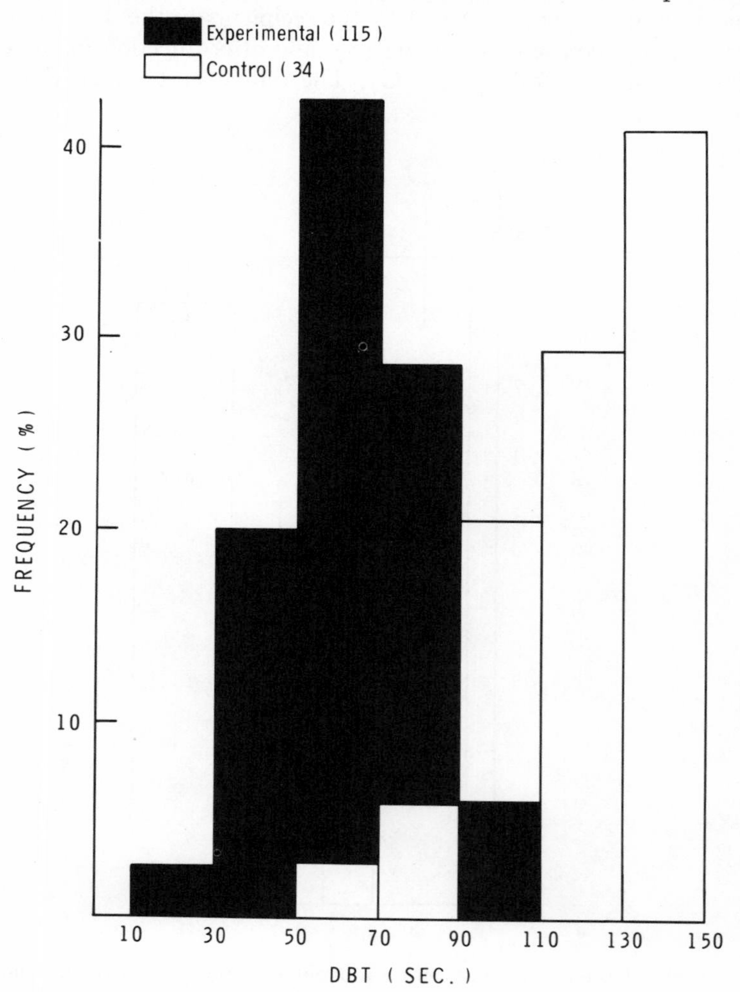

FIG. 2. Distribution of activity in 149 rat brain extracts. Abscissa = time in dark box (DBT); ordinate = frequency in percent among groups of 6 to 12 mice injected with 115 extracts from trained rats (black columns) and 34 extracts from untrained rats (white columns).

extracts of brains taken from trained rats induced a significant behavioral change.

Specificity of the Method

The problem that remains to be examined is whether the substances we are assaying represent the code for acquired information or are merely stimulating certain types of learning. This is a very important question to which we do not yet have a complete answer.

Several experiments indicate an intermodal stimulus specificity. It was mentioned above that material from sound-habituated donors induces in the recipients only habituation to sound and not to airpuff, the viceversa (Ungar, 1967a). I have also tested the specificity of the passive avoidance assay and found that extracts from dark-avoiding donors have no effect on the avoidance of stepdown, and viceversa (Ungar, 1970b).

At present, the only experiments that suggest specificity within the same sensory modalities are those of Zippel and Domagl (1969), who trained goldfish in color and taste discrimination paradigms and found that the recipients followed the preferences learned by the donors.

Chemistry of the Information-Carrying Material

From the outset, the essential purpose of the work done in this laboratory has been the isolation and identification of the active material extracted from the brain of the donors and the bioassay method has been only the means to this end.

On the ground of Hydén's and McConnell's work, it was assumed in most of the laboratories that the active substances were RNA sequences. Some workers made no assumptions about the nature of the material (Reinis, Byrne and Samuel, Dyal et al., Chapouthier, Wolthuis) and four groups found that it had the chemical properties of peptides (Ungar, Rosenblatt, Giurgea, Domagk).

In view of the effects observed with both RNA and peptide preparations, we tried to solve the problem by making extracts from the same pool of rat brain trained for dark-avoidance by using the peptide extraction technique (Ungar, 1970a) and by the RNA method described by Røigaard-Petersen et al. (1968). Both preparations were active after intraperitoneal injection, as shown in Table 3. They were then submitted to the action of trypsin, which was known to destroy the material isolated by the peptide extraction procedure, and it was found that the RNA extract was also inactivated by this same enzyme while it was left intact by RNase. Inactivation by proteases was observed also with two other partially purified active substances: the material that induces habituation to sound is de-

TABLE 3. Comparison between "RNA" and "peptide" preparations treated with enzymes

Enzyme	DBT (sec) ± S.D. "RNA"	"Peptide"
None	67 ± 31	76 ± 26
Trypsin	109 ± 29	112 ± 28
Chymotrypsin	68 ± 34	58 ± 22
RNAse	72 ± 30	65 ± 26

DBT = time spent in dark box; mean values ± standard deviation obtained from 12 animals tested 24 and 48 hr after intraperitoneal injection.

stroyed by chymotrypsin and the one that causes avoidance of step-down in inactivated by both trypsin and chymotrypsin.

In view of these findings, it seemed probable that the active material present in the "RNA" preparations was a peptide more or less loosely attached to RNA (Ungar and Fjerdingstad, 1970). This hypothesis was confirmed when at low pH (<4.0) the active substance was dialyzed out, leaving behind the RNA, which showed no behavior-inducing effect. This material reacted on gel filtration and thin-layer chromatography like the peptide isolated by the original procedure, and it contained the same amino acids. It should be noted that untrained brain submitted to the same extraction procedure lacks the chromatographic spot which contains the active material.

In view of the findings of Faiszt and Adam (1968), it is most probable that the peptide forms a complex with ribosomal RNA. It is not known at present whether the complex exists as such in the cell or is formed during the extraction procedure.

The only substance that has been definitely identified is the dark-avoidance-inducing factor. It has been obtained in a highly purified state and showed significant activity at the dose of 50 to 100 ng per mouse. It is a pentadecapeptide containing four residues of glutamic acid or glutamine, three of aspartic acid or asparagine, three of glycine, two of serine, and one of alanine, lysine and tyrosine. The N-terminal group is serine, and trypsin splits the peptide at a lysine-serine bond.

It is hoped that isolation of the first of the "code words" will be followed by identification of several others in this laboratory and in others. This will allow us to gain some insight into the "rules" under which the code operates: its vocabulary, grammar, and syntax.

Tentative Interpretation

There is probably a large measure of agreement that the processing of acquired information in the brain is associated with some sort of a chemical change. The controversy chiefly concerns the interpretation of this chemical change. Of the three approaches to the problem (mentioned in the introduction to this chapter), the first two merely indicate that retention of information is accompanied by increasing RNA and protein synthesis and that it may be impaired if the increase is prevented. The results of the third approach, the bioassay method, go a step further; they suggest that the chemical changes represent an actual coding of the information.

Such a molecular coding can be interpreted in the framework of the two opposing theories on higher nervous function: the field or mass hypothesis and the connectionist view. The field hypothesis is based essentially on the finding of Lashley, although its origins go back to the debates of the last century over cerebral localization. It assumes that the brain is "equipotential" and, if it has any organization at all, it is not relevant to higher nervous function. As Hebb (1949) put it, the field theory attributes to the brain as much differentiation as can be found in a "bowlful of porridge." Those who subscribe to the field theory see the code substances as records of individual experience, "tape-recorder molecules" (McConnell, 1965), more or less floating in the porridge and storing information in a random filing system. Many variants of this hypothesis have been published and for details the reader is referred to reviews by Ungar (1970d, e).

It is important to emphasize that the hypothesis of molecular coding is in no way linked to the field theory. On the contrary, it fits in much better with the connectionist idea. The hypotheses of Szilard (1964), Rosenblatt (1967), Ungar (1968), and Best (1968) are all based on highly organized neural connection that are controlled by molecular "switches." In my interpretation, the coding system operating the switches derives from the chemical recognition process by which the nervous system becomes organized during embryonic life. This process is based on the chemospecificity of neural pathways proposed by Sperry (1963) to explain the mechanism by which neurons belonging to the same pathway identify each other and establish synaptic connections. This may be the refinement of an even more basic process by which cells recognize other cells of the same type. Moscona and Moscona (1965) have shown that this is due to a specific coating material, probably a glycoprotein. The synthesis of this material is genetically determined since administration of puromycin inhibits recognition and selective aggregation of cells.

Similarly, the neural coding system is probably genetically determined and controls the inborn stimulus-response patterns. The problem is the way in which this same molecular code can operate in the processing of acquired information. Learning and other forms of information processing can be conceived as the formation of new connections between existing pathways. The main mechanism by which such new connections can be created is the concomitant firing of adjacent neurons. Neural activity is associated with changes in the synaptic areas that would favor the passage of material from the presynaptic to the postsynaptic neuron. If, for example, the molecular Label *a* of Pathway A combines with Label *b* of Pathway B, the complex *ab* can be the code name for the newly formed synapse. In a given task, such a process takes place probably at thousands of disseminated sites wherever the two pathways or cell assemblies (Hebb, 1949) are in close enough proximity. During training, the synthesis of *ab* may be increased tenfold or 100-fold so that the material can be extracted from the brain in a large excess. When it is injected into the recipients, the molecular complex attaches itself to the homologous neurons and reconstitutes the synaptic connections symbolized by it. The probability of reproducing in the recipients the behaviour of the donors is related to the number of sites reached by the coded molecule.

For details of the hypothesis, the reader is referred to several papers (Ungar, 1968, 1970c, d, e; Best, 1968). Future progress in this area will depend on the identification of a large enough number of coded molecules to permit a study of the code: the rules which determine the formation of the code words and the mechanism by which information is retained, stored, and retrieved. By identifying the first of the code names, we hope to stimulate other workers to continue and extend the exploration of this important and rewarding area of knowledge.

Acknowledgment

Experimental work from the author's laboratory referred to in this chapter supported by U.S. Public Health Service Grant MH 13361.

References

Adam, G., and Faiszt, J. (1967). Nature (London), 216:198.
Albert, D. J. (1966). Neuropsychologia, 4:49.
Babich, F. R., Jacobson, A. L., and Bubash, S. (1965). Proc. Nat. Acad. Sci. U.S.A., 54:1229.
Bennett, E. L., and Calvin, M. (1964). Neurosci. Res. Prog. Bull., July–August.
Best, R. M. (1968). Psychol. Rep., 22:107.
Block, R. A., and McConnell, J. V. (1967). Nature (London), 215:1465.
Booth, D. A. (1967). Psychol. Bull., 68:149. ——— (1970). In: Molecular Mechanisms in Memory and Learning, Ungar, G., ed. New York: Plenum Press.
Bozhko, G. Kh. (1968). Pavlov J. Higher Nervous Activity, 18:1085.
Byrne, W. L., and Hughes, A. (1967). Fed. Proc., 26:676. ——— and Samuel, D. (1966). Science, 154:418. ——— Samuel, D., Bennett, E. L., Rosenzweig, M. R., Wasserman, E., Wagner, A. R., Gardner, R., Galambos, R., Berger, B. D., Margules, D. L., Fenichel, R. L., Stein, L., Corson,

J. A., Enesco, H. E., Chorover, S. L., Holt, C. E., III, Schiller, P. H., Chiapetta, L., Jarvik, M. E., Leaf, R. C., Dutcher, J. D., Horowitz, Z. P., and Carlson, P. L. (1966). Science, 153:658.
Chapouthier, G., and Ungerer, A. (1968). C. R. Acad. Sci. (D) (Paris), 267:769. ——— and Ungerer, A. (1969). Rev. Comportement Animal, 3:64.
Cohen, H. D. (1970). In Molecular Mechanisms in Memory and Learning, Ungar, G., ed. New York: Plenum Press.
Corson, J. A., and Enesco, H. E. (1968). J. Biol. Psychol., 10:10.
Daliers, J., and Rigaux-Motquin, M. L. (1968). Arch. Int. Pharmacodyn., 176:461.
Dyal, J. A., and Golub, A. M. (1968). Psychon. Sci., 11:13. ——— Golub, A. M., and Marrone, R. L. (1967). Nature (London), 214:720.
Essman, W. B., and Lehrer, G. M. (1967). Fed. Proc., 26:263.
Faiszt, J., and Adam, G. (1968). Nature (London), 220:367.
Fjerdingstad, E. J. (1969). Scand. J. Psychol., 10:220. ——— (1970). In: Chemical Transfer of Learned Information, Fjerdingstad, E. J., ed. North Holland Publishing Company (in preparation). ——— Nissen, Th., and Roigaard-Peterson, H. H. (1965). Scand. J. Psychol., 6:1.
Gay, R., and Raphelson, A. (1967). Psychon. Sci,. 8:369.
Gibby, R. G., and Crough, D. G. (1967). Psychon. Sci., 9:413. ——— Crough, D. G., and Thios, S. J. (1968). Psychon. Sci., 12:295.
Giurgea, C., Daliers, J., and Mouravieff, F. (1969). Abstracts, 4th Internat. Cong. Pharmacol. (Basel, Switzerland, July 14–18, 1969). Basel: Schwabe & Co.
Golub., A. M., and McConnell, J. V. (1968): Psychon. Sci., 11:1. ——— Epstein, L., and McConnell, J. V. (1969). J. Biol. Psychol., 11:44.
Gordon, M. W., Deanin, G. G., Leonhardt, H. L., and Gwynn, R. H. (1966). Amer. J. Psychiat., 122:1174.
Gross, C. G., and Carey, F. M. (1965). Science, 150:1749.
Gurowitz, E. M. (1969). The Molecular Basis of Memory. Englewood Cliffs, N.J.: Prentice-Hall.
Hebb, D. O. (1949). The Organization of Behavior. New York: Wiley.
Hoffman, R. F., Steward, C. N., and Bhogavan, H. N. (1967). Psychon. Sci., 9:151.
Hydén, H. (1959). 4th Internat. Cong. Biochem. (Vienna). New York: Pergamon Press.
Jacobson, A. L., Babich, F. R., Bubash, S., and Goren, C. (1966a). Psychon. Sci., 4:3. ——— Fried, C., and Horowitz, S. D. (1966b). Nature (London), 209:599. ——— Fried, C., and Horowitz, S. D. (1966c). Nature (London), 209:601. ——— Fried, C., and Horowitz, S. D. (1967). J. Comp. Physiol. Psychol., 64:73.
Janković, B., Rakic, L., Veskov, R., and Horvat, J. (1968). Nature (London), 218:270.
Jensen, D. (1965). Animal Behav., 13:9.
John, E. R. (1964). In: Brain Function, Brazier, M. A., ed., Vol. 2. Berkeley, California: Univ. of California Press.
Kabat, L. (1964). Worm Runner's Digest, 6:23.
Kimble, R. J., and Kimble, D. P. (1966). Worm Runner's Digest, 8:32.
Kleban, M. H., Altschuler, H., Lawton, M. P., Parris, J. L., and Lorde, C. A. (1968). Psychol. Rep., 23:51.
Krylov, O. A., Kalyuzhnaya, P. I., and Tongur, V. S. (1969). Pavlov J. Higher Nervous Activity, 19:286.

Lagerspetz, K. M. J., Raitis, P., Tirri, R., and Lagerspetz, K. Y. H. (1968). Scand. J. Psychol., 9:225.

Lambert, R., and Saurat, M. (1967). Bull. C.E.R.P., 16:435.

Luttges, M., Johnson, T., Buck, C., Holland, J., and McCaugh, J. (1966). Science, 151:834.

Machlus, B., and Gaito, J. (1969). Nature (London), 222:573.

McConnell, J. V. (1962). J. Neuropsychiat., 3(Suppl. 1):s42. ——— (1965). Worm Runner's Digest, 7:3. ——— and Shelby, J. M. (1970). In: Molecular Mechanisms in Memory and Learning, Ungar, G., ed. New York: Plenum Press. ——— Shigehisa, T., and Salive, H. (1968). J. Biol. Psychol., 10:32.

Moscona, M. H., and Moscona, A. A. (1963). Science, 142:1070.

Moos, W. S., Le Van, H., Mason, B. T., Mason, C. S., and Hebron, D. L. (1969). Experientia, 25:1215.

Nissen, Th., Røigaard-Petersen, H. H., and Fjerdingstad, E. J. (1965). Scand. J. Psychol., 6:265.

Reinis, S. (1965). Activ. Nerv. Super., 7:167. ——— (1966). Worm Runner's Digest, 8:7. ——— (1968). Nature (London), 220:177.

Revusky, S. H., and DeVenuto, F. (1967). J. Biol. Psychol., 9:18.

Røigaard-Petersen, H. H., Nissen, Th., and Fjerdingstad, E. J. (1968). Scand. J. Psychol., 9:1.

Rosenblatt, F. (1967). In: Computer and Information Sciences, Vol. II, Tou, J., ed. Washington, D.C.: Spartan Books. ——— (1970). In: Molecular Mechanisms in Memory and Learning, Ungar, G., ed. New York: Plenum Press. ——— Farrow, J. T., and Herblin, W. F. (1966). Nature (London), 209:46.

Rosenthal, E., and Sparber, S. B. (1968). Pharmacologist, 10:168.

Sperry, R. W. (1963). Proc. Nat. Acad. Sci. U.S.A., 50:703.

Szilard, L. (1964). Proc. Nat. Acad. Sci. U.S.A., 51:1092.

Ungar, G. (1967a). In: Proc. 5th Internat. Cong. C.I.N.P. (Washington, March, 1966). Amsterdam: Excerpta Medica. ——— (1967b). J. Biol. Psychol., 9:12. ——— (1968). Perspectives Biol. Med., 11:217. ——— (1971a). In: Methods in Pharmacology, Schwartz, A. ed. New York: Appleton-Century-Crofts. ——— (1970b). In: Symposium on Protein Metabolism in the Nervous System, Lajtha, A., ed. New York: Plenum Press. ——— (1970c). In: Handbook of Neurochemistry, Lajtha, A., ed. New York: Plenum Press (in press). ——— (1970d). In: Molecular Mechanisms in Memory and Learning, Ungar, G., ed. New York: Plenum Press. ——— (1970e). Int. Rev. Neurobiol., 13:223. ——— and Fjerdingstad, E. J. (1970). In: Symposium on Biology of Memory (Tihany, Hungary, 1969). Hungarian Academy of Science (in press). ——— and Irwin, L. N. (1967). Nature (London), 214:453. ——— and Oceguera-Navarro, C. (1965). Nature (London), 207:301. ——— Galvan, L., and Clark, R. H. (1968). Nature (London), 217:1259.

Westerman, R. A. (1963). Science, 140:676.

Wolthuis, O. (1969). Arch. Int. Pharmacodyn. 182:439. ——— Anthoni, J., and Stevens, W. (1969). Acta Physiol. Pharmacol., 15:93.

Zelman, A., Kabat, L., Jacobson, R., and McConnell, J. V. (1963). Worm Runner's Digest, 5:14.

Zippel, H. P., and Domagk, G. F. (1969). Experentia, 25:938.

Chapter 7

Biological Activity of Antibrain Antibody—An Introduction to Immunoneurology

BRANISLAV D. JANKOVIĆ

Microbiological Institute, University of Belgrade; Department of Experimental Immunology, Institute of Immunology and Virology, and Immunology Unit, Institute for Biological Research, Belgrade, Yugoslavia

The term "antibrain antibody" used in the title designates any antibody which reacts with brain tissue antigen regardless of the properties of the antibody's combining groups. Antibrain antibody also refers to different experimental situations in which the whole brain, brain regions, populations of neurons, or neuronal components are employed as antigen to prepare antibrain serum. Since antibrain antibody cannot be precisely defined in terms of its immunochemical specificity, its use in this paper can be justified only by convention.

Immunoneurology, like any other interdisciplinary biological science in its infancy, is primarily concerned with the development of new experimental and theoretical models in order to prepare the ground and provide more elements for scientific activity and creative imagination rather than to seek immediate results which have the exactitude of chemical and physical findings. Being a biological science, immunoneurology is faced with the materialistic fact that in nature everything is chemistry and physics, but it is also confronted with the subtle and essential truth that a number of macromolecules, particularly proteins, and associations of these molecules have a unique quality: *the quality of life*. The high instability of proteins and the extremely intricate conditions needed for their existence make very difficult and complex any research directed toward structural and functional characterization of these macromolecules, as well as other macromolecules involved in the operational composition of a living cell. Nevertheless, this specific position of biological sciences, including immunoneurology, by no means grants the privilege of, or excuse for ignorance of any kind, since in natural sciences "ignorantia non est argumentum" (Baruch Spinoza).

From the immunologic point of view, immunoneurology may be regarded as a new chapter in the immunology of biologically active molecules. On the other hand, for neurosciences immunoneurology may be a powerful and promising extension of the interdisciplinary approach to the study of biological galaxies organized and functioning in the form of the neuron and assemblages of neurons.

Introductory Comments

The history of immunoneurology, to my knowledge, began in 1900 when Delezenne used a "neurotoxic" serum in experiments in vivo. In order to obtain antibrain sera, he immunized ducks with different parts of the dog brain. Antisera thus prepared were used for injection (0.5 to 0.6 ml) into the frontal lobes of the dog's brain. Most of the treated animals became paralyżed, and some showed epileptic salivation and clonic-tonic convulsions. On the other hand, control dogs given intrafrontal injection of normal duck serum did not exhibit any apparent behavioral abnormalities. In 1906 Armand-Delille described histologic changes in the brain of dogs that died following intracerebral injections of antidog brain serum produced in guinea pigs. It should be mentioned here that the experiments of Besredka (1919), who used the intracerebral route in studying the nervous origin of anaphylactic symptoms, belong to a more comprehensive review of immunoneurological attempts made at the beginning of this century. Unfortunately, in spite of the brilliant originality of the experimental approach of these French scientists, the level of immunoscience at that time did not allow a full understanding of the results.

In the 40 years that followed, interest in antibrain antibody declined until the work on experimental allergic encephalomyelitis, in which antibrain antibody was used to transfer disease passively (Kabat et al., 1948), once more attracted the attention of investigators to antibrain antibody. Although Kabat and his associates failed to produce characteristic lesions in the brain and spinal cord by means of antibody, they did not completely abandon the possibility that antibrain antibody was involved in the pathogenesis of allergic encephalomyelitis. In a series of experiments, Hurst (1955a,b) studied the spectrum of pathological changes in the brain following intra-arterial and intracisternal injections of goat antimonkey brain sera and antimonkey spleen sera. He reported meningitic and parenchymal lesions which exhibited a morphology similar to an Arthus reaction. In almost all experiments dealing with the local effect of antibrain antibody following subarachnoid, intracisternal or intracerebral injection, the histopathological changes of nervous tissue were the focus of interest, whereas neurological and behavioral descriptions were of an incidental nature (Waksman, 1961).

Unlike these and later investigators, our primary concern was not to gather information related to the pathogenesis of immunologic disorders of the central nervous system but rather to approach a fundamental problem—the in-vivo activity which an antibrain antibody may exert on an antigen situated in the intact nerve cell or populations of nerve cells.

As a matter of fact, 13 years ago I was very much intrigued by the relationship between the antigenic structure and functional complexity of the tissue cells, and in searching for a suitable experimental model which would enable the study of biological activity of anti-tissue antibodies, I discussed with Dr. M. Draškoci, a pharmacologist, the problem of immunologic, physiologic, and pharmacologic events which may have their origin in the in-vivo and in-vitro (tissue culture) collision between antitissue antibody and corresponding antigen. Several disadvantages were mentioned with respect to the injection of antitissue antibody into the circulation, *inter alia,* the considerable dilution of antibody in the blood stream, nonspecific consuming of antibody by other tissues due to cross-reactivity, the problem of crossing the blood-tissue barrier, and the relatively long distance that antibody has to travel to reach the antigenic sites in the tissue. On the other hand, the existence of the brain cavity, technical advances in neurophysiology and behavioral sciences, refined methods used in recording the activity of the central nervous system, and the specificity of immune systems were the starting premises which led us to construct immunoneurological models that could be used in studying the biological activity of antibrain antibody. We considered immunologic, neurophysiologic, behavioral, and other techniques as necessary instruments in serving the ultimate goal: the understanding of neural correlates of behavior.

We believed that the introduction of immunologic techniques into neurosciences would yield completely new information about the connection between neurons and between brain regions and that this kind of research would provide further details of the structure and function of the central nervous system. However, as usually happens in biological research, what appeared to be a simple and clear experimental undertaking soon proved to be much more complicated.

Some Methodologic and Physiologic Correlates of Immunoneurology

When mammalian species are used, several different experimental findings—as well as morphologic, physiologic, and technologic circumstances—contribute to immunoneurological modelling. Some of them are considered below.

Immunologic Specificity

The extreme specificity of immune reactions is a great advantage in studying the structural and functional correlates of the neuron. The anti-brain antibody seems to provide a suitable tool for this kind of research. Although relatively small portions of the antigen molecule are involved in an antigen-antibody reaction, the in vivo contact between antigen and antibody is presumably capable of initiating a series of structural and biochemical changes related to the antigen and its environment, and that may have important repercussions on the function of the neuron to which the antigen belongs. The immunologic specificity of brain antigens will be discussed later in more detail.

Intraventricular Injection

The cellular structure of the "blood-brain barrier," which involves at least the connective tissue that ensheaths even the smallest blood vessels and separates them from the nervous tissue (Nakajima et al., 1965) and the glial component, particularly astroglia (Bariati, 1958), prevents antibodies circulating in the blood from reaching the antigenic sites in the brain. This barrier can be bypassed by administering antibodies directly into the brain cavity.

It has been stated that intraventricularly injected radioglobulins are distributed in the brain and other organs "by normal processes" and that the needle puncture does not significantly affect the pattern of distribution of globulins (Day et al., 1967). This may be correct from the immunologic point of view, depending on the aim of experimental design, but this way of injecting antibodies nevertheless causes local damage to the nervous tissue and breaks the brain-blood barrier, since even microscopic injuries to blood vessels in the brain cause a leakage of antibodies from the site of the injection. In our experience, the application of even a minute amount of serum (0.002 ml) to the brain tissue by means of a microinfusion technique produces injuries to blood vessels and small hemorrhages at the site of the infusion. All these undesirable consequences of direct injection of proteinic material into brain tissue can be avoided by the insertion of a permanent cannula into the lateral ventricle of the brain. The operation itself, of course, causes damage to the brain tissue and elicits a sterile inflammatory reaction around the cannula, but this is soon followed by the proliferation of connective tissue and glia and by the healing of the brain wound (Clemente, 1955). If properly treated, an animal with a cannula permanently implanted in the lateral ventricle of the brain according to the method of Feldberg and Sherwood (1953) does not exhibit any apparent behavioral and other abnormalities for months. There is thus no serious reason to regard the administration of antibrain antibody into the cerebral cavity through a permanent cannula more artificial than injections given intravenously or intraperitoneally.

The delicately balanced internal milieu of the brain is probably disturbed by intraventricular injection of antibody or any other protein. The injected material may slightly increase the cerebrospinal fluid pressure and this may help the antibody penetrate from the cerebrospinal fluid into the brain tissue (Lee and Olszewski, 1960). Pertinent to this is the fact that small variations in the cerebrospinal fluid pressure are physiologic and may be recorded under normal conditions (Davson, 1967). In our experience, the injection of a volume larger than 0.3 ml into the lateral ventricle of the cat, monkey, and rabbit induces in some instances a prompt change in the animal's behavior, characterized by passivity and complete loss of interest in the environment. However, these animals tolerate 0.2 ml injections of saline very well, without displaying apparent alterations in behavior and bioelectrical pattern.

Contact between Antigen and Antibody

The injection of antibrain antibody into the cerebral cavity allows more or less direct contact between antibody and antigen, but does not ensure an ideal experimental situation in which immunologically defined determinant groups of antigen and combining sites of the antibody are brought into juxtaposition.

The small volume of the cerebral cavity, when compared with the volume of the blood vessel system, certainly contributes to a higher concentration of antibody in situ. The surface of cerebral cavities and communications among them enable antibrain antibodies to gain access to antigens situated in topographically distant regions. The slow rate of clearance of antibody from the cerebrospinal fluid also enhances the reaction between antigen and antibody.

The direction and speed of cerebrospinal fluid circulation may influence the localization of injected antibrain antibody (Lee and Olszewski, 1960). Thus the slow flow of cerebrospinal fluid in a given portion of the brain helps to maintain a higher concentration of antibody molecules in that sector.

Complement

Cerebrospinal fluid contains the complement, which is an important component of antigen-antibody reaction occurring in-vivo. It seems that the complement itself is capable of assisting vascular permeability and some other features related to the inflammatory reaction (Cinader and Lepow, 1967). The involvement of complement in an immunologically induced injury to the cell membrane under tissue culture conditions results in a leakage of ribonucleotides and protein (Green et al., 1959). Consequently, the presence of complement in cerebrospinal fluid is one of the most valuable requisites for the biological activity of antibrain antibody.

Choroid Plexus

The physiology of choroid plexus should be taken into consideration when heterologous or homologous antibrain antibodies are introduced into the lateral ventricle of the brain. Bowsher (1957) detected some radioactivity in the choroid plexus of the third and lateral ventricles following intraventricular injection of radioactive homologous serum proteins. Pappenheimer et al. (1961) supported the view that choroid plexus could be the site of active transport of foreign organic molecules from cerebrospinal fluid to blood. It has been claimed that the uptake of both fluorescein-labeled bovine albumin and gamma globulin by the epithelial cells of incubated choroid plexus is an active process (Smith et al., 1964). The presence of microvesicles in the choroidal epithelium following intraventricular injection of thorium dioxide particles indicates that the pinocytosis takes part in the clearance of material from the cerebrospinal fluid (Tennyson and Pappas, 1961).

Macrophages

There is a possibility that the macrophages may participate in the mechanism which helps antibrain antibody to reach corresponding antigens in deep structures of the brain. Bowsher (1960) stated that the absorption of material from the craniospinal subarachnoid space occurs through leptomeningovascular and along perineurolymphatic ways. That macrophages may indeed play a role in transporting foreign particles through nervous tissue has been shown by Klatzo (1964). These workers described numerous scattered macrophages containing fluorescein-labeled bovine serum albumin in the subarachnoid space.

Relevant to this subject, although indirectly, is the function of macrophages in immune processes. It seems that in an immune response triggered by radiolabeled proteins, the macrophages exert both a catabolic and an immunogenic function. If this immunogenic role really operates, then macrophages transport proteinic antigens (Unanue and Askonas, 1967).

Brain Cell Membrane

The cell membrane may represent an important barrier to the antibody, the specificity of which is directed against an intracellular antigen. Since there are no convincing data demonstrating the active transport of macromolecules through biological membranes (Heinz, 1967), it seems unlikely that antibrain antibody is capable of entering the nerve cell (Levine, 1967). Nevertheless, it may be assumed that an antibody is capable of penetrating into a cell by altering the structure and composition of a cell membrane, by activating or inhibiting biologically active substances, or by disturbing the mechanisms which control membrane permeability (Charnock and Opit, 1968). The molecular biological view of the function of brain cell membrane and the relationship of cell membrane to its microenvironment has recently been described in an essay by Schmitt (1969).

It is pertinent to mention here the existence of antinuclear antibodies, particularly those reacting to deoxyribonucleic acid, which are involved in the pathogenesis of systemic lupus erythematosus (Miescher and Peronetto, 1969). Leaving aside the question of the specificity of lupus erythematosus cell phenomenon, the important facts remain that this phenomenon occurs in vivo and that the activity of the antibody in circulation is directed against nucleoproteins. Immune hemolysis is an example showing the damaging effect of antibody on cell membrane; biochemical events before the lysis of erythrocytes lead to the appearance of holes in the erythrocyte membrane (Cinader and Lepow, 1967).

Brain-Blood Barrier

An antibrain antibody introduced into the cerebral space is not only faced with the problem of how to enter the nerve cell but also has to reach the microenvironment of brain cells by crossing *the brain component of the brain-blood barrier*. It is not within the scope of this chapter to discuss the morphology of intercellular space and its relation to the complex of anatomical, physiological, and biochemical phenomena called the brain-blood barrier (Roth and Barlow, 1961). However, the mention of some basic data directly related to the subject of this chapter may be found useful in understanding the pathways used by foreign macromolecules following their introduction into the cerebral cavity.

The distribution of radioactive iodinated bovine albumin after injection into the subarachnoid space of the rabbit, cat and dog has revealed the presence of labeled protein in the subpial brain tissue. It has been suggested that the "physiological" path of radioactive proteins is along the perivascular space (Lee and Olszewski, 1960). The diffuse radioactivity observed in the subpial layer was probably the result of direct penetration of tracer from the subarachnoid space into the brain parenchyma. An interesting point is the lack of radioactivity in the ependyma and subependymal area, which is probably due to the direction of cerebrospinal fluid flow. Bowsher (1957) claimed that radioactive homologous serum protein injected into the subarachnoid space was absorbed by leptomeningovascular, perineurolymphatic, and ependymal routes. This investigator advanced a hypothesis that the leptomeninx is a cerebral part of the reticuloendothelial system, the immunologic capacity of which is unknown.

As suggested by Draškoci et al. (1960), the mechanism of uptake of foreign proteins in the tissue surrounding the ventricles is not necessarily identical to that operating in the area of pial surfaces. Differences between ependymal cells in various regions and variations in properties of neuroglia could account for differences in penetration of a substance into the periventricular zones (Feldberg and Fleischhauer, 1960). Lajtha (1961) pointed out that the exchange rates between plasma and brain differ in various parts of the brain. The size of molecules—e.g., sizes of albumin and immunoglobulins—can also influence the ability of a protein to enter the periventricular tissue. Lipid-soluble molecules with a molecular weight up to 50,000 readily pass the brain-blood barrier (Prockop et al., 1961). The same is true of bovine serum albumin (Sherwin et al., 1963).

The dynamic state of the brain-blood barrier and the capacity of cellular and soluble antigens to cross this barrier and reach the antibody-synthesizing sites in lymphoid tissue has been demonstrated in experiments in which the immune machinery of the body was stimulated by sheep red blood cells or human gamma globulin administered via cannula directly into the lateral ventricle of the rabbit brain (Janković et al., 1961; Mitrović et al., 1964).

Results presented in Table 1 show that foreign erythrocytes may cross the brain-blood barrier and that the intraventricular administration of antigen represents a powerful way of stimulating the immune apparatus. It should be stressed that in spite of high titers of hemolysins in the serum, these antibodies were absent or present only in traces in the cerebrospinal fluid. In another experiment, cats immunized intraventricularly with bovine gamma globulin produced a fair amount of antibody in the serum, but all attempts to detect antibody in cerebrospinal fluid by means of im-

TABLE 1. Peak Hemolysin Titler ($\log_2$) in the Cerebrospinal Fluid (CSF) of Rabbits Receiving Sheep Erythrocytes into the Lateral Ventricle of the Brain

Rabbit No.	Before injection of sheep red cells		Day after the last injection of sheep red cells						
			2	4	6	8	10	12	12
	Serum	CSF	Serum						CSF
4	5	0	11	13	14	12	10	9	1
6	5	0	11	13	14	15	14	12	1
7	8	0	12	14	12	10	10	9	1
8	8	0	9	9	11	11	11	10	0

Modified from Janković et al., 1967.

munoelectrophoresis and passive cutaneous anaphylaxis utterly failed (Janković et al., 1969).

The mechanism by which foreign erythrocytes are removed from the cerebrospinal fluid is not clear. Dupont et al. (1961) stated that red blood cells from the cerebral cavity are phagocitized by mesothelial-lining cells. It has been shown that subarachnoid administration of human serum globulin into rabbits is just as effective in stimulating antibody production as is intravenous injection (Sherwin et al., 1963). High agglutinin titers were obtained in rabbits following injection of canine red blood cells into the cisterna magna (Panda et al., 1965). This high antibody-producing rate was thought to be the result of a "slow leak" of antigen into the blood.

It should be emphasized that our immunoneurological experiments, in which antibrain antibodies were injected into the lateral ventricle of the brain, differ from those in which the intraventricular route was used to introduce materials into the brain cavity in order to study either the clearance mechanism of particles from cerebrospinal fluid or the production and appearance of antibody in the blood. First of all, material employed in the above-mentioned experiments did not exhibit a specificity for brain cells or their constituents, whereas in our immunoneurological experiments we used macromolecules the specificity of which was directed toward brain antigens. It is much more relevant for an antibrain antibody to reach corresponding antigenic sites in the brain than to get into the blood circulation. Such an immunologically induced destination of macromolecules probably plays an important role in determining the pathway and velocity of their migration from the cerebrospinal fluid, and this may influence factors that function in the brain-blood barrier system (Bakay, 1956; Schmitt, 1969).

Correlates of Brain Antigen Specificity

The brain impresses one not only by its immense cytoarchitectonical and functional complexity but also by its fantastic antigenic heterogeneity. Morphologically, the brain is divided into a variety of regions and structures which are connected by means of a widespread net of communications such as, for example, fibers in the cerebral cortex which come from very remote parts of the central nervous system. Further, the dynamic state of neurons and glia certainly involves the synthesis of macromolecules that may have antigenic properties. These circumstances thus further increase the range of antigenic potentialities of the brain. In this section I should simply like to introduce to those who are not very familiar with immunology some data relevant to the antigenic mosaic of the brain, with no intention of providing the reader with a list of brain antigens, their immunologic specificity, or their localization in the nervous system. Immunoneurology still waits, as do many other related sciences, to see the spectacular achievements of immunochemistry applied in neurosciences—assuming that the biological aspects of the antigen and its antibody will not be neglected because of an exclusive interest in the chemical structure of these functional partners which nature has created in one of her experiments.

Organ-Specificity of Brain Lipids

According to Landsteiner (1945), organ-specificity means "that a serum reacts with homologous organs of unrelated species or differentiates one organ from others within a species" (p. 80). The brain is usually considered to be a typical representative of organ-specificity. It has been accepted that this kind of specificity is shared by lipids extracted from whole-brain tissue (Witebsky and Steinfeld, 1928). Lewis (1933) used alcoholic brain extracts from various species and found that all preparations were antigenic in rabbits. Antibrain sera thus produced were organ-specific, and only a weak cross-reactivity was observed with antigens extracted from testicles of a variety of species. Reichner and Witebsky (1934) demonstrated regional brain specificity. Using antisera prepared to boiled suspensions of white and gray matter, they showed by complement fixation reaction that gray matter could be distinguished from white matter. An interesting point is that anticaudate nucleus and antihippocampus antisera produced positive reactions with gray matter antigen, although they reacted weakly with the white matter.

In 1958 we started our immunoneurological experiments on the central nervous system by investigating the serological specificity of lipids ex-

TABLE 2. Reactivity of Rabbit Anticat Brain Region Sera with Lipids Extracted from Total Brain of Different Species

Anticat serum for	Antigen (brain lipid) Cat	Guinea pig	Rat	Rabbit	Pheasant	Pigeon	Duck	Chicken
Frontal cortex	9	7	8	7	7	7	7	7
Occipital cortex	8	7	7	7	7	7	7	7
Temporal cortex	8	6	6	5	5	5	5	5
Cerebellar cortex	8	6	5	6	6	6	5	5
Caudate nucleus	9	7	7	7	6	6	6	6
Thalamus	9	7	6	7	6	6	6	7
Cerebral white matter	10	8	7	7	7	7	7	7
Medulla	9	6	6	7	6	6	6	7
Spinal cord	8	6	5	5	5	5	5	5

The peak antibody titers (complement-fixation) of anticat brain sera are expressed in term of logarithms to the base 2.
Modified from Janković et al., 1960.

tracted from various parts of the cat brain (Janković et al., 1960). For this purpose, alcoholic extracts from the frontal cortex, occipital cortex, temporal cortex, cerebellar cortex, caudate nucleus, thalamus, cerebral white matter, medulla, and spinal cord were used as antigens to prepare antisera in rabbits. The examination of rabbit anticat brain region sera by means of complement fixation reaction revealed that none of the antisera differentiated among the lipids isolated from different regions of the central nervous system of the cat. Furthermore, the cross-reactivity of these antisera with lipids obtained from the same regions of heterologous brains (dog, ox, monkey, and man) was very pronounced. These antisera also reacted with lipids from the whole brain of cat, guinea pig, rat, rabbit, pheasant, pigeon, duck, and chicken (Table 2).

In an attempt to make anticat brain region sera more specific, antisera were absorbed with lipids from various regions of the cat brain, and only the anticaudate serum exhibited a slight regional specificity. In fact, serological tests confirmed that lipids from the various regions of the cat brain lack individual specificity but that all are highly organ-specific. However, the inability of serological reactions to demonstrate the region-specificity of antibrain antibodies did not entirely exclude the possibility that such a specificity might be recorded in an in-vivo experimental situation.

Brain Proteins as Antigens

Proteins of the neuron, populations of neurons, and regions of the central nervous system offer the most challenging and fascinating problems to molecular biologists, neurophysiologists, biopsychiatrists, immunologists, and other investigators interested in neurosciences. Neural proteins (i.e., antigens) represent a wide range of molecular weights and chemical compositions, which situation corresponds to the complex function of the nervous system (Waelsch and Lajtha, 1961). The glial component cannot be disregarded since the glia constitutes the great majority of brain cell population, and because of the possibility that the "true unit of the nervous system is not the neuron, but the neuron-glia complex" (Schmitt, 1967).

A short survey of the basic data on the isolation and identification of protein antigens from the brain of various species will be given here. Using starch-gel electrophoresis, Bernsohn et al. (1961) succeeded in demonstrating 11 to 12 protein bands in a water-soluble preparation of the rat brain. Immunodiffusion and immunoelectrophoresis of saline extracts of rat brain revealed about 12 antigens, and one antigen exerted a specificity limited to the species brain (MacPherson and Liakopoulou, 1966). These investigations claimed that the majority of the proteins of rat brain is shared with other organs.

In this connection, I should like to mention our experiments on cats trained to perform defensive conditioned responses (Janković et al., 1968). The immunodiffusion analysis of rabbit anticat liver protein serum revealed a faint precipitin line when brain extra was used in the test, thus pointing to the cross-reactivity component of the antiliver serum. However, the same antiserum did not exert any notable effect on conditioned reflexes following injection into the lateral ventricle of trained cats. On the other hand, conditioned responses were strikingly affected by intraventricular injection of anticat brain protein serum. The observed serological cross-reactivity between cat liver and brain seems to be very instructive, since it indicates at least three possibilities: (1) the brain antigen related to the process of learning is *not accessible* to the cross-reacting antibody from antiliver serum; (2) the brain antigen which is also present in the liver is *not related* to the learning process; and/or (3) *serological reactions do not necessarily reflect the biological activity of the antibrain antibody.* This again emphasizes the discrepancy between the results of serological tests and those observed in the in-vivo reaction between antigen and antibody.

As for the human brain, 14 to 15 different proteins were demonstrated in water-soluble material of white matter (Bernsohn et al., 1961). Thirteen immunoelectrophoretically distinct proteins were isolated (by means of chromatography on DEAE-cellulose) from gray matter (Rajam and Bogoch, 1966), and, with the use of similar techniques, three basic antigens specific to the human brain were identified (Rajam et al., 1966). It is asserted that the human brain contains over 100 proteins of different size and charge (Bogoch et al., 1964). And this is only the beginning of the efforts devoted to mapping protein antigens of the brain.

S-100 Protein

The structural complexity of the brain makes it difficult to obtain particulate and soluble fractions of sufficient purity for use in studying the protein synthesis in the brain (Suzuki et al., 1964) and the antigenic composition of nervous tissue. The S-100 protein, isolated by Moore and McGregor (1965) from beef brain, seems to meet the immunochemical requirements. This protein exists in a number of isomeric forms (McEwen and Hydén, 1966) and is mainly localized in the glia, since anti-S-100 sera precipitate with glia protein in a microcapillary gel medium but not with neuronal proteins. However, fluorescent antibody analysis shows specific fluorescence in the big Deiter's nerve cells (Hydén and McEwen, 1966). The S-100 protein makes up 0.5 percent of the soluble protein fraction of the brain, which is only a tiny portion of the brain proteins. It is still not clear whether S-100 is synthesized in the neuron or glia or both. The S-100 protein is an example of an antigen which is unique for the brain but of unknown function.

Myelin Protein (Basic Protein)

Water-soluble fractions were isolated from bovine (Roboz and Henderson, 1959) and guinea pig nervous tissue (Kies and Alvord, 1959). These preparations proved to be encephalitogenic since they produced experimental allergic encephalomyelitis when injected into animals. Myelin protein isolated from brain tissue of different species exhibited similar chemical structures and immunologic cross-reactivity (Kies, 1965). All preparations of myelin protein are highly susceptible to proteolitic digestion, unlike Folch-Pi's (1963) proteolipids. The molecular weights of basic proteins range from 10,000 to 40,000 (Kibler et al., 1964; Nakao and Roboz-Einstein, 1965; Kies et al., 1965), probably because of variations in procedures employed for the isolation. Raunch and Raffel (1964) have provided evidence that the basic protein is localized in the myelin sheath.

A basic protein of low molecular weight, having alanine as N-terminal amino acid, was found in the myelin sheaths of the central and peripheral nervous tissue of a variety of mammalian and nonmammalian species. It has been suggested that this cationic protein may play a role in the protein synthesis of the neuron (Kornguth and Anderson, 1965). An interesting finding is that this protein reacts with corresponding antibody only after previous damage to the myelin sheath by virus, bacteria, or trauma. For a more detailed review of proteins and other nervous tissue antigens related to demyelinating disease, the reader is referred to Paterson (1966).

Brain Proteins and Learning

According to Hydén (1967) the process of learning includes modifications of the protein pattern of the nerve cell that are associated with DNA-RNA-protein in macromolecule systems (Gaito, 1966, 1967; Griffith and Mahler, 1969). The situation becomes more complex if memory results from joint functions of neurons and adjacent glia. In spite of conceptual differences, it is supposed that the memory specificity is coded in a protein (Bogoch, 1968).

The isolation of an already existing, modified, or de novo synthesized protein, which is the "backbone" of processes underlying the storage of information, would be of fundamental significance and would permit a more specific and precise experimental approach to the study of memory. Bogoch (1965), working on pigeons, demonstrated the relationship between the amount of 11B-11A cerebroproteins and the amount of learning, and suggested that the encoding processes of memory occur in the glucoproteins of the brain. So far, this has been the only experimental finding that memory-linked proteins do exist.

In connection with the transfer of memory from one animal to another (Bogoch, 1968), I should like to stress here that some investigations used in their transfer experiments a rather immunologically incompatible pairing of donors and recipients; trained rats served as donors of a saline brain extract, designated as "clear supernatant," obtained by centrifugation at 26,000 g and injected into recipient naive mice. Positive results were ascribed to the transfer of brain peptides or proteins but not to nucleic acids (Ungar and Irwin, 1967).

If one accepts the idea that any information is capable of inducing at least some configurational changes in the molecule of brain protein, it follows that the sum of information equals the number of modified (or newly formed) macromolecules. In immunologic terms, a change of steric configuration of the protein molecule may induce a new immunologic specificity. In light of this, let us consider, for example, the brain of an "ordinary" man who is not obliged to exercise his intellect to any great extent. Even the brain of such a man contains a quantity of information so enormous as to astound an astronomer.

On the other hand, intellectual work, which involves not only a high rate of information storage but also an intensive and continuous process of analyzing and synthesizing the recorded information, increases the number of learning-linked proteins, i.e., the number of proteins which have different immunologic specificities. It follows from this that the brain of an intellectual is chemically and immunologically different from the brain of an "ordinary" man as well as from the brain of any other individual. This indicates that *the quantitative changes* in the brain, probably in the form of an increased number of learning-linked proteins, bring about the appearance of *a new quality:* knowledge. This assumption is based on the present evidence and concepts related to the chemistry of memory; it has nothing to do with science fiction.

A Remark on Specificity

Not only the brain as a whole but also its parts and basic structural units show a high antigenic complexity. Synaptic relations, for example, are different anatomically (morphologic relationship between neurons), physiologically (electrical or chemical transmission), and pharmacologically (type of chemical transmitter), but they also differ immunologically (antigens involved in synaptic relations). It would be unwise, therefore, to insist on prompt answers to questions about the specificity of brain antigens and antibrain antibodies and to consider the phenomena that are produced by an antigen-antibody reaction a matter of secondary importance. Some investigators who are not trained in immunology are overimpressed by the fascinating possibilities offered by immunologic techniques and simply exaggerate potentialities of serological tests. It should not be forgotten, in this connection, that in spite of a formidable amount of data your knowledge of antibody specificity still remains highly speculative (Edelman, 1967).

To conclude, a scientific approach to the study of any biological problem is incompatible with the neglect of observed phenomena just because of our inadequate experience with regard to the characterization of components that are involved in these phenomena. Certainly, the use of immunochemically undefined materials in immunoneurological experiments represents a crude method of studying the central nervous system. Nevertheless, the immunoneurological phenomena that will be described here are in fact specific, because *they all were produced only by preparations containing the antibrain antibody.*

Experimental Approaches in Immunoneurology

At present, immunoneurology may be regarded as a kind of "dissection" of the central nervous system at different levels of its morphology and physiology by means of immunologic techniques. The introduction of immunology into neurosciences has created not only a new manner of studying neurophysiologic and behavioral phenomena but also a new way of thinking about them. I shall concentrate here on several immunoneurological models used at various times and places by a number of investigators whose experiments were motivated by ideas different from ours.

Experiments on Tissue Culture

Kimura (1928) was the first to use antibrain sera in cultures of nerve cells and to report the cytotoxic effect of antibrain antibody. About 20 years later, Grunwalt (1949) used three antigens from newly hatched chicken brain to prepare antibrain sera in rabbits. The addition of antibody against the whole brain, antibody against alcohol-insoluble fraction, or antibody against alcohol-soluble fraction of the chicken brain to the culture of chicken embryonic spinal cord induced the cessation of outgrowth of fibers and cells.

Bornstein and Appel (1961, 1965) and Appel and Bornstein (1964), using the myelinated cultures of rat cerebellum, described specific demyelination which occurred when serum from animals with experimental allergic encephalomyelitis or serum from multiple sclerosis patients was added to nervous tissue explants. The demyelinating activity was complement-dependent and restricted to the myelin sheath and glial membrane, as demonstrated by fluorescent antibody technique. Winkler and Aranson (1966) found that intact lymph node cells from animals sensitized with sciatic nerve exerted a specific demyelinating effect on trigeminal ganglion culture. This cytodestructive action was blocked by a rabbit mono-specific antiserum directed against rat immunoglobulin A. The above experiments were primarily concerned with the pathogenesis of autoallergic disorders of nervous tissue.

After our first experiments on biological activity of antibrain antibody in 1960–1965, it became clear that cultured nervous tissue provides a suitable object for the exploration of antibrain antibody activity, since tissue culture permits the study of well-defined populations of neurons under controlled experimental conditions. In 1965 Bornstein and Crain applied serum from animals with allergic encephalomyelitis and humans with multiple sclerosis to the myelinated cultures of the cerebral neocortex and spinal cord of the mouse. The allergic encephalomyelitis serum, as well as the serum from multiple sclerosis patients, induced alterations in the bioelectrical properties of cultured nervous tissue characterized by partial or complete block of electrical activity within an hour of the addition of serum. Electronmicroscopic examination, however, did not reveal morphologic changes in synaptic structures. These investigations stated that the blocking of neuronal transmission was caused by a circulating factor in the serum which acts in the presence of complement. Prieto et al. (1967), by recording the electrical activity of meningioma cells, showed that the addition of antiserum with complement to human meningioma tissue culture produces deplorarization and a decrease in resistance.

Experiments on Isolated Axon

Perfusion of squid giant axon with antibodies against intraaxonal proteins abolished the action potential while the effect on membrane potential was less expressed (Huneeus-Cox, 1964). Antiaxon antibodies may block the propagation of action potentials by combining with one or more of the 14 identified axonal proteins. Partial purification of both low (Schmitt and Davison, 1961) and high (Huneeus-Cox, 1964) molecular constituents of the squid axoplasm did not provide evidence about their function in the axon.

Another "simple" biological system was used by our group (Mihailović et al., 1965) in a series of experiments devoted to studying the effect of antilobster nerve antibody on membrane potentials of the giant axon of the Adriatic lobster (*Palinurus vulgaris*). The antibody activity was tested on giant axons of the lobster ventral cord in which KCl microelectrodes were inserted. Membrane potentials of axons immersed in normal rabbit gamma globulin, like those dipped in the artificial sea water containing guinea pig complement, could hold for more than 10 hr. On the other hand, potentials of axons immersed in the solution containing antilobster nerve antibody and complement could be recorded for only 2 to 6 hr. The changes in action potentials induced by immune globulin were characterized by an initial increase in the amplitude which was soon followed by gradually smaller action potentials. We supposed that changes in membrane potentials probably originated from an interaction between antibody and axonal macromolecules. It has been suggested that "simple" systems (for example, Aplysia preparations) are suitable not only for the study of electrical propagation but also for the elucidation of structural changes in macromolecules related to memory (Applewhite, 1967).

Experiments on Insects

The nervous system of the insects, because of its "simple" structural organization, may represent a convenient model for exploring the bioelectrical changes induced by antibrain antibody. Recently, we described the electrical activity of the cockroach (*Blatta orientalis L.*) brain following application of antilobster brain antibody (Janković and Rakić, 1969; Janković et al., 1969).

The brain from Adriatic lobster (*Palinurus vulgaris*) was used as antigen to prepare antibrain serum in rabbits, as it was not possible to obtain a sufficient amount of cockroach brain necessary for immunization and

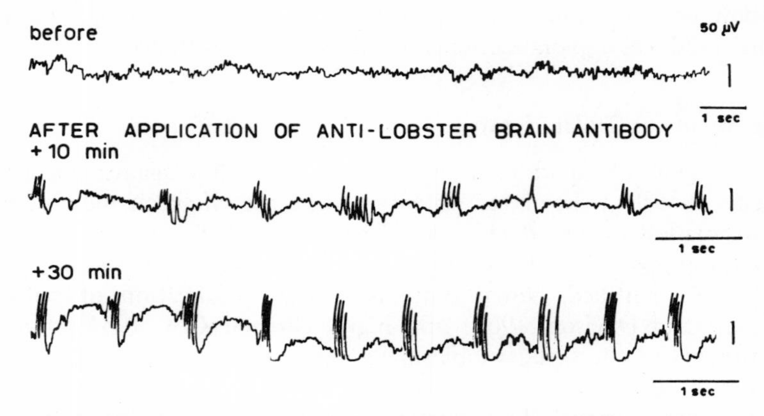

FIG. 1. Electrical activity of the cockroach brain before and after application of anti-lobster brain antibody. Note the bursts of high voltage spikes in rather synchronous time intervals. (From Janković, Rakić, and Šestović. 1969. *Experientia,* 25:1049.)

because many morphologic characteristics are common to the nervous system of arthropodes (Bullock and Horridge, 1965). Physiological solution for insects, normal rabbit gamma globulin, and immune rabbit gamma globulin containing antibrain antibody were applied on the surface of the cockroach brain, in which electrodes were inserted. The great majority of cockroaches tested with antibrain antibody developed bioelectrical changes in the brain in the form of continuous spike activity or bursts of high-voltage spikes (Fig. 1). On the other hand, no changes in the electrical activity of the brain were observed in insects treated with saline or normal gamma globulin.

These results suggest that an in-vivo contact between antigens of the cockroach brain and antibrain antibodies causes the repetitive firing. We advanced the possibility, among others, that the in-vivo reaction between brain antigen and antibody may enhance or sustain the activity of some pharmacologically active substances (Nachmansohn, 1959).

Experiments on Mammals

In an experiment initially designed to study the role of antibrain antibody in the pathogenesis of experimental allergic encephalomyelitis (Janković et al., 1966a, b), we used as the source of antibrain antibody the immune gamma globulin isolated from rabbits that developed allergic encephalomyelitis. This preparation contained antibrain antibodies, while normal rabbit gamma globulin was serologically inactive against brain antigens. Normal rabbits with implanted electrodes and cannulae were injected intraventricularly either with homologous antibrain antibody or with normal gamma globulin and saline.

The first injection of antibrain antibody was followed by high-voltage slow activity, which in some cases first occurred in the caudate nucleus and soon spread to the frontal and occipital cortex, and to the hippocampus (Fig. 2). This activity was combined with spindles and interspersed with short periods of faster rhythm similar to that seen before the injection. On the other hand, animals injected into the lateral ventricle with normal rabbit globulin showed no apparent alterations in electrical activity (Fig. 3). These results proved that antibrain antibodies are capable of inducing bioelectrical abnormalities in various brain structures.

In a previous work we demonstrated that the injection of anticaudate nucleus antibody into the cerebral cavity of the cat brain caused a modification in electrical activity which was particularly expressed in the caudate nucleus (Mihailović and Janković, 1961, 1965). Antibody against hippocampus did not affect the electrographic pattern of the caudate nucleus. Although these results suggested that antibrain antibody may exert a regional specificity, they still await confirmation by other investigators.

In another experiment rabbits were immunized with thalamus tissue of the cat brain, and antithalamus antibody thus obtained was injected through a cannula into the lateral ventricle of the cat brain (Radulovački and Janković, 1966). The injection of antithalamus antibody was followed by long-lasting desynchronization and by single spikes in thalamic leads. These spikes were followed by a generalized strong discharge in all cortical and subcortical structures. Injections of saline or normal rabbit gamma globulin did not induce electrical abnormalities in the thalamus and other structures.

Experiments on Paradoxical Sleep

Paradoxical sleep and slow sleep originate probably from two different processes and different structures (Jouvet, 1967). The nucleus reticularis pontis caudalis was described by Jouvet (1965) as the center of paradoxical sleep, while Candia et al. (1967), stated that the nucleus reticularis pontis oralis performs the same function. We assumed that the application of antibodies prepared in rabbits against pons, against nucleus reticularis pontis caudalis (NRPC), and against midbrain reticular formation (MRF) of the cat brain would provide some information about the sleep-wakefulness structures (Janković et al., 1970). Therefore, cats with implanted electrodes and cannula were deprived of paradoxical sleep (Jouvet et al., 1964), and injected intraventricularly with antipons, antiNRPC or anti-MRF antibody. The sleep-wakefulness patterns of three representative cats are shown in Table 3.

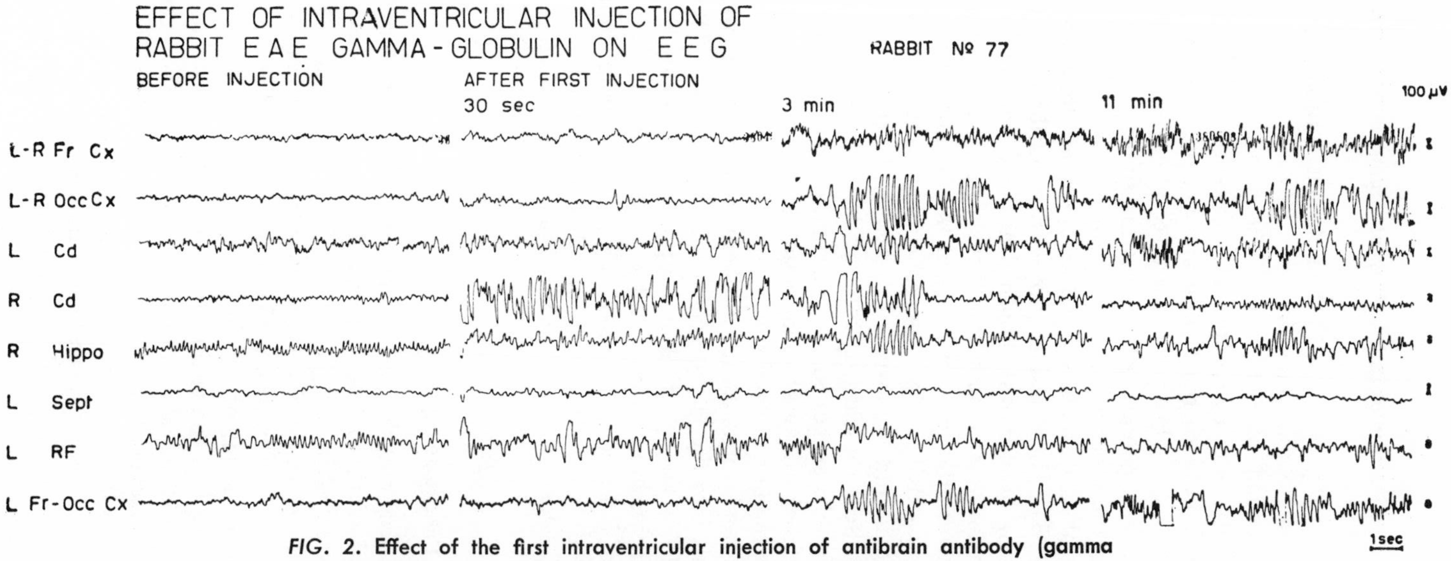

EFFECT OF INTRAVENTRICULAR INJECTION OF RABBIT EAE GAMMA-GLOBULIN ON EEG

RABBIT № 77

BEFORE INJECTION

AFTER FIRST INJECTION

30 sec

3 min

11 min

100 µV

L-R Fr Cx

L-R Occ Cx

L Cd

R Cd

R Hippo

L Sept

L RF

L Fr-Occ Cx

1 sec

FIG. 2. Effect of the first intraventricular injection of antibrain antibody (gamma globulin from rabbits which developed experimental allergic encephalomyelitis) on EEG activity. Note the high voltage slow waves in caudate nucleus and reticular formation, spread of this activity to other brain structures, and appearance of faster rhythms. (From Janković et al. 1966. *Path. Europ.*, 2:87.)

Abbreviations for EEG figures: L—left; R—right; Fr Cx—frontal cortex; Occ Cx—occipital cortex; Cd—caudate nucleus; Hippo—dorsal hippocampus; Sept—septum pellucidum; RF—midbrain reticular formation.

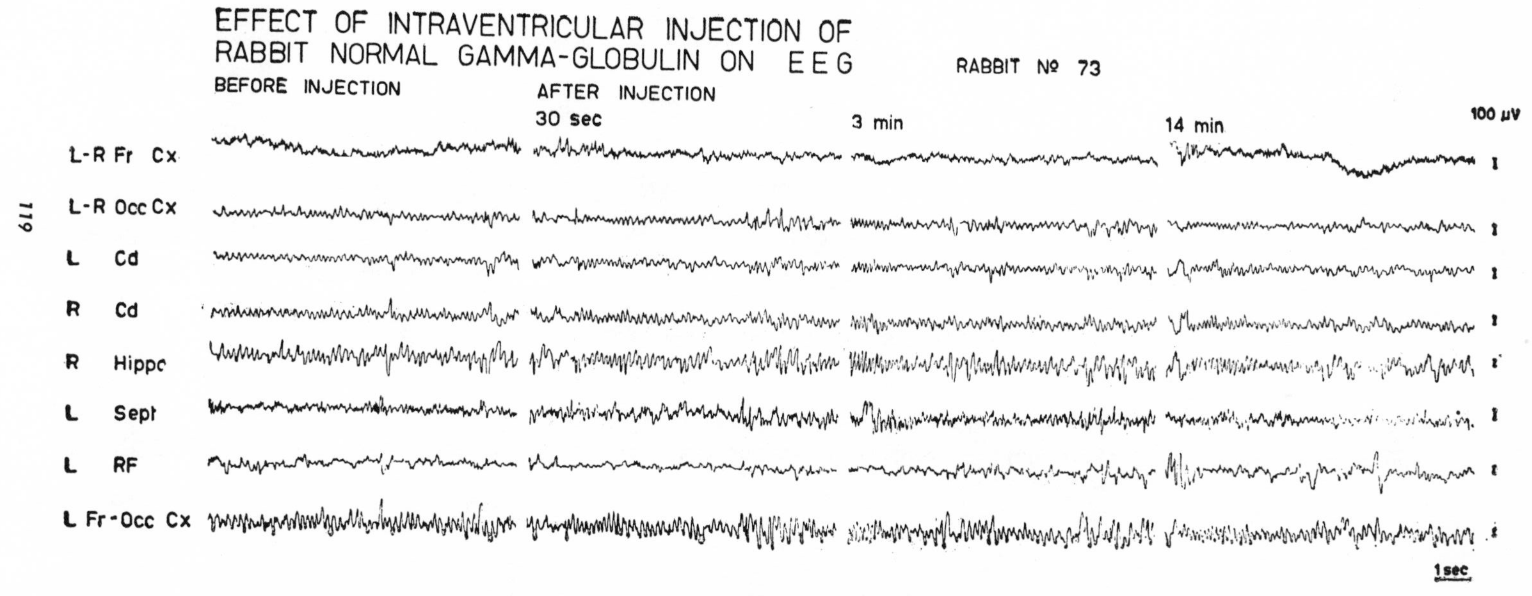

FIG. 3. Effect of intraventricular injection of normal rabbit gamma globulin on EEG activity. Lack of EEG changes. (From Janković et al. 1966. *Path. Europ.*, 2:87.)

TABLE 3. Percentage of Waking State (WS), Slow Sleep (SS), and Paradoxical Sleep (PS) following Injection of Antibrain Antibodies into the Lateral Ventricle of the Brain of Cats Deprived of Paradoxical Sleep

Kind of sleep	Anticat brain region antibody injected intraventricularly								
	Antipons			AntiNRPC			AntiMRF		
	Cat No.	Before injection	After injection	Cat No.	Before injection	After injection	Cat No.	Before injection	After injection
WS	29	10.1	32.4 (3.41)	22	18.4	27.2 (1.48)	14	15.6	20.1 (1.29)
SS		56.5	65.2 (1.15)		62.0	65.0 (1.05)		71.2	68.2 (0.96)
PS		33.4	2.4 (0.07)		19.6	7.8 (0.39)		13.2	11.7 (0.87)

NRPC, nucleus reticularis pontis caudalis; MRF, midbrain reticular formation.
Numbers in parenthesis indicate the ratios of the experimental values (after intraventricular injection of antibrain antibody) to the control values (before injection).

The most pronounced reduction of paradoxical sleep and significant increase in the waking state were induced by antipons antibody. The slow sleep, however, was not affected by antibrain antibodies.

The results summarized in Table 3 indicate that antibrain antibody may be used in the study of sleep and wakefulness, as well as electrical and chemical stimulations of hypnogenic structures or alterations of the sleep-wakefulness pattern by cerebral lesions. It is, however, a difficult task to offer an acceptable explanation for the effects exerted by antibrain antibody on paradoxical sleep. The possibility remains that antibody affects the catecholaminergic mechanisms which are presumed to be related to this kind of sleep (Gori, 1958). Another possibility would be that antibrain antibodies do not influence the activity of monoamine-oxidase inhibitors since no increase in slow sleep occurred following administration of antibodies, and monoamine-oxidase inhibitors are known to depress paradoxical sleep and increase slow sleep (Jouvet, 1967).

The above-described experiment also provided some important information: almost all cats that received a single intraventricular injection of antipons antibody died between 4 and 72 hr following the injection. We interpreted this high mortality rate as a kind of "immunologically induced death," since it was supposed that antibodies from antipons serum exerted their effect on vital centers in the lower structures of the central nervous system. The antipons-antibody-injected animal on which information is included in Table 3 did not show behavioral and other abnormalities for at least 48 hr.

Experiments on Learning

The delineation, in terms of neuronal macromolecular composition and ultrastructural activity, between "normal" behaviour and changes induced by processes of learning and memory is a difficult task, since all the structural and functional properties of the central nervous system are still unknown. Several mechanisms that underlie learned behavior (Nelson, 1967) can be blocked by antimetabolites acting against protein synthesis (Agranoff, 1967). For example, puromycin exerts its antilearning effect most probably through its structural similarity with RNA (Yarmolinsky and Haba, 1959) and thus blocks memory of training (Flexner et al., 1963). Hydén and Egyhási (1962) found an increase in the amount of neuronal RNA during the process of learning. If learning and memory are dependent on particular endocellular arrangements that include the DNA-RNA-protein chain, then we assumed that the interference of antibrain antibody with this biochemical series should affect neurons, their synaptic networks, and other components related to learning. This reasoning led our immunoneurophysiology study team to investigate the relation-

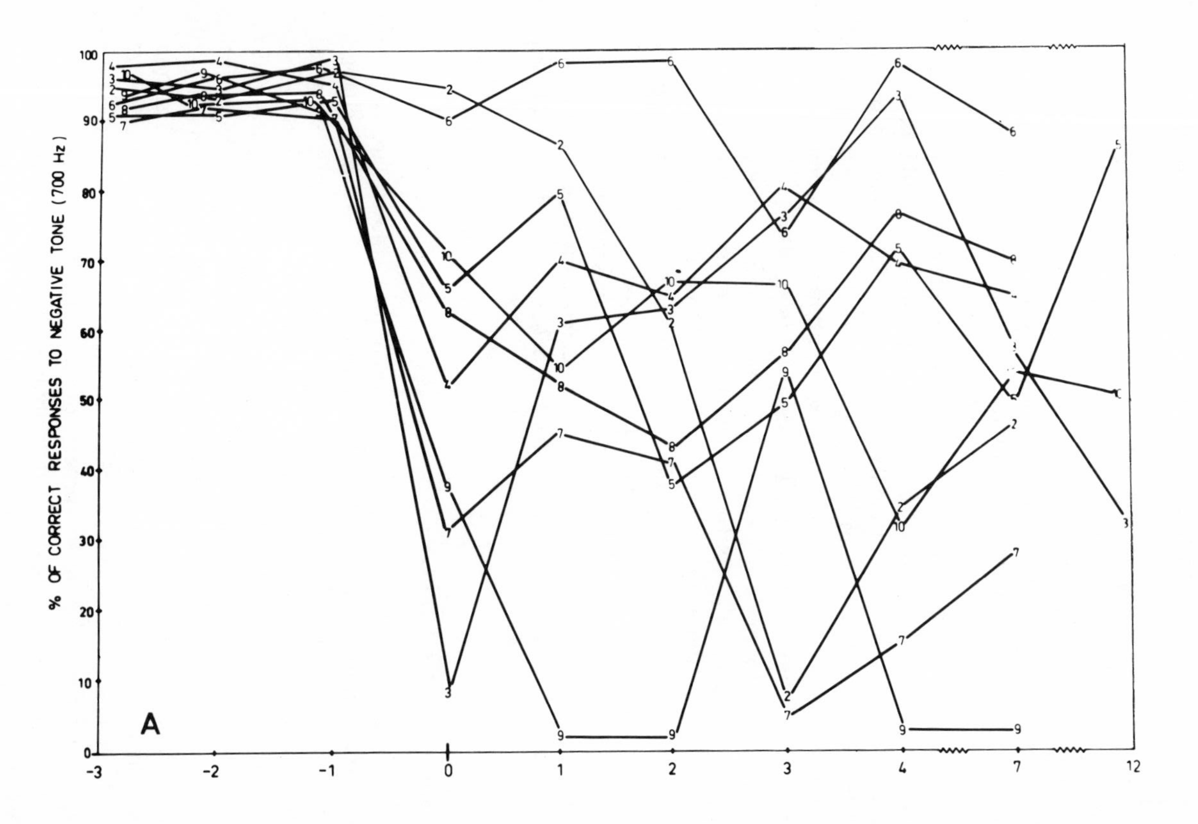
A
% OF CORRECT RESPONSES TO NEGATIVE TONE (700 Hz)
-3
-2
-1
0
1
2
3
4
7
12
100
90
80
70
60
50
40
30
20
10
0

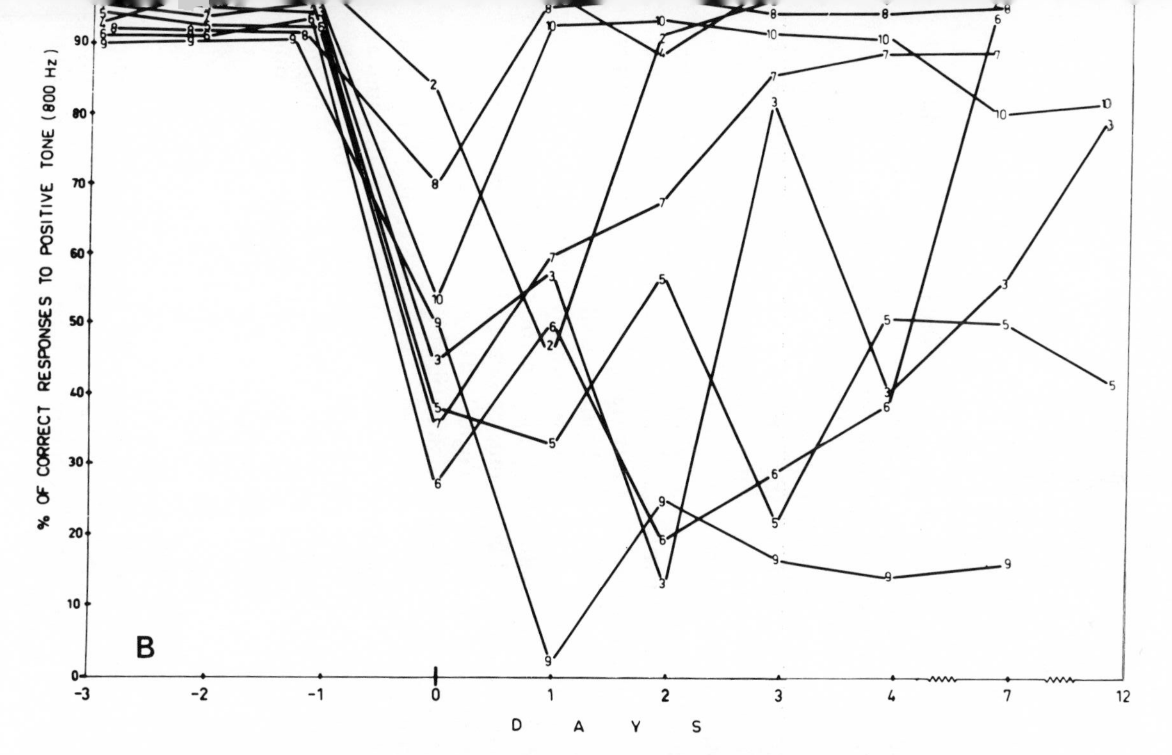

FIG. 4. Effect of intraventricular injection of antibrain antibody on defensive conditioned reflexes in nine cats (Nos. 2–10). The injection was given on day 0. A, response to negative tone; B, response to positive tone. Calculations of correct and incorrect responses to positive and negative tone are based on leg flexion and electromyographic activity. (From Janković et al. 1968. *Nature,* 218:270.)

ship between proteins and the learning process by means of antibrain antibody prepared in rabbits with a mixture of proteins extracted from the brain of naive cats (Janković et al., 1968).

Cats trained to respond by leg flexion to a "positive" tone of 800 Hz, and not to a "negative" tone of 700 Hz, received intraventricularly a single 0.2-ml injection of antibrain antibody, anticat liver antibody, or normal rabbit globulin. Injection of antibrain antibody produced significant changes in conditioned responses immediately after the application of antibrain antibody and on subsequent days (Fig. 4). On the other hand, injection of antiliver antibody, normal rabbit globulin or saline did not affect defensive conditional reflexes. These results suggest that brain proteins and metabolically related compounds probably represent the substrate for antibrain antibody activity and that an antigen-antibody reaction on neuronal level is responsible for altered behavioral phenomenology.

Pursuing these kinds of investigations, we isolated and immunologically characterized ribosomal proteins (Bigley et al., 1963) from the cat brain. Ribosomal proteins, containing 54 to 57 percent of RNA and 43 to 46 percent of protein, were injected into rabbits in order to prepare antibrain serum. Anticat brain ribosomal protein serum thus obtained was injected into the cerebral cavity of cats trained to perform an alimentary task, barpressing for food reward. Preliminary trials demonstrated beyond doubt that the process of learning was markedly affected by antibrain antibody but not by intraventricular injection of anticat liver ribosomal protein antibody, normal rabbit serum, or saline (Janković et al., 1970). These results again point to the biological effect on learning phenomena exerted by materials containing antibrain antibody.

In addition, we recently demonstrated (Mihailović et al., 1969) that in monkeys trained to perform delayed alternation and visual discrimination tasks, the injection of anticaudate nucleus and antihippocampus antibodies into the lateral ventricle produced significant impairment in the delayed alternation performance. The results of this experiment are complementary to those in the previous work done along a similar line.

Experiments on Schizophrenia

A globulin fraction, which electrophoretically migrated with immunoglobulin G, was isolated from the sera of schizophrenic patients and called "taraxein" (Heath, 1966). This "antibody" combines in vivo with neural cell nuclei of the septum and basal caudate nucleus. The injection of taraxein into normal subjects induced psychosis, and it was supposed, therefore, that schizophrenia is an immunologic disorder (Heath and Krupp, 1967; et al., 1967a, b) in which antibrain antibody plays an important role. However, other authors (Fessel, 1963; Vulchanov and

Hadjieva, 1964; Jensen et al., 1964; Rubin, 1965) were unable to demonstrate serological differences between sera from schizophrenic and nonschizophrenic subjects.

Although the concept that a typical mental disease may be of immunologic origin was quite interesting and attracted considerable attention, there are nevertheless several questions that should be answered prior to the classification of schizophrenia among autoimmune diseases. The crucial question is whether the antibrain antibody in the serum of a schizophrenic patient is *the cause* of disease or *the consequence* of a process which was initiated by some nonimmunologic factors and which started before antibrain antibody occurred in the blood.

Schizophrenia was mentioned here simply to present a more or less complete picture of antibrain antibody activity. Of course, antibodies directed against nervous antigens may be involved in the pathogenesis of some other disorders of nervous tissue, such as allergic neuritis, encephalomyelitis, and multiple sclerosis. These syndromes are also within the scope of immunoneurology, but they are not directly relevant to the subject of this book.

Concluding Remarks on Antibrain Antibody

By analogy with what was observed in an enzyme-antibody system (Cinader, 1967), the effect produced by antibrain antibody may depend on the steric configuration of immunogenic determinants and biologically active groups of nervous tissue antigen. In almost all experiments on enzyme-antibody systems extracellular antigens were employed, whereas in a neuron-antibody system antibrain antibody acts on an antigen which functions in the cell and which is a part of the cellular integrity. Therefore, the study of a neuron-antibody system is morphologically, functionally, and immunologically more complicated. Whereas an immunoenzymologic experiment involves the enzyme (as antigen), antienzyme antibody, and substrate, an immunoneurological experiment is composed of the nerve cell and its constituents (as antigen), antibrain antibody, and innumerable behavioral and nonbehavioral manifestations related to the function of the neurons and their associates.

Generally speaking, an antibody is capable of locating and identifying an antigen in its in-vivo constellation. Fluorescent antibody technique, for example, revealed that antibody prepared against purified ganglioside localizes in the body of the nerve cell (Bogoch, 1960). There is no reason why antibrain antibody should not induce chemical and structural changes in the radius of neuronal morphology and activity. Glia, because of its functional and morphologic intimacy with neurons (Hydén, 1962), may also be of importance for molecular events initiated by antibrain antibody. If proteins play a vital role in the function of nerve cells, then any kind of blockade or modification of protein molecules and protein synthesis leads to the functional disharmony and excitation or collapse of nervous tissue. This effect may be exerted on major molecular pathways such as ribosomes (Spirin and Gavrilova, 1969), or be accomplished, as suggested by Levine (1967), by eliminating sterically and biologically active sites of neuronal antigen, by interfering with conformational changes taking place in antigen, and by protecting neuronal antigen and thus allowing its biological action.

Several other mechanisms are possible when one attempts to explain the described parade of neurological and behavioral phenomena conducted by immunoneurological methodology. (1) Antibrain antibody may act on mitochondria, particularly in the terminal axons where they are concentrated, and thus influence the principal points which provide energy for a variety of neuronal activity. (2) Antibrain antibody may affect membrane-associated "electrogenic proteins" of a neuron-glia system (Schmitt and Davison, 1965). (3) Antibrain antibody may interfere with synaptic and transmissive relations between neurons by acting on cell membrane, the transmitting portion of the neuron, or the zone of neuron communication (neuropils). (4) Antibrain antibody may influence the release of chemically defined or undefined mediators in synapses. It has been suggested that biogenic amines have a neural transmitter role and also correlate with behavior (Kety, 1967). The biological potential of antibrain antibody is so enormous that in searching for the explanations of its activity—for example, its activity on paradoxical sleep—even some of the very old theories, such as Piéron's (1913) theory of hypnotoxins, should be taken into account. Whatever the true mechanism of antibrain antibody action may be, there is little doubt that antibody is capable of changing the functional equilibrium of the neuron.

The immunoneurological approaches to the study of the central nervous system and its cellular, subcellular, and molecular components, which form the basis of behavior, raise many methodological problems, in particular the purification and identification of brain antigens and antibrain antibodies. Ideally, what one would like to have is a list of antibodies the specificity of which is defined and which react with precisely located nervous antigens. It is now evident that in the course of immunization several classes of immunoglobulins appear, and that, in addition to physicochemical diversity, they exhibit a striking chemical heterogeneity based mainly on differences in amino acid sequences (Fahey, 1962). The dissection of the antibody molecule, in order to evaluate capacity for recognizing the antigenic sites, is the job of an immunochemist. However, even the most precise definition of antigen and antibody cannot predict the sequence of molecular, functional, and behavioral events which may be induced by in-vivo reaction between brain antigen and antibrain antibody. This emphasizes again that in biology structure and function should always be considered as parts of a single dynamic system.

Acknowledgments

This work was supported by the U.S. Public Health Service Grant 6X9803 from the National Institutes of Health, Bethesda, and by grants from the Federal Scientific Fund, Belgrade.

References

Agranoff, B. W. (1967). In: The Neurosciences, Quarton, G. C., Melnechuk, T., and Schmitt, F. O., eds. New York: Rockefeller Univ. Press.
Appel, S., and Bornstein, M. B. (1964). J. Exp. Med., 119:303.
Applewhite, P. B. (1967). Yale Univ. J. Biol. Med., 40:205.
Armand-Delille, P. F. (1906). Ann. Inst. Pasteur, 20:838.
Bairati, A. (1958). In: Biology of Neuroglia, Windle, W. F., ed. Springfield: Thomas.
Bakay, L. (1956). The Blood-Brain Barrier. Springfield: Thomas.
Bernsohn, J., Barron, K. D., and Hess, A. R. (1961). Proc. Soc. Exp. Biol. Med., 107:773.
Besredka, A. (1919). Anaphylaxis and Anti-Anaphylaxis and Their Experimental Foundations. St. Louis: Mosby.
Bigley, N. J., Dodd, M. C., and Geyer, V. B. (1963). J. Immunol., 90:416.
Bogoch, S. (1960). Nature (London), 185:392. ——— (1965). Neurosci. Res. Prog. Bull., 3:38. ——— (1968). The Biochemistry of Memory. London: Oxford Univ. Press. ——— Rajam, P. C., and Belval, P. C. (1964). Nature (London), 204:73.
Bornstein, M. B., and Appel, S. (1961). J. Neuropath. Exp. Neurol., 20:141. ——— and Appel, S. (1965). Ann. N.Y. Acad. Sci., 122:280. ——— and Crain, S. M. (1965). Science, 148:1242.
Bowsher, D. (1957). Anat. Rec., 128:23. ——— (1960). Cerebrospinal Fluid Dynamics in Health and Disease. Springfield: Thomas.
Bullock, T. H., and Horridge, G. A. (1965). Structure and Function in the Nervous Systems of Invertebrates. San Francisco: W. H. Freeman.
Candia, O., Rossi, G. F., and Sekino, T. (1967). Science, 155:720.
Charnock, J. S. and Opit, L. J. (1968). In: The Biological Basis of Medicine, Bittar, E. E., and Bittar, N., eds., Vol. 1. New York: Academic Press.
Cinader, B. (1967). In: Antibodies to Biologically Active Molecules, Cinader, B., ed. Oxford: Pergamon Press. ——— and Lepow, I. H. (1967). In: Antibodies to Biologically Active Molecules, Cinader, B., ed. Oxford: Pergamon Press.
Clemente, C. D. (1955). In: Regeneration in the Central Nervous System, Windle, W. F., ed. Springfield: Thomas.
Davson, H. (1967). Physiology of the Cerebrospinal Fluid. London: Churchill.
Day, E. D., Rigsbee, L., Wilkins, R., and Mahaley, Jr., M. S. (1967). J. Immunol., 98:62.

Delezènne, C. (1900). Ann. Inst. Pasteur, 14:686.
Draškoci, M., Feldberg, W., Fleischhauer, K., and Haranath, P. S. R. (1960). J. Physiol., 150:50.
Dupont, J. R., Wart, C. A. van, and Kraintz, L. (1961). J. Neuropath. Exp. Neurol., 20:450.
Edelman, G. M. (1967). In: The Neurosciences, Quarton, G. C., Melnechuk, T., and Schmitt, F. O., eds. New York: Rockefeller Univ. Press.
Fahey, J. L. (1962). Adv. Immunol., 2:41.
Feldberg, W., and Fleischhauer, K. (1960). J. Physiol., 150:451. ——— and Sherwood, S. L. (1953). J. Physiol., 120:3 P.
Fessel, W. J. (1963). Arch. Gen. Psychiat., 8:614.
Flexner, J. B., Flexner, L. B., and Stellar, E. (1963). Science, 141:57.
Folch-Pi, J. (1963). In: Brain Lipids and Lipoproteins, and the Leucodystrophies, Folch-Pi, J., and Bauer, H., eds. Amsterdam: Elsevier.
Gaito, J. (1966). In: Macromolecules and Behavior, 1st ed., Gaito, J., ed. New York: Appleton-Century-Crofts. ——— (1967). In: Chemistry of Learning, Corning, W. C., and Ratner, S. C., eds. New York: Plenum Press.
Gori, E. (1958). Atti. Soc. Lombarda Sci. Med. Biol., 13:26.
Green, H., Fleicher, R. A., Bazzow, R., and Goldberg, B. (1959). J. Exp. Med., 109:511.
Griffith, J. S., and Mahler, H. R. (1969). Nature (London), 223:580.
Grunwalt, E. (1949). Texas Rep. Biol. Med., 7:270.
Heath, R. G. (1966). Int. J. Neuropsychiat., 2:597. ——— and Krupp, I. M. (1967). Arch. Gen. Psychiat., 16:1. ——— Krupp, I. M., Byers, L. W., and Liljekvist, J. I. (1967a). Arch. Gen. Psychiat., 16:10. ——— Krupp, I. M., Byers, L. W., and Liljekvist, J. I. (1967b). Arch. Gen. Psychiat., 16:24.
Heinz, E. (1967). Ann. Rev. Physiol., 29:21.
Huneeus-Cox, F. (1964). Science, 143:1036.
Hurst, E. W. (1955a). J. Neurol. Neurosurg. Psychiat., 18:174. ——— (1955b). J. Neurol. Psychiat., 18:260.
Hydén, H. (1962). Endeavor, 21:144. ——— (1967). In: The Neurosciences, Quarton, G. C., Melnechuk, T., and Schmitt, F. O., eds. New York: Rockefeller Univ. Press. ——— and Egyhási, E. (1962). Proc. Nat. Acad. Sci. U.S.A., 48:1366. ——— and McEwen, B. S. (1966). Proc. Nat. Acad. Sci. U.S.A., 55:354.
Janković, B. D., Draškoci, M., and Isaković, K. (1961). Nature (London), 191: 288. ——— Isaković, K., and Mihailović, Lj. (1960). Int. Arch. Allergy, 17: 211. ——— Radulovaćki, M., and Mitrović, K. (1970). In preparation. ——— and Rakić, Lj. (1969). Proc. Acad. Med. Sci., U.S.S.R., 4:44. ——— Rakić, Lj., Horvat, J., and Veskov, R. (1970). In preparation. ——— Rakić, Lj., Janjić, M., Ivanus, J., and Mitrović, K., (1966a). Experientia, 22:459. ——— Rakić, Lj., Janjić, M., Mitrović, K., and Ivanuš, J. (1966). Path. Europ., 2:87. ——— Rakić, Lj., and Šestović, M. (1969). Experientia, 25:1049. ——— Rakić, Lj., Veskov, R., and Horvat, J. (1968). Nature (London), 218:270. ——— Rakić, Lj., Veskov, R., and Horvat, J. (1969). Experientia, 25:864.
Jensen, K., Clausen, J., and Osterman, E. (1964). Acta Psychiat. Scand., 40:280.
Jouvet, D., Vimont, P., Delorme, J. F., and Jouvet, M. (1964). C.R. Soc. Biol., Paris, 158:756. ——— (1965). Prog. Brain. Res., 18:20. ———

(1967). In: The Neurosciences, Quarton, G. C., Melnechuk, T., and Schmitt, F. O., eds. New York: Rockefeller Univ. Press.
Kabat, E. A., Wolf, A., and Bezer, A. E. (1948). J. Exp. Med., 88:417.
Kety, S. S. (1967). In: The Neurosciences, Quarton, G. C. Melnechuk, T., and Schmitt, F. O., eds. New York: Rockefeller Univ. Press.
Kibler, R. F., Fox, R. H., and Shapiro, R. (1964). Nature (London), 204: 1273.
Kies, M. W. (1965). Ann. N.Y. Acad. Sci., 122:242. ——— and Alvord, E. C., Jr. (1959). In: "Allergic" Encephalomyelitis, Kies, M. W., and Alvord, E. C., Jr., eds. Springfield: Thomas. ——— Nakao, A., Roboz-Einstein, E., Caspary, E. A., Field, E. J., and Honegger, C. G. (1965). Ann. N.Y. Acad. Sci., 122:161.
Kimura, R. (1928). Z. Immunitätsforsch., 55:501.
Klatzo, I., Miquel, J., Ferris, P. J., Prokop, J. D., and Smith, D. E. (1964). J. Neuropath. Exp. Neurol., 23:18.
Kornguth, S. E., and Anderson, J. W. (1965). J. Cell Biol., 26:157.
Lajtha, A. (1961). In: Regional Neurochemistry, Kety, S. S., and Elkes, J., eds. Oxford: Pergamon Press.
Landsteiner, K. (1945). The Specificity of Serological Reactions. Cambridge: Harvard Univ. Press.
Lee, J. C., and Olszewski, R. (1960). Neurology, 10:814.
Levine, L. (1967). In: The Neurosciences, Quarton, G. C., Melnechuk, T., and Schmitt, F. O., eds. New York: Rockefeller Univ. Press.
Lewis, J. H. (1933). J. Immunol., 24:193.
MacPherson, C. F. C., and Liakopoulou, A. (1966). J. Immunol., 97:450.
McEwen, B. S., and Hydén, H. (1966). J. Neurochem., 13:823.
Miescher, P. A., and Peronetto, F. (1969). In: Textbook of Immunopathology, Miescher, P. A., and Müller-Eberhard, H. J., eds. New York: Grune & Stratton.
Mihailović, Lj., Divac, I., Mitrović, K., Milošević, D., and Janković, B. D. (1969). Exp. Neurol., 24:325. ——— and Janković, B. D. (1961). Nature (London), 192:665. ——— and Janković, B. D., (1965). Neurosci. Res. Prog. Bull., 3:8. ——— Janković, B. D., Beleslin, B., Milošević, D., and Ćupić, D. (1965). Nature (London), 206:904.
Mitrović, K., Draškoci, M., and Janković, B. D. (1964). Experientia, 20:700.
Moore, B. W., and McGregor, D. (1965). J. Biol. Chem., 240:1647.
Nachmansohn, D. (1959). Chemical and Molecular Basis of Nerve Activity. New York: Academic Press.
Nakajima, Y., Pappas, G. D., and Bennett, M. V. L. (1965). Am. J. Anat., 116:471.
Nakao, A., and Roboz-Einstein, E. (1965). Fed. Proc., 24:242.
Nelson, P. G. (1967). In: The Neurosciences, Quarton, G. C., Melnechuk, T., and Schmitt, F. O., eds. New York: Rockefeller Univ. Press.
Panda, J. N., Dale, H. E., Loan, R. W., and Davis, L. E. (1965). J. Immunol., 94:760.
Pappenheimer, J. R., Heisey, S. R., and Jordan, E. F. (1961). Am. J. Physiol., 200:1.
Paterson, P. Y. (1966). Adv. Immunol., 5:131.
Piéron, H. (1913). Le Problème Physiologique du Sommeil. Paris: Masson.
Prieto, A., Kornblith, P. L., and Pollen, D. A. (1967). Science, 157:1185.
Prockop, L. D., Schanker, L. S., and Brodie, B. B. (1961). Science, 134:1424.

Radulovački, M., and Janković, B. D. (1966). Biological and Physiological Problems of Psychology, 247, Moscow (abstract).
Rajam, P. C., and Bogoch, S. (1966). Immunology, 11:211. ——— Bogoch, S., Rushworth, M. A., and Forrester, P. C. (1966). Immunology, 11:217.
Rauch, H. C., and Raffel, S. (1964). J. Immunol., 92:452.
Reichner, H., and Witebsky, E. (1934). Z. Immunitätsforsch., 81:410.
Roboz, E., and Henderson, N. (1959). In: "Allergic" Encephalomyelitis, Kies, M. W., and Alvord, E. C., Jr., eds. Springfield: Thomas.
Roth, L. J., and Barlow, C. F. (1961). Science, 134:22.
Rubin, R. T. (1965). Brit. J. Psychiat., 111:1003.
Schmitt, F. A. (1967). In: The Neurosciences. Quarton, G. C., Melnechuk, T., and Schmitt, F. O., eds. New York: Rockefeller Univ. Press.
Schmitt, F. O. (1969). Neurosci. Res. Prog. Bull., 7:281. ——— and Davison, P. F. (1961). In: Actualities Neurophysiologiques, Monnier, A. M., ed., 3 ième série. Paris: Masson. ——— and Davison, P. F. (1965). Neurosci. Res. Prog. Bull., 3:355.
Sherwin, A. L., O'Brien, G. J., Richter, M., Cosgrove, J. B. R., and Rose, B. (1963). Neurology, 13:703. ——— Richter, M., Cosgrove, J. B. R., and Rose, B. (1963). Neurology, 13:113.
Smith, D. E., Streicher, E., Milković, K., and Klatzo, I. (1964). Acta Neuropath., 3:372.
Spirin, A. S., and Gavrilova, L. P. (1969). The Ribosome. Berlin: Springer.
Suzuki, K., Korey, S. R. and Terry, R. D. (1964). J. Neurochem., 11:403.
Tennyson, W. M., and Pappas, G. D. (1961). In: Disorders of the Developing Nervous System, Fields, W. S., and Desmond, M. M., eds. Springfield: Thomas.
Unanue, E. R., and Askonas, B. A. (1967). J. Reticuloendothelial Soc., 4:440.
Ungar, G., and Irwin, L. N. (1967). Nature (London), 214:453.
Vulchanov, V. H., and Hadjieva, Y. (1964). Z. Immun. Allergieforsch., 127:138.
Waelsch, H., and Lajtha, A. (1961), Physiol. Rev., 41:709.
Waksman, B. H. (1961). In: Immunopathologie in Klinik und Forschung, Miescher, P., and Vorländer, K. O., eds. Stuttgart: G. Thieme.
Winkler, G. F., and Aranson, B. G. (1966). Science, 153:75.
Witebsky, E., and Steinfeld, J. (1928). Z. Immunitätsforsch., 58:271.
Yarmolinsky, M. B., and Haba, G. L. de la (1959). Proc. Nat. Acad. Sci. U.S.A., 45:1721.

Chapter 8

Correlation of the S-100 Brain Protein with Behavior

HOLGER HYDÉN and PAUL W. LANGE

Institute of Neurobiology, Faculty of Medicine,
University of Göteborg, Göteborg, Sweden

During the last few years we have investigated the acidic brain protein S-100 in hippocampal nerve cells during a behavioral test in rats. We wish to report that the amount of nerve cell S-100 protein increases in trained animals and that the S-100 protein is specifically correlated to learning. This linkage was demonstrated by the use of antiserum against the S-100 protein, which was injected intraventricularly during the course of the training and localized in the hippocampus by specific fluorescence. The presence of antiserum against the S-100 protein in the hippocampus prevents further learning during continued training.

The S-100 protein is a defined and brain-specific protein and its correlation to learning seems important since brain specific protein can be supposed to mediate neural functions. This protein, described in 1965 by Moore et al., has a molecular weight of 21,000, a high content of glutamatic and aspartic acid and, therefore, moves closest to the anodal front in electrophoresis at $pH>8$. The S-100 protein is mainly a glial protein but occurs also in the nerve cells (Hydén and McEwen, 1966), and constitutes about 0.2 percent of the total brain proteins. The anodal band containing S-100 can be separated into at least three components, two of which precipitate with antiserum against S-100 and have a high turnover (McEwen and Hydén, 1966). The S-100 protein seems to be composed of three subunits of 7,000 molecular weight (Dannies and Levine, 1969). Its appearance in the human frontal cortex parallels the onset of neurophysiological function (Zuckerman et al., 1970).

Moore and Perez (1968) have described another acidic brain-specific protein (14-3-2) localized in nerve cells. Still another acidic protein ("antigen α") unique to the brain has been characterized by Bennett and Edelman (1968). In addition, evidence for the existence of other brain-specific soluble proteins has been presented by Bogoch (1968), MacPherson and Liakopolou (1965), Kosinski and Grabar (1967), and Warecka and Bauer (1967).

The training of animals involves a number of variables, such as motor and sensory activity, motivation, orientation reflexes, stress, and the learning processes, per se. Active controls are, therefore, essential to the experimental animals in which—except for learning—these factors have become equated.

In a well-planned behavioral test surgical, mechanical, or electrical measures to the body should be avoided and the stress factor should be small. For these reasons, we have chosen reversal of handedness in rats as a behavior experiment (Hydén and Egyhazi, 1964). The active controls perceive and act similarly to the experimental animals.

Eighty-one Sprague-Dawley rats weighing 150 to 175 g were used. The experimental set-up has been described previously in detail (Hydén and Egyhazi, 1964). It need only be pointed out that the rat retrieved one food pill at a time by reaching into the glass tube housing the pills. The rats were induced to use the nonpreferred paw by arranging a wall parallel and close to the glass tube on the opposite side of the preferred paw. The controls used the preferred paw and received the same amount of reward as the experimental animals. The rats were trained during two sessions of 25 min per day. The performance, defined as number of reaches per day, was linear up to the eighth day (Fig. 1). The rats used in our experiments all showed performance curves similar to those in Figure 1. Once learned, this new behavior will remain for a long time (Wentworth, 1942).

The three persons handling the training of the rats and registering the performance did not know with what serum the rat was injected. Neither did the three persons carrying out the injections of sera and chemical analysis know about the performance of the rats.

In the present experiments, fresh pyramidal nerve cells of the CA3 region of the hippocampus were used. The method for dissection has been described elsewhere (Hydén, 1959). The cell sample, weighing around 1 μg, was homogenized in a microhomogenizer in the following solution: 20 μM sodium thioglycolate, 0.25 M sucrose with 0.1 percent Triton X-100 solution, buffered to pH 6.7 by a solution containing 2.85 g Tris and 1 M H_3PO_4 and H_2O to 50 ml. After centrifugation, the protein sample was separated electrophoretically on a 400-μ diameter polyacrylamide gel in glass capillaries, according to a previously described technique (Hydén and Lange, 1968).

Increase of the S-100 protein. When the electrophoretic pattern of the samples was studied, we observed the presence of a double anodal protein band in trained rats. The electrophoretical pattern from the sam-

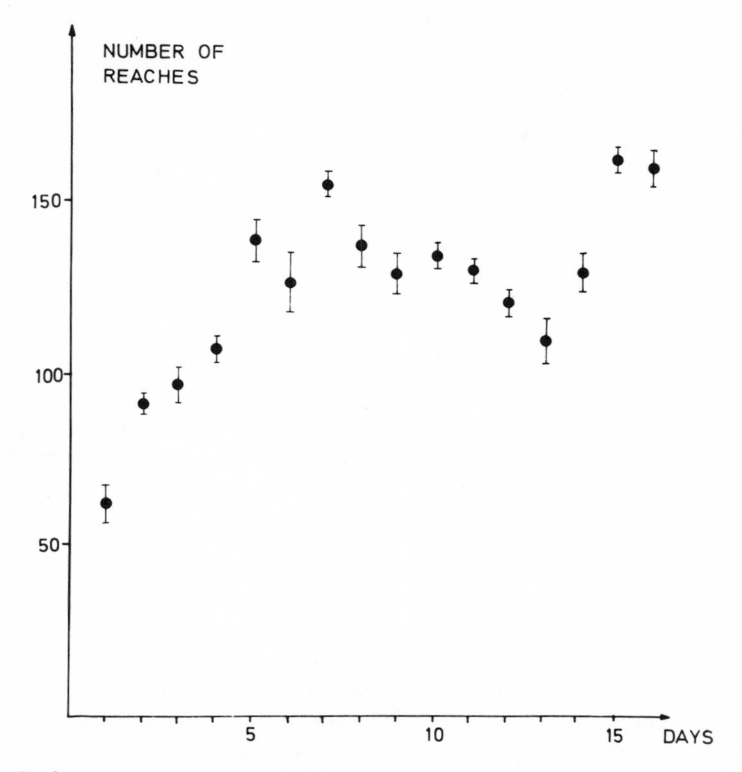

FIG. 1. **Performance curve of a group of twelve rats given as the average of the number of reaches as function of number of training sessions (2 × 25 min per day).**

ples of the control rats, however, showed one single anodal protein band. This observation was noted in a recent paper on brain protein synthesis (Hydén and Lange, 1970). Densitometric recordings were made of 75 electrophoretic patterns from 23 rats (Fig. 2). Table 1 shows the presence of two frontal protein bands in 30 recordings out of 55 (16 rats) from trained animals. One front band only was observed in all 20 recordings from controls (seven rats).

The following two tests were performed to see whether both or only one of these frontal protein bands contained S-100 protein. Protein extracted from pyramidal nerve cells of the CA3 hippocampal region from five trained rats was separated. After the separation, the gel cylinders were placed for 15 min in saturated ammonium sulphate solution and briefly rinsed. The protein was fixed in sulfosalicylic acid and stained with brilliant blue. This treatment with ammonium sulphate caused the protein band closest to the anodal front to disappear, which is a characteristic

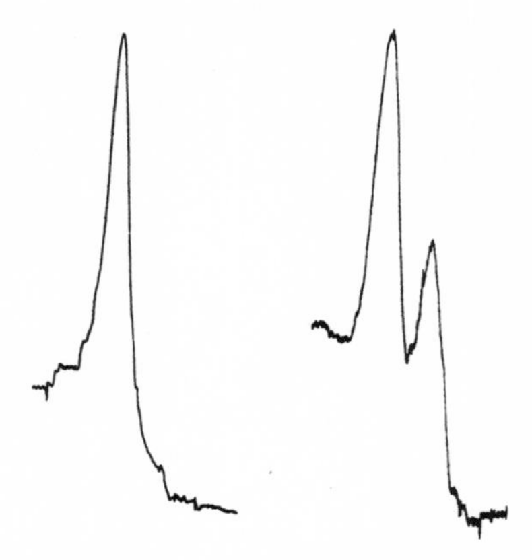

FIG. 2. **Two recordings of the front anodal protein band from the electrophoretic pattern of the soluble hippocampal protein from a control rat (left) and a trained rat (5 days). See text.**

solubility property of the S-100 protein. The band immediately behind that diminished in size but did not disappear completely. Hippocampal nerve cell protein from another group of five trained rats was electrophoretically separated on 400-μ diameter gels and precipitated with 80 percent alcohol for 3 min. The gel cylinders were placed in fluorescein-conjugated antiserum against S-100 (dilution 1:4) for 24 hr and then examined in a fluorescence microscope and photographed. Both front protein bands showed specific fluorescence and had thus reacted positively with the anti-S-100 antiserum.

The localization of the extra protein band closest to the anodal front, the reaction of both protein fractions to ammonium sulphate and anti-

TABLE 1. Frequency of Single and Double Front Anodal Protein Fractions in Electrophoretic patterns*

Controls		Resumed training on Day 14		Resumed training on Day 30	
One fraction	Two fractions	One fraction	Two fractions	One fraction	Two fractions
20	0	5	10	20	20

**75 polyacrylamide gels from 23 rats (7 controls, 4 resumed training on Day 14, 12 resumed training on Day 14 and on Day 30).*

serum against the S-100 protein merit the following conclusion. A second protein band in front of the S-100 protein complex had emerged in the nerve cell protein from the trained rats. This band contains S-100 protein. The result also indicates that the original S-100 band contains proteins other than the S-100 component.

The question then arose if the amount of S-100 protein had increased in the nerve cells of the trained animals. Measurements of the absorbance of the single protein band from controls and the two bands of the trained animals were performed and compared with an integrating microphotometer. The same amount of protein from trained and control animals was used for the electrophoresis and the procedure was identical in all experiments. The protein was stained with brilliant blue. The electrophoretic patterns were photographed together with a step wedge, and the areas under the curves were calculated. It was found that the amount of protein contained in the two anodal bands of the trained rats was ten percent greater than the amount of protein contained in the one band of the controls. Figure 2 gives an example of the recordings of the S-100 bands. The integrated value of the largest peak of the bands from the trained rats (right in Fig. 2) did not differ from that of the single protein band of the controls (left in Fig. 2). In addition, there is the new band exclusively containing S-100 protein in the samples from the trained rats (right in Fig. 2).

THE EFFECT ON ANTISERUM AGAINST S-100 PROTEIN. The next question was whether the increase of the S-100 protein reversal of handedness specifically relates the S-100 protein to learning processes occurring in the hippocampal nerve cells. As we pointed out above, training involves several factors not related to learning per se. In the reversal of handedness experiments, such unspecific factors have been eliminated or reduced to a minimum. The motor and sensory activity, attention, motivation, and reward are equated between the experimental and control animals, and the stress involved in reversal of handedness is minimal. In the following experiments, designed to test the specificity of the S-100 protein increase, even the stress factors have been equated between controls and experimental rats. A group of six rats were trained during two 25-min sessions per day for three days. Between the first and second training session on the fourth day, the rats were injected intraventricularly on both sides with 2×30 μg of antiserum against S-100 in 2×30 μl. During further training for three days after the injection, the rats did not increase in performance, i.e., the number of reaches per day remained at the same values as those immediately before the injection (Fig. 3). It should be noted that the rats were not affected by the S-100 antiserum with respect to motor function and sensory responses.

To demonstrate the specific effect on the antiserum against the S-100

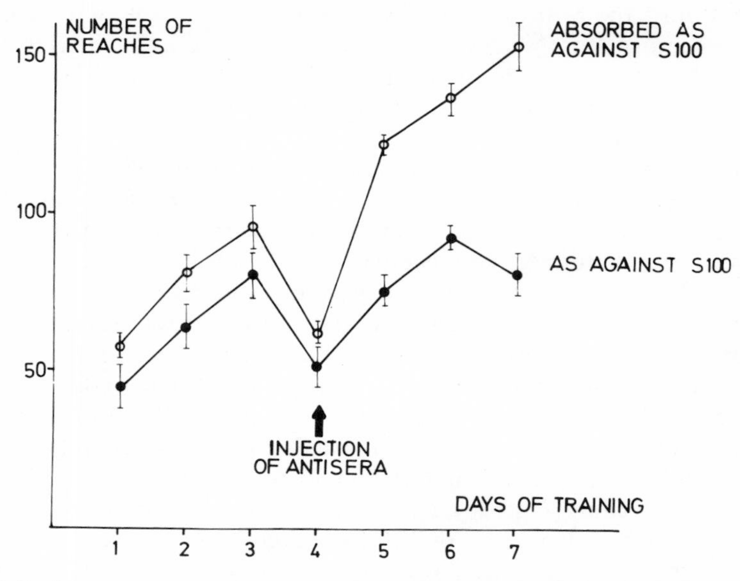

FIG. 3. **Performance curves of 6 rats injected intraventricularly on day 4 with 2 × 30 μg of antiserum against S-100 protein and 8 rats injected with 2 × 30 μg of antiserum against S-100 protein absorbed with S-100 protein.**

protein, the following experiment was carried out. Antiserum against S-100 was absorbed with S-100 protein according to a procedure previously used (Mihailović and Hydén, 1969).

The mixture of S-100 antiserum and the S-100 protein was left for 6 hr at room temperature and transferred to +4°C for a further 65 hr. Three times a day the mixture was agitated for 30 min by using a magnetic stirrer, and finally centrifuged. The supernatant was tested for effect of antibodies against the S-100 protein by Coons' double-layer method, with sheep antirabbit γ-globulin conjugated with fluorescein (Flow Lab., Irvine, Scotland). The test material was cryostat sections through the Deiters' nucleus from rats that were fixed in cold acetone after which both absorbed and unabsorbed antiserum against S-100 were applied to the sections and finally treated with the fluorescein conjugated antirabbit γ-globulin. When the sections were observed in the fluorescence microscope, a bright specific fluorescence was found in glial cell bodies in the sections treated with unabsorbed antiserum. On the other hand, with the use of the absorbed antiserum, only a weak unspecific fluorescence was found in the glia cells, not stronger than that in the nerve cell cytoplasm. Since the S-100 protein is mainly a glia protein and localized to glial cell bodies and to nuclei only of nerve cells, the result showed that the absorption with glia and nerve cell homogenate had removed the antibodies against the S-100 protein in the antiserum.

Eight rats were trained for four days and injected on Day 4 with 2×30 μg of S-100 antiserum absorbed with S-100, as described. The rats were then trained for further three days. The performance of these rats, as number of reaches per training session (Fig. 3), increased in the same way as did the performance of the noninjected control rats shown in Figure 1.

Since the active molecules of an antiserum have a large molecular weight, it is important to know if the antibodies injected intraventricularly will reach and can be localized to the hippocampal structures. Therefore, rats were injected intraventricularly with 2×30 μg of antiserum against S-100, and other rats with 2×30 μg S-100 antiserum absorbed with S-100 protein. At 1 hr (two rats) and 18 hr (four rats) after the injection, the rats were decapitated and cryostat sections were made of the hippocampus. Coons' double-layer method was applied on the two types of material to demonstrate the possible localization of antibodies to cell structures, by using a rabbit-antirat-γ-globulin conjugated with fluorescein isothiocyanate (Behring Werke AG, Marburg-Lahn, Germany). Figure 4 demonstrates specific fluorescence localized to nerve cells in the hippocampus of rats injected with antiserum against S-100 (Fig. 4a). Evidently, the hippocampus differs from the brain stem, insofar as the astro- and oligodendroglia do not contain the S-100 protein in amounts sufficient to give an immunofluorescent reaction. No such nerve cell fluorescence can be observed in the material from rats injected with S-100 antiserum absorbed with S-100 (Fig. 4b).

A pertinent question is whether the effect of the S-100 antiserum on behavior is due to a S-100 antibody-antigen reaction in limbic structures, as Klatzo et al. (1964) reported that fluorescein-labeled globulin does not penetrate through the ependyma into the brain tissue. When a small, local, cold lesion was produced on the surface of the brain cortex, Steinwall and Klatzo (1964) could show, on the other hand, that fluorescein-labeled globulin entered through the minute surface lesion and spread rapidly through the underlying subcortical area. In our present experiments, the antisera are injected into the narrow lateral ventricles. It can be suspected that the thin needle (20 gauge) may slightly damage the walls of the lateral ventricles when inserted and thus give free passage to globulins to enter. Sham injections were therefore made with NaCl administered intraventricularly in rats, in our standard way. The ventricles were then carefully exposed, flooded for 5 sec with a 0.1 percent erythrocin solution in 0.9 percent NaCl and after washing examined under low-power magnification in UV light. From the remaining staining it was seen that the walls of the lateral ventricles were superficially damaged where the needle had been touching the walls.

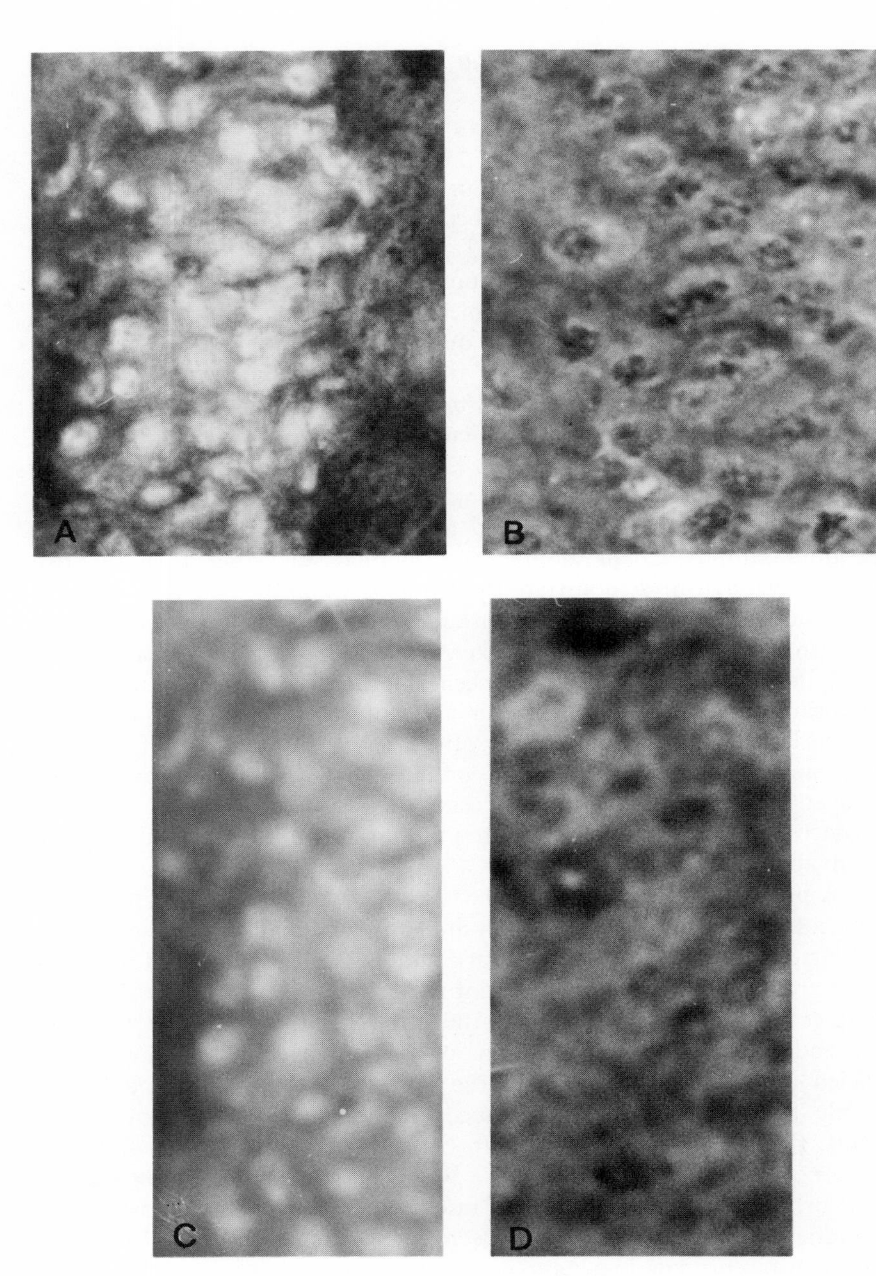

FIG. 4. Specific fluorescence showing the presence of antibodies against the S-100 protein localized to the nerve cell nucleus and cytoplasm in the CA3 region of the hippocampus (a) and in granular cells of gyrus dentatus of rats injected with 60 μg of antiserum against S-100 protein (c). There is no specific fluorescence to be seen in the corresponding structures of rats injected with antiserum against S-100 protein absorbed with S-100 protein (b and d).

In conclusion, it can be said that the S-100 protein is specifically correlated with learning processes within training. It does not mean, of course, that injection of S-100 protein should give rise to a spontaneous change of handedness.

It was then of interest to study a possible effect of injected antiserum containing antibodies not directed against S-100 protein. Therefore, six rats were injected with 2×25 μg of antiserum against rat γ-globulin from goat; four rats with the same amount of antiserum against rat γ-globulin from rabbit; and four rats with rabbit γ-globulin from goat. As is seen from the curves in Fig. 5, this has no impeding effect on the animals' performance. Before injection of antisera, all rats followed an identical performance curve. After the injection of antisera, these rats followed a performance curve which was an extrapolation of the performance curve before the injection. The same was the case with a rat which was injected with the same volume of physiological NaCl solution.

Another way to present the results is the following. For each rat, the sum of reaches for the first three training days is calculated, as is also the sum of reaches for the last three training days. The number of reaches during the day of injection are thus not included in these sums. The difference between the second and first sum is calculated. The averages of

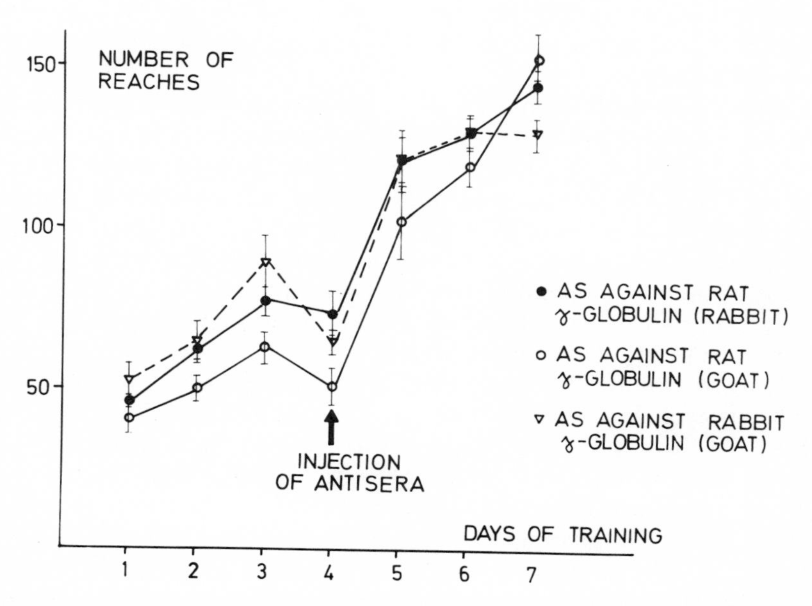

FIG. 5. Performance curves of rats injected intraventricularly on the 4th day with 2×30 μg of antiserum against rat gamma-globulin from rabbit (4 rats), and from goat (4 rats), and rabbit gamma-globulin from goat (4 rats).

this difference are 60 ± 14 for the rats injected with S-100 antiserum and 178 ± 14 for the control rats. The difference between these numbers is highly significant ($P<0.001$, 10 d.f.). It is clear that the experimental rats show a decrease in learning capacity.

Discussion

There are so far no experiments reported which have aimed at correlating brain protein components to learning processes in mammals.

In the present experiments, we have used a uniform nerve cell population of the hippocampus in rats. This brain region was chosen because of its importance for learning (Hydén and Lange, 1970). In order to relate behavior to biochemical correlates, correct control experiments are essential. In an ideal learning experiment, all factors involved in the training should be identical for experimental and control animals—with the exception of learning. The above experiments with four types of antisera fulfill these requirements. The increase of the S-100 protein in hippocampal nerve cells, its alteration in physical properties, and the blocking effect of the S-100 antiserum on behavior point to the S-100 protein as a true biochemical correlate to behavior. This conclusion is backed up by the localization of the antibodies against the S-100 by specific immunofluorescence and by the lack of effect of antisera not specifically directed to the S-100 protein.

Presumably the S-100 protein is not the only protein of importance for behavior; hence the role of the rest of the brain proteins in this experiment should be discussed. A young adult rat's brain contains about 200 mg protein. The S-100 protein constitutes around 0.4 mg thereof. Antisera were administered in amounts of 0.05 mg per rat, i.e., 0.025 percent of the total protein. The amount of antiserum against S-100 constituted 12.5 percent of the total amount of the S-100 protein of the whole brain. It is to be noted, though, that the antisera were injected intraventricularly and thus in the vicinity of the hippocampus.

Even so, it must be realized that the amount of anti-γ-globulin serum is not only exceedingly small relative to the total amount of brain protein but also relative to the total hippocampal protein. It is therefore questionable that the antiserum would influence the total brain protein synthesis in a measurable way. The antiserum against the S-100 protein is selective for the 0.2 percent of the total brain proteins. The same conclusion can therefore be made about the total protein synthesis. To confirm this assumption, the incorporation of ^{3}H-leucine into the hippocampal nerve cell protein was studied after the last training session on the seventh day. No difference in incorporation of ^{3}H-leucine in the hippocampal protein was found between rats receiving antiserum against S-100 and rats receiving rat γ-globulin antiserum intraventricularly. The effect observed of the S-100 protein antiserum on behavior can thus not be explained by an influence from a change total protein synthesis in the hippocampus.

In a recent paper (Hydén and Lange, 1970), we reported the response during reversal of handedness of two protein fractions of hippocampal nerve cell (CA3). These proteins move on the acidic side at electrophoresis at pH 8.3. During the course of one month of intermittent training (Days 1–4, 14, and 30–33), the values for the incorporation of ^{3}H-leucine into the protein were increased after the first and second training period, but not after the last training. The rats performed well at each training period. This temporal link between behavior and protein synthesis indicated that the protein response was linked to learning processes of the neurons and was not an expression of increased neural function in general. The two acidic protein fractions mentioned above are most probably not pure single protein components. It is therefore of special interest to observe that in the experiments presented in this paper we have been able to correlate a change in behavior to a single, defined nerve cell protein which is, furthermore, brain specific.

Janković et al. (1968) reported the in-vivo effect of antibrain protein antibodies on defensive conditioned reflexes in the cat. Their results show that the intraventricular injection of these antibodies produces significant changes in conditioned responses immediately after the administration and on subsequent days. Injection of antiliver antibody, normal γ-globulin, and saline produced no effect.

Mihailović et al. (1969) studied the effects of intraventricularly injected antibrain antibodies on visual discrimination tests performance in Rhesus monkeys. The animals were injected with anticaudate nucleus, antihippocampal and normal γ-globulin, respectively. The anticaudate and antihippocampal animals were significantly impaired in the performance as compared to the normal γ-globulin animals. The impairment was temporary.

It is also interesting to note that a high level of environmental stimulation leads to a thicker hippocampus, with a higher density in both oligo- and astroglia compared to those of control rats living in isolation (Walsh et al., 1969).

Summary

The brain-specific acidic protein S-100 in the pyramidal nerve cells of the hippocampus was investigated as a possible correlate to learning during transfer of handedness in rats. The amount of S-100 increased during training. Intraventricular injection of antiserum against the S-100 protein during the course of training prevented the rats from further learning but did not affect motor function in the animals. Antibodies against the S-100 protein could be localized after the injection to hippocampal structures by immunofluorescence; presumably they penetrated through slight ependymal lesions caused by the injection. By contrast, control animals subjected to the same training and injected with S-100 antiserum absorbed with S-100 protein or with other antisera against γ-globulins showed no decrease in their ability to learn. The conclusion is that the brain-specific protein S-100 is linked to the learning process within the training used.

Acknowledgments

We thank Dr. L. Levine, Brandeis University, Waltham, Mass., who kindly provided the antiserum against the S-100 protein, and Dr. Blake Moore, Department of Psychiatry, Washington University, School of Medicine, St. Louis, Mo., who generously gave us S-100 protein.

This study has been supported by the Swedish Medical Research Council, Grant B69-11X-86-05B, and a grant from Riksbankens Jubileumsfond.

The main content of this article is accepted for publication in proceedings of the National Academy of Sciences of USA, Vol. 67, March 10, 1970.

References

Bennett, G. S., and Edelman, G. M. (1968). J. Biol. Chem., 243:6234.

Bogoch, S. (1968). The Biochemistry of Memory. London: Oxford University Press.

Dannies, P. S., and Levine, L. (1969). Biochem. Biophys. Res. Commun., 37:587.

Hydén, H. (1959). Nature, 184:433. ——— and Egyhazi, E. (1964). Proc. Nat. Acad. Sci. U.S.A., 52:1030. ——— and Lange, P. W. (1968). J. Chromat., 35:336. ——— and Lange, P. W. (1970). Proc. Nat. Acad. Sci. U.S.A., 65:898. ——— and McEwen, B. (1966). Proc. Nat. Acad. Sci. U.S.A., 55:354.

Janković, B. D., Rakic, L., Veskov, R., and Horvat, J. (1968). Nature, 218:270.

Klatzo, I., Miquel, J., Ferris, P. J., Prokop, J. D., and Smith, D. E. (1964). J. Neuropath. Exp. Neurol., 23:18.

Kosinski, E., and Grabar, P. (1967). J. Neurochem., 14:273.

MacPherson, C. F. C., and Liakopolou, A. (1965). Fed. Proc., 24:Part 1, Abstr. 272.

McEwen, B. S., and Hydén, H. (1966). J. Neurochem., 13:823.

Mihailović, L., Divac, I., Mitrovic, K., Milosevic, D., and Janković, B. D.

(1969). Exp. Neurol., 24:325. ——— and Hydén, H. (1969). Brain Res., 16:243.
Moore, B. W., and McGregor, D. (1965). J. Biol. Chem., 240:1647. ——— and Perez, V. J. (1968). Physiological and Biochemical Aspects of Nervous Integration, Carlson, F. D., ed. Englewood Cliffs, N.J.: Prentice-Hall.
Steinwall, O., and Klatzo, I. (1964). Acta Neurol. Scand., 41:Suppl. 13.
Walsh, R. N., Budtz-Olsen, O. E., Penny, J. E., and Cummins, R. A. (1969). J. Comp. Neurol., 137:361.
Wareck, A. K., and Bauer, H. (1967). J. Neurochem., 14:783.
Wentworth, K. L. (1942). Genet. Psychol. Monogr., 26:55.
Zuckerman, J., Herschman, H., and Levine, L. (1970). J. Neurochem., 17:247.

Section III

MACROMOLECULES AND INTRACELLULAR, INTERCELLULAR, AND SYNAPTIC EVENTS

If one assumes that some macromolecules such as proteins exert their effects at the synapse, then there is the requirement of getting the protein which is synthesized in the soma to the synapse, or having an independent protein synthesis system in the axon. The first three chapters are concerned with basic aspects of the axon and synapse. Chapter 9 (Ochs) discusses the axoplasmic flow of substances whereas Chapter 10 (Koenig) examines evidence for an axonic synthesizing system. A detailed treatment of biochemical aspects of the synapse is provided in Chapter 11 (Appel). Chapter 12 (Rosenzweig, Bennett, and Diamond) describes a research program using the direct approach which began by evaluating synaptic chemicals during behavior. In the course of the investigations, however, numerous interesting neuroanatomical findings were uncovered. Chapter 13 (Deutsch) provides the results of experiments using drugs which affect synaptic chemicals and notes the effects on behavior (an indirect approach). The next chapter (Altman) discusses an autoradiographic procedure used to determine brain macromolecular changes and the growth of brain cells. It is suggested that development and migration of some undifferentiated cells into the synapse during behavior may provide new synaptic connections. Chapter 15 (Pevzner) is a comprehensive review of nucleic acid and protein changes in the neuron-neuroglia unit during behavior.

Chapter 9

AXOPLASMIC FLOW—THE FAST TRANSPORT SYSTEM IN MAMMALIAN NERVE FIBERS

SIDNEY OCHS

Indiana University Medical Center
Indianapolis, Indiana

In the relatively short time since a review of axoplasmic flow appeared in the first edition of this book (Ochs, 1966), study of the subject has undergone several remarkable shifts in emphasis. Whereas it was previously believed necessary to present evidence in support of the concept that materials synthesized in the neuron somas are continuously transported down inside the axons, a number of studies have since then fully confirmed the phenomenon (e.g., Friede, 1966; Barondes, 1967; Ochs, 1969; Grafstein, 1969; Droz, 1969). Another change was the realization that slow and fast-moving components of axoplasmic flow are present in the fibers. Most of the previous evidence obtained with isotope-labeling techniques had supported a rate of axoplasmic flow of several millimeters per day, a value close to an earlier one arrived at from morphological evidence of damming proximal to a constriction of the nerve trunk (Weiss and Hiscoe, 1948). The increase in the volume of the fibers in the dammed portions above the constriction was interpreted as reflecting a growth of the axoplasmic contents down inside the fibers. However, evidence obtained by Dahlström and Häggendal (1966), Karlsson and Sjostrand (1968), Kerkut et al. (1967), Burdwood (1965), Lasek (1967, 1968), Livett et al. (1968), Ochs et al. (1967), Ochs and Johnson (1969), and Sjöstrand (1969) indicated the presence of a much faster moving component additional to the slow-moving component.

By making use of the longer lengths of fiber present in the cat sciatic nerve, and injecting the precursor ^{3}H-leucine into the lumbar seventh (L7) ganglion, definitive evidence of a fast transport system was given by a crest of activity in the sciatic nerve, with its displacement showing a movement down the fibers at a rate close to 400 mm/day (Ochs et al.,

1969). Our recent work on this system and its differentiation from the slow transport system will be presented. The rapidity and the regularity with which materials are carried down the fibers by the fast transport system suggest that this system plays a crucial role in nerve function. For example, Wallerian degeneration occurs in a few days distal to a transection. The molecular nature of the transport system, however, is not yet known. Recently, a new advance has been made which gives us hope of uncovering the mechanism. Fast transport is locally present all along the fibers and can be maintained in vitro by oxygen (Ochs and Ranish, 1970b). A close dependence on oxidative metabolism was found, and this will be described in a later section of this review.

Fast and Slow Axoplasmic Transport Systems

The distribution of activity found in cat L7 ventral roots removed from animals at various times after injection of ^{3}H-leucine into the cord ventral horn region near L7 motoneurons showed, for the most part, an outward moving axoplasmic flow similar to that previously reported with ^{32}P orthophosphate (^{32}P) as the precursor (Ochs et al., 1962). There was, however, a difference seen early after injection with a break in the distribution of activity indicative of a second fast-moving component having a rate of flow of at least several hundred mm per day (Ochs et al., 1967). Lasek (1967, 1968) had, on similar grounds, also estimated a faster component as being present in sensory nerve fibers. When the interval between ventral horn injection and root removal was shortened to several hours, the first component of axoplasmic flow was more clearly evident (Ochs and Johnson, 1969). Reasons were given, however, for not placing to much credence for an estimation of the rate of fast transport on the basis of the furthest extent of the spread of activity in a short time interval. Fortunately, our hopes for a better method of determining the rate of the fast transport component were realized. In studies made by using the longer length of fiber available in the sciatic nerve, the L7 dorsal root ganglia were injected with ^{3}H-leucine and the sciatic nerves removed at times from 2 to 8 hr after injection (Ochs et al., 1969). The distribution of activity found present in the sciatic nerves clearly showed the presence of a fast axoplasmic flow by the crest of activity present in the nerves (Fig. 1). The pairs of curves in this illustration represent the distribution of activities in the two sciatic nerves taken from each of the animals sacrificed at the indicated times after injection of their ganglia. In the nerves removed 2 hr after injection, fast transport is suggested only by the outward slope of activity spreading from the ganglia into the dorsal roots and into the sciatic nerves. However, at 3 hr and thereafter, a plateau of

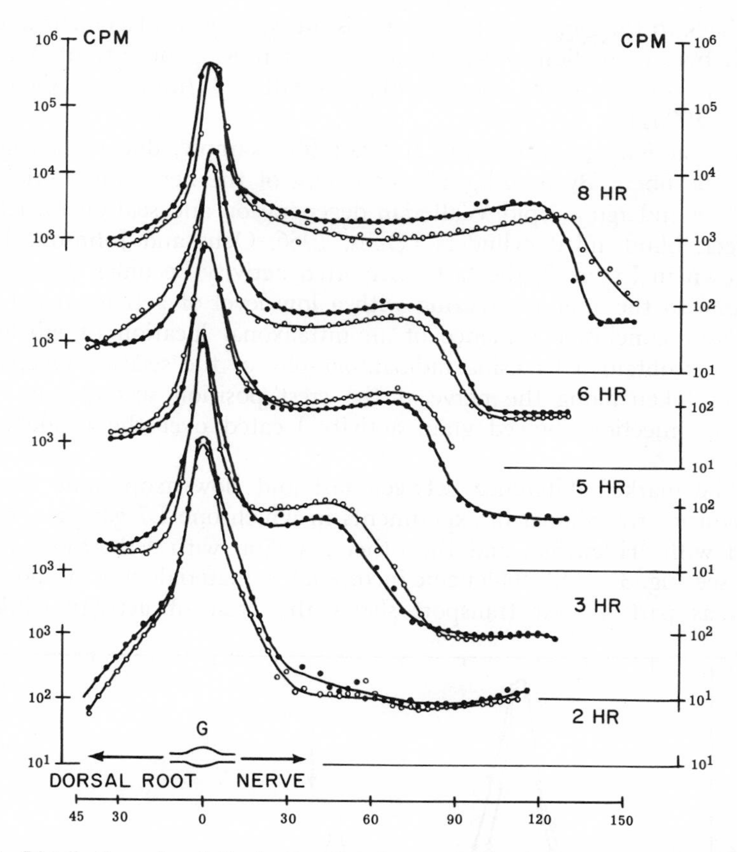

FIG. 1. Distribution of activity in the dorsal roots and sciatic nerves of five cats at times from 2 to 8 hours after injection of ^{3}H-leucine into the L7 ganglia (G). The symbols (O, ●) represent the activity in 3-mm segments taken from roots, ganglia and nerves of each side. The ordinate scales are logarithmic and shown only in part. That for the 2-hour nerves is given at the bottom left with divisions in counts per minute (cpm). At bottom right, only the lowest level, 10 cpm is shown. Above at the top on the right, a full scale is given for the 8-hour nerves. Abscissa gives distance in mm. (From Ochs, Sabri, Johnson. 1969. *Science,* 163:686.)

activity was found distal to the ganglia, with a rise to a crest more distally in the nerve and a sharp fall of the crest to background levels. This crest of activity represents a wave of labeled materials being moved down the fibers, with the crest appearing at increasingly greater distances from the ganglion as more time elapses between L7 ganglion injection and the removal of the sciatic nerve. Because the crests have essentially the same shape from experiment to experiment, an estimation of the rate of axoplasmic transport can be made from a line drawn through the forward slope back to the ganglion. By using this measure of displacement and the time between ganglion injection and nerve removal, the rate of transport found in large series of experiments was 401 ± 35/mm day (Ochs and Ranish, 1970a).

The activity is intraaxonic and not for example, due to leakage between the fibers. By freezing a short length of the nerve for a brief time the fibers undergo a rapid Wallerian degeneration and seal off to become, in effect, blind-ended cylinders (Ochs, 1966; Ochs and Johnson, 1969). As shown in Figure 2, the fast-transported activity becomes dammed up proximal to the zone of freezing with a low level of activity distal to it. The most conclusive evidence of an intraaxonal location of labeled activity was obtained by using radioautography of the sciatic nerve. Nerve sections taken along the nerve at the crest position several hours after ganglion injection showed grain activity located over the inside of the fibers.

The marked difference between fast and slow axoplasmic transport mechanisms was shown in experiments in which one L7 ganglion was injected with ^{3}H-leucine, and the other ganglion with ^{32}P-orthophosphate (^{32}P; see Fig. 3). The ^{3}H-leucine-incorporated materials moving down the axons as part of fast transport shows the crest of activity while the

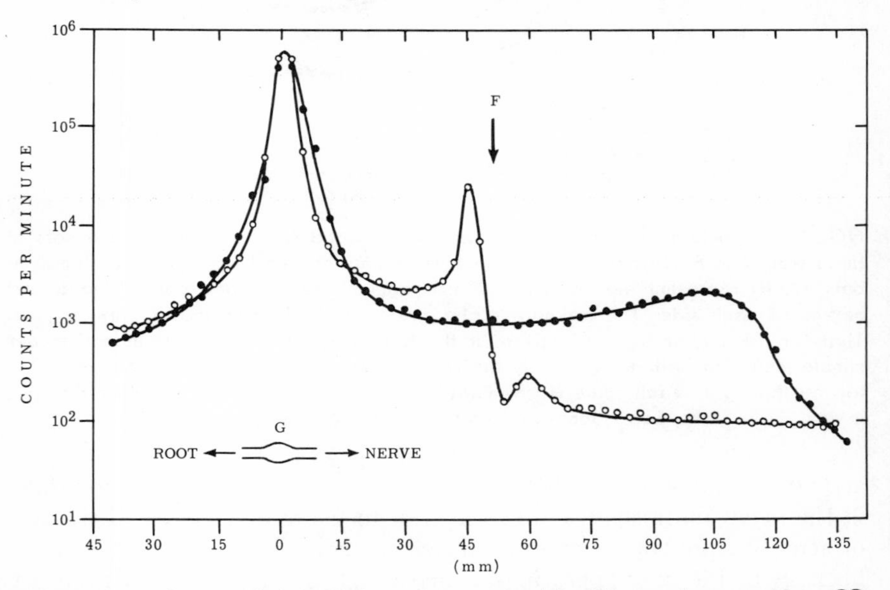

FIG. 2. A small part of the sciatic nerve on one side (O) was frozen with a CO_2 cooled bar 16 hr before injection of the L7 ganglion. The other side (●) was injected with ^{3}H-leucine as a control. A damming of activity is present above the frozen zone (F) with a fall to base-line levels distally. (From Ochs and Ranish. 1970a. *J. Neurobiol.,* 1:247.)

^{32}P-incorporated activity remains close to the ganglion, with only a suggestion of slow axoplasmic transport (indicated by the broadness of the activity sloping from the ganglion). The slow phase of transport becomes readily evident when sciatic nerves were taken several days after L7 ganglion injection with ^{32}P (Ochs and Ranish, 1970a). In that case an outward, declining slope of activity was seen similar to that previously described for the motor fibers of the L7 ventral roots. The results confirm our earlier studies in which ^{32}P-incorporated materials were found to move down the motor nerve fibers at a slow rate (Ochs et al., 1962). It is likely that these components enter into the distribution of the phosphorous compounds, described by Miani (1963).

Fast transport is also present in motor fibers, as shown by a similar crest of activity in the sciatic nerve when ^{3}H-leucine is injected into the ventral horn region of the spinal cord near the L7 motoneurons. A technical difficulty is that the precursor can spread to the motoneurons of the adjoining L6 and S1 segments and upon incorporation flow down the fibers of the L6 and S1 ventral roots (as well as the L7 roots) to produce an irregular pattern of activity in the sciatic nerve (Ochs and Johnson, published experiments). To prevent this, the L6 and S1 roots were first ligated and then the L7 ventral horn on one side of the spinal cord was

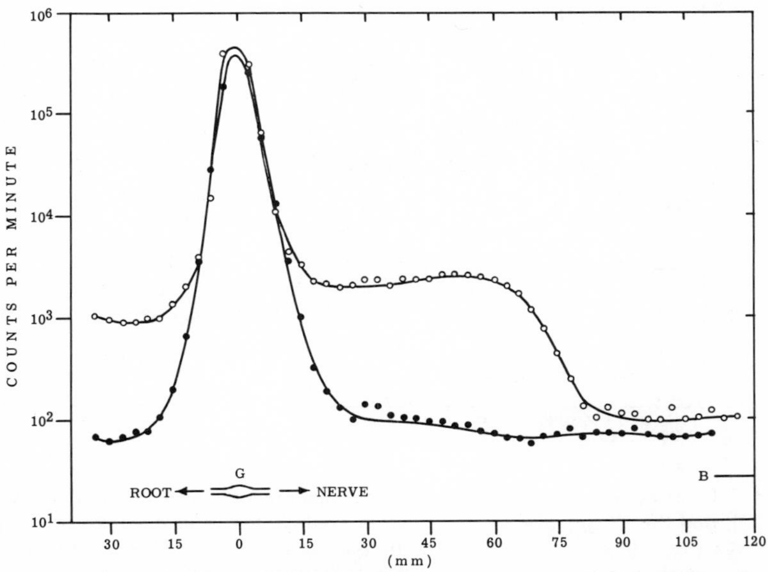

FIG. 3. A peak of activity due to fast transport is present in the nerve on the ^{3}H-leucine injected side (O) with little evidence of fast transport present in the nerve or root on the ^{32}P injected side (●). Nerves removed for sampling 5 hr after injections. (From Ochs and Ranish. 1970a. *J. Neurobiol.*, 1:247.)

injected with ^{3}H-leucine. The L7 dorsal root ganglion on the opposite side was also injected in such experiments so that a comparison with fast transport in sensory nerve fibers could be made. Six hours after the injections, the sciatic nerves on the two sides were removed and the distributions of activity determined. The usual crest of activity indicative of fast transport was present in the sciatic nerve on the L7 ganglion injected side, and a similar though more proximally positioned crest of activity was found in the sciatic nerve on the spinal cord injected side (Fig. 4). The proximo-distal placement of the two crests could be adequately accounted for by more distal position of the L7 ganglion relative to the position of the L7 motoneuron cell bodies in the spinal cord and a similar rate of fast transport for motor and sensory fibers (Ochs and Ranish, 1970a).

It is not known as yet whether the differing rates given in the literature for fast transport represents species differences or whether they reflect different methods of measurement of axoplasmic flow. Some reported fast transport rates are not too far different from those we have found for the cat. Miledi and Slater (1970) determined a flow rate of 360 mm/day for the rat by using the failure of MEPP activity in diaphragm muscle upon

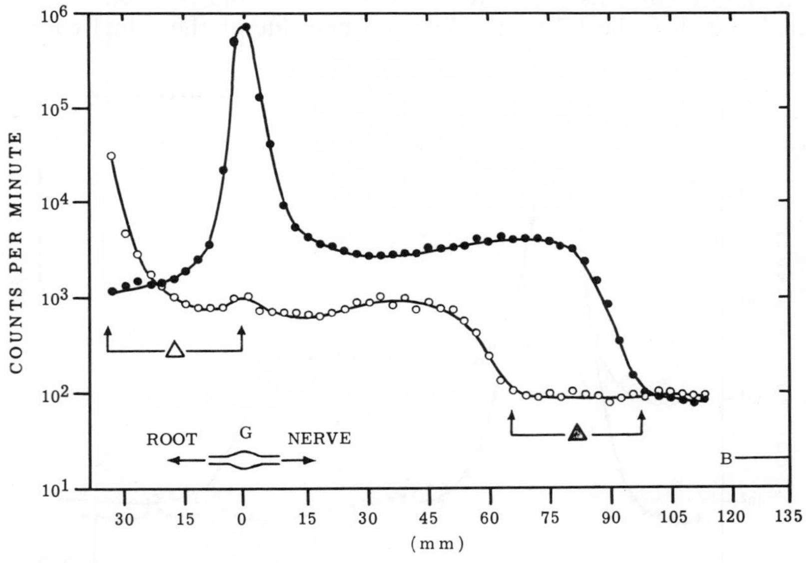

FIG. 4. ^{3}H-leucine was injected into the L7 motoneuron region on the left side of the cord (O), and the L7 dorsal root ganglion injected on the right side (●). Six hr later the nerves were removed for sampling. The antero-posterior displacement (△) of the peaks is comparable to the anatomical antero-posterior positions of the cord motoneurons and ganglion cell bodies (△). (From Ochs and Ranish. 1970a. *J. Neurobiol.,* 1:247.)

cutting phrenic nerves close to and far from the muscle. Bray and Austin (1969) estimated a rate of transport in chicken sciatic nerve of 250 to 350 mm/day. Sjöstrand (1969) reported a rate of 383 to 408 mm/day for the rabbit vagus and 240 to 380 mm/day for the hypoglossal nerve. Using the crest displacement as the measure of fast transport in their sciatic nerves, a rate of 360 mm/day was found for the goat and rabbit and a rate of 400 mm/day for the rhesus monkey (Ochs, unpublished data).

Labeled Components Carried by Fast and Slow Axoplasmic Transport Mechanisms

In addition to the transport differences seen when using ^{32}P and ^{3}H-leucine as precursors (Fig. 3), evidence that different materials incorporated after ^{3}H-leucine injection are carried down by the fast and slow transport systems was obtained in studies of ventral roots. The roots taken at different times after injection into the L7 ventral horn region were homogenized and subjected to subcellular centrifugation (Ochs, et al., 1967; Kidwai and Ochs, 1969). Activity was found in the "nuclear" (N), "mitochondrial" (M), "particulate" (P), and high-speed supernatant or soluble (S) fraction. The soluble fraction had the highest relative level of activity, with most of the labeling present in the soluble protein as shown by TCA precipitation. After hydrolysis of the TCA precipitate, the leucine identified with paper chromatography was found to contain the label. Somewhat similar results were later reported by Bray and Austin (1968) and by McEwen and Grafstein (1968).

The components of the soluble fraction were further separated with Sephadex G-100 gel filtration. Labeled higher-molecular-weight soluble proteins appeared in the first emerging peak while the second peak contained lower-molecular-weight polypeptides and free leucine (Ochs, 1967; Kidwai and Ochs, 1969). Changes in the relative degree of labeling of the two peaks were related to the fast and slow phases of axoplasmic flow. More activity was present in the polypeptides and free leucine during fast flow at early times, and after a day slow transport carried down more labeled activity in the higher-molecular-weight proteins.

A still further analysis of the soluble protein was made possible with Sephadex G-200 columns (Fig. 5). As shown by suitable markers the soluble proteins were separable into several components having molecular weights of approximately 450,000 and 68,000 and labeled Peaks Ia and Ib, respectively. The lower-weight polypeptides in the molecular weight range of 5 to 10,000 constitutes Peak II (Fig. 5). Less activity is found in the protein peaks at early times (5 hr) while later (21 hr) more soluble protein is transported (Ochs et al., 1969a). A characteristic of the fast transport mechanism is the high specific activity present in the small particulate (P) fraction. Using the goldfish optic nerve system, McEwen and Grafstein (1968) placed more emphasis on the labeling of particulate materials. In their study, all the sediment of a high-speed centrifugation comprised their "particulate" fraction. In our studies, a high specific activity was found in the small particulate fraction and a smaller degree of labeling of the "N" and "M" fractions. Some of the P fraction could have a transmitter function, but the high specific activity found for the P fraction of

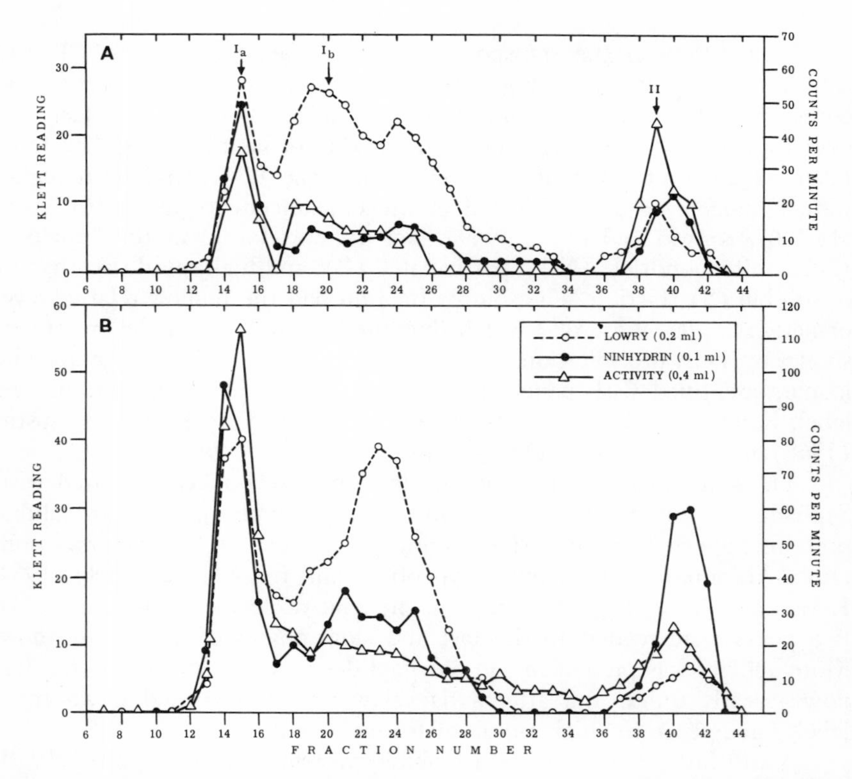

FIG. 5. Sephadex G-200 gel filtration eluates of the high speed supernatant (fraction S) of nerves (two combined) taken 6 hr (A) and 21 hr (B) after injection of the L7 ganglia with ^{3}H-leucine. Peaks Ia and Ib represent higher weight proteins and peak II polypeptides. Lowry (O) and ninhydrin (●) determinations of aliquots from the fraction tubes on the abscissa are represented on the ordinate with 100 Klett units equal to 100 μg Bovine serum albumin for Lowry and 10 μg of L-leucine for ninhydrin. Activity (△) in cpm. (From Ochs, Sabri, and Ranish. 1970a. *J. Neurobiol.,* 1:329.)

sensory nerves indicates that this component has a function in nerve which is not yet understood (Ochs et al., 1969b).

In addition to a movement of labeled materials into the terminals by fast transport (Droz and Barondes, 1969), there is also a possibility that a local synthesis of protein can take place in the terminals (Austin and Morgan, 1967; Autilio et al., 1968; Morgan and Austin, 1969). Possibly such a local synthesis in the terminal is related to an uptake of transmitter substances or their precursors, a phenomenon which has been extensively studied for autonomic nerve (Iversen, 1967). The relationship of an uptake of transmitter precursors by the terminals to the evidence of an axoplasmic transport of transmitter materials (Dahlström and Haggendal, 1966; Livett et al., 1968; Bank et al., 1969; Johnson, 1970) is not fully understood.

Somal Origin of Labeled Components in the Axons and Synthesis Time

Our earlier evidence for a somal synthesis of the axonally transported materials in ventral roots was the effect of puromycin injected near the cell bodies ½ to 1 hr before ^{3}H-leucine to block the downflow of activity (Ochs et al., 1967; Kidwai and Ochs, 1969). The mechanism of puromycin's action is likely due to its molecular similarity to tRNA acting on the somal ribosomes, where a complex array of coupling agents (Lipmann, 1969) are required for synthesis of protein chains. Similar results were obtained with acetocycloheximide as the protein-blocking agent with use of the retinal neurons and optic nerve outflow of labeled incorporated components by McEwen and Grafstein (1968) and with puromycin by Sjöstrand and Karlsson (1969). However, some amount of RNA is present in axons (Miani et al., 1966); Koenig, 1967; Edström et al., 1969) and protein synthesis was reported in axons separated from Schwann cells (Koenig, 1967; Edström and Sjöstrand, 1969). There is also the possibility that some materials synthesized locally in the Schwann cells can gain access to the axons (Singer and Salpeter, 1968).

In recent studies of the effect of protein-blocking agents on fast-transported substances, puromycin when injected into the L7 ganglion in a concentration of 10 mM before the injection of ^{3}H-leucine into that same region was effective in blocking the downflow of labeled material into the sciatic nerve (Ochs et al., 1970a). Puromycin blocked when injected at times from approximately 3 hr to just before the time of injection of ^{3}H-leucine. As can be seen in Figure 6, puromycin injected 1 hr before ^{3}H-leucine was effective in blocking the activity expected in the plateau

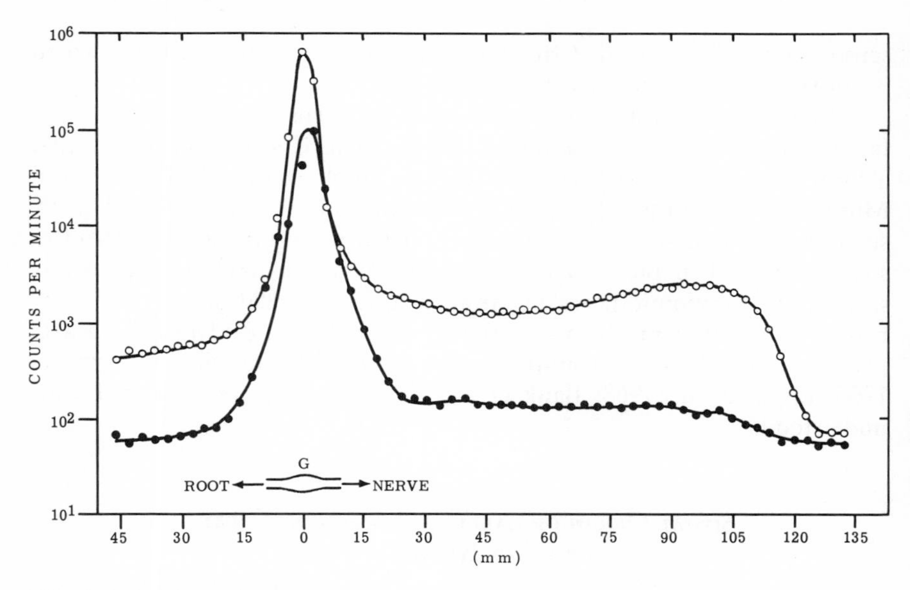

FIG. 6. Left L7 ganglion (●) was injected with 10 μl of a 4.3 mM of puromycin, the right (O) with an equal volume of Ringer solution. One hour later both ganglia injected with 5 μl of ^{3}H-leucine. Seven hrs afterwards the nerves, L7 ganglia and dorsal roots were removed and sectioned into 3 mm pieces for determination of activity distribution. (From Ochs, Sabri, and Ranish. 1970a. *J. Neurobiol.,* 1:329.)

and the crest regions. A reduction in the amount of activity present in the ganglia was also seen. Cycloheximide in a concentration of 18 mM also produced a block when it was injected into the ganglion 20 to 30 min before ^{3}H-leucine (see Fig. 7A). In homogenates of the nerve taken from the side injected with puromycin or cycloheximide, a relative increase in the supernatant fraction was found compared to other components; this suggests a downflow of incomplete polypeptides chains (Ochs et al., 1970a).

The above experiments with puromycin and cycloheximide are in accord with a somal blocking action on protein synthesis, except for the possibility that these agents enter the soma and are then transported down into the axons to effect a block of protein synthesis locally in the fibers. In order to test this hypothesis ^{3}H-puromycin was injected into the ganglion on one side while the contralateral ganglion was injected with ^{3}H-leucine (Ochs and Ranish, 1970a). The experiments showed that puromycin is not moved down the fibers by the fast transport system and therefore it does not block protein synthesis locally in the fibers. Puromycin and cycloheximide have a similar time course of action, thus providing further support for their point of action on the ribosomes in the soma, although their mechanism of action differs. Puromycin acts by blocking tRNA on ribosomes (Yarmolinsky and de Haba, 1969) while

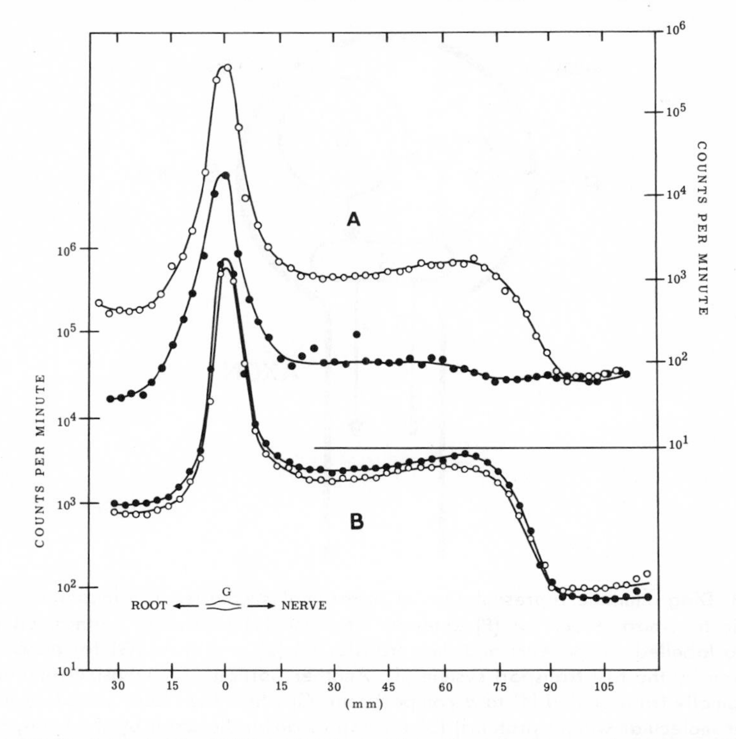

FIG. 7. (A) Cycloheximide (17.8 mM) injected into the left L7 ganglion (O) and Ringer solution into the right (●) 20 min before injection of ³H-leucine into each ganglia. Nerves removed 4 hr later. (B) Cycloheximide (17.8 mM) injected into the right L7 ganglion (O) and Ringer solution into the left (●) 30 min after ³H-leucine was injected into each ganglion. Nerves removed 5 hr after ³H-leucine injection. (From Ochs, Sabri, and Ranish. 1970a. *J. Neurobiol.,* 1:329.)

cycloheximide effects a block by an interruption of the enzymatic transfer action of mRNA along the ribosomes (Wettstein et al., 1964).

When the order of administration was reversed and puromycin as cycloheximide was injected only 5 min after ³H-leucine, an incomplete block was found. Very little effect was seen when these agents were injected 20 to 30 min after ³H-leucine (see Fig. 7B). The failure of blocking effect when puromycin and cycloheximide was injected so soon after the precursor suggests that protein synthesis in the somas occurs quickly after precursor entry, a result in conformity with the rapid turnover found for free leucine in the CNS (Lajtha, 1964). It also indicates that the later outflow of a high level of labeled soluble proteins into the axons comes from a somal compartment which is insensitive to the action of the protein synthesis blocking agents, a concept pictured in Figure 8. A portion of the materials synthesized is transferred to the postulated compartment in the

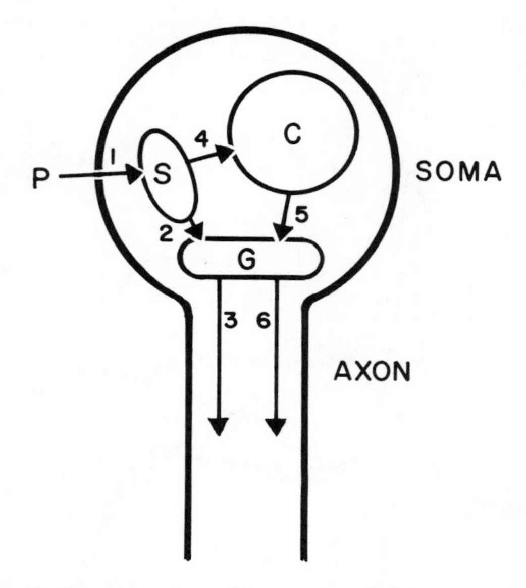

FIG. 8. Diagrammatic representation of intrasomal synthesis, translocation and axoplasmic transport. Precursor (P) entering the soma (1) is rapidly synthesized at site (S) into labelled components and then transferred (2) to a gate (G) for export down the axon by the fast transport system (3). Another portion of synthesized materials is intrasomally translocated (4) to a compartment (C) which then later releases materials (higher molecular weight proteins) (5) for export down the axon by the transport systems (6). (From Ochs, Sabri, and Johnson. 1969a. *Science,* 163:686.)

soma (C) and then, later, proteins from the compartment are exported down into the axon to be carried outward by the transport mechanisms.

An alternative possibility to the above schema is that the free leucine and polypeptides carried down into the axon by the fast transport system are locally synthesized into higher-molecular-weight soluble proteins in the axons. To examine this point the L7 ganglion was injected and labeled materials were allowed to enter the nerve fibers via fast transport for a period of 5 hr. Then, further entry into the axons was prevented by ligating the nerve just distal to the ganglion. The other ganglion was left intact so that a normal unhindered downflow of labeled components could take place in its nerve fibers. A total of 21 hr elapsed before the animals were sacrificed and the nerves removed for comparison. If a local synthesis of protein had occurred in the fibers during the 16 hr when the nerves remained in the animal after ligation, the composition of labeled materials in those nerves should be similar to that present in control unligated nerves taken from the opposite side. The results were that the ligated nerves had actually less activity in the protein than did the control nerves taken 5 to 6 hr after ganglion injection (Ochs et al., 1969b). Not only did a local incorporation not take place but the fast transport system moved labeled compounds distally out of the sampled region.

The Presence of the Fast Transport Mechanism Locally in Axons

The form of the peak activity moved down the nerve by the fast transport system suggests that diffusion or a force of propulsion exerted by the soma is not the cause of this phenomenon. Definitive evidence that the mechanism of fast axoplasmic transport is locally present along all the fiber itself was obtained by "exclusion" experiments. Ganglia were injected with ^{3}H-leucine and a period of time allowed for labeled materials to gain entry into the axons. Further entry of materials from the somas was prevented by means of a ligature tied just distal to the ganglion. An additional period of time was then allowed for fast transport to carry labeled materials farther down the nerve fibers to a more distal position. With ligatures placed 1 to 3 hr after ^{3}H-leucine injection, and an additional time of 4 to 6 hr allowed for transport, a peak of activity was seen distally in the nerve at the same position as the crest on the control side (Fig. 9). When more time was allowed for somal outflow before the ligation was made, there was more activity present in the peak moved down to a position comparable to the crest in the control nerve. These results show that fast transport continues at the usual rate in ligated nerves, even though the continuity with the cell bodies in the ganglion had been destroyed so that no somal force could be exerted on the axons.

The Dependence of Fast Transport on Oxidative Metabolism

The local mechanism of transport was found to require oxidation for its maintenance (Ochs and Ranish, 1970; Ochs et al., 1970a; Ochs and Hollingsworth, 1970). The L7 ganglia in a group of animals were injected with either ^{3}H-leucine or ^{3}H-lysine and then 3 hr later the animals were bled to death and the nerve on one side was left in situ for an additional 3 hr in the animal while the temperature was kept close to 38°C.

There was no further distal displacement of the crest position in those nerves as compared to the 3-hr control nerve (Fig. 10). This suggested that in the nerve in situ, fast transport failed because of asphyxiation. To test this point in a subsequent group of experiments, nerves were removed 3 hr after ganglion injection with ^{3}H-leucine and then placed into a chamber containing either O_2 or 5 percent CO_2 + 95 percent O_2 for

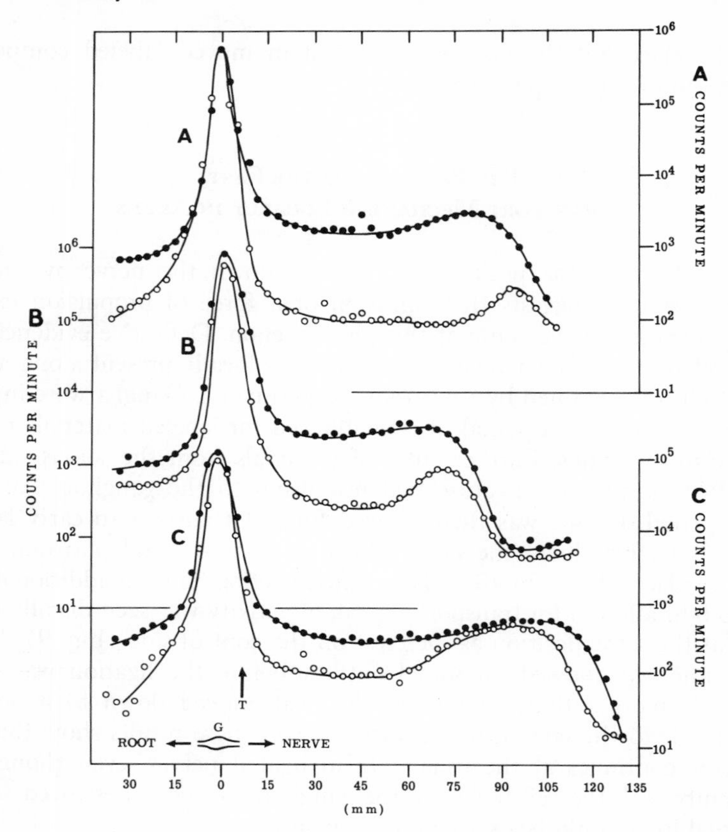

FIG. 9. (A) The L7 ganglia on the two sides were injected with ^{3}H-leucine and one hour later the nerve on one side (O) was ligated just distal to the ganglion (T). Six hours later control and ligated nerves were removed. The control side (●) shows a crest typical of 7 hours of downflow; the ligated nerve shows a peak at this same position. (B) As in A, except that the S1 ganglion was injected and the nerve (O) was ligated just below the S1 ganglion 2 hr after ^{3}H-leucine injection. It and control (●) nerves were removed 6 hr after injection. (C) Nerve ligation (O) was made just distal to the L7 ganglion on one side 3 hr after injection. Control (●) not ligated. (From Ochs and Ranish. 1970a. *J. Neurobiol.*, 1:247.)

an additional 3 hr. The nerve was kept moist with Ringer solution at a temperature of 38°C. In that case an additional distal movement of the crest of activity occurred, indicating that fast transport was maintained in vitro under those conditions (Fig. 11). The rate of fast transport found for the preparation was close to 401 ± 35 mm/day, the rate found in the living animal (Ochs and Ranish, 1970a).

When in similar such in vitro experiments the nerves were exposed in the chamber to N_2, fast transport was blocked as shown by the lack of

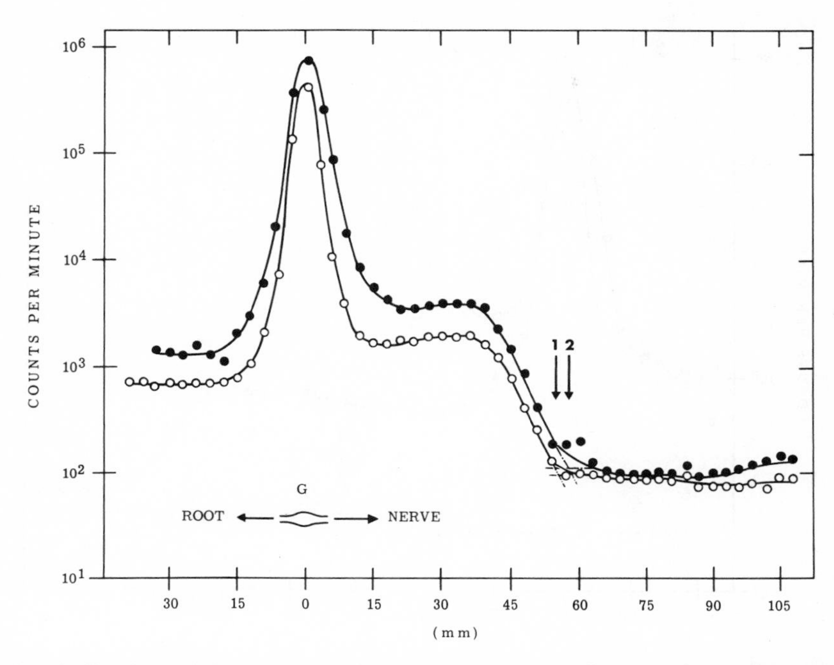

FIG. 10. **The L7 ganglia on each side were injected with ^{3}H-leucine and 3 hr later the sciatic nerve, ganglion and dorsal root from the left (O) side was removed from the animal. The extent of transport is shown at the foot of the crest by the dashed lines by Arrow 1. The nerve on the right side (●) was left in situ in the sacrificed animal for an additional 3 hr and very little additional displacement of its crest position is shown by Arrow 2. (From Ochs and Ranish. 1970b. *Science,* 167:878.)**

a further movement of the crest of activity. It was particularly striking how little the crest of activity had moved in the N_2 asphyxiated nerves beyond that of the controls, indicating that the failure of fast transport occurred well within 15 min of the time of asphyxiation. A similar rapid block of fast transport was seen when nerves in vitro were exposed to a solution of 20 mM NaCN (Ochs et al., 1970b; Ochs and Hollingsworth, 1970).

The time of failure of fast transport is similar to the time of failure of mammalian nerves to conduct action potentials upon their asphyxiation (Gerard, 1932; Lehman, 1937; Wright, 1946, 1947; Maruhashi and Wright, 1967). A time of 20 to 30 min was given for the survival time of action potentials on N_2 asphyxiation. Somewhat shorter survival time of approximately 10 min were found in our studies (Ochs et al., 1970b). In any case, the similarity of the survival time of action potentials on asphyxiation to the failure time of fast transport suggests an exhaustion of a common source of energy supply. Most likely this is due to an interrup-

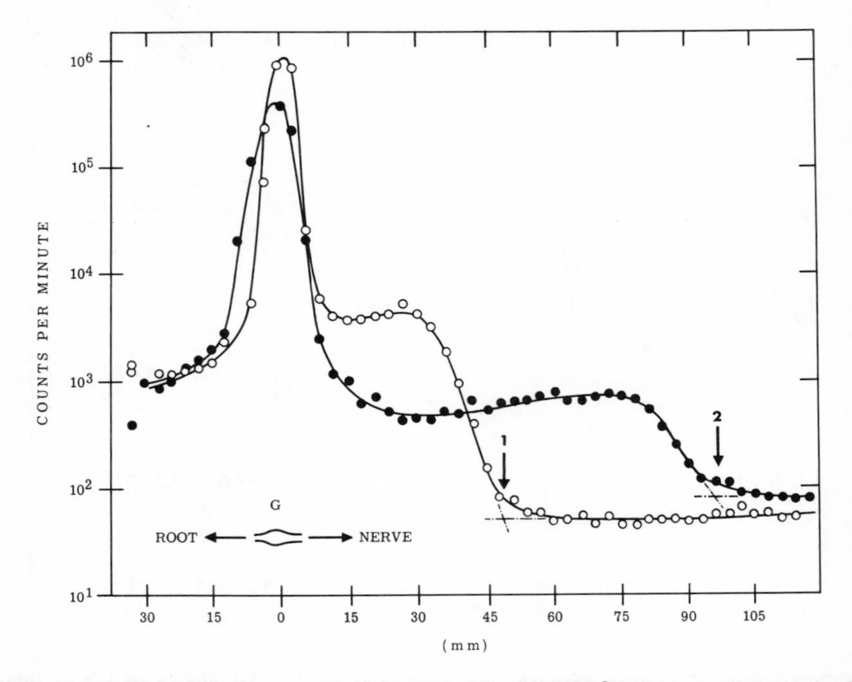

FIG. 11. The L7 ganglia on each side were injected with ^{3}H-lysine and the animal sacrificed 3 hr later. The control nerve (O) removed from the animal at that time shows a crest with as indicated by Arrow 1 at the foot at a distance expected of a fast transport lasting 3 hrs. The opposite nerve (●) was removed from the animal placed in a chamber containing O_2 and kept moist with Ringer solution at a temperature of 38° C for an additional 3 hr. The displacement of the crest in this in vitro preparation shown by Arrow 2 is that expected of a fast axoplasmic transport lasting for a period of 6 hr. (From Ochs and Ranish. 1970b. *Science*, 167:878.)

tion of the supply of ATP to the ionic mechanisms underlying the excitability of the nerve membrane and as well to the mechanisms subserving fast transport.

Hypothesis Regarding the Fast Transport Mechanism

Earlier the hypothesis that best seemed to explain axoplasmic flow was a contraction of the axoplasm, which in exaggerated form was seen as the phenomenon of beading (Weiss and Pillai, 1965; Ochs, 1966). This mechanism now seems more likely to explain transport (Ochs and Johnson, 1969). Attention turned to the possibility that microtubules (neurotubules) and/or neurofilaments were somehow involved (Burdwood, 1965; Ochs, 1967; Schmitt, 1968). The beading phenomenon had shown that the neurofilaments and/or neurotubules were not themselves labeled (Ochs, 1966). This was done by stretching nerves to produce beading, freeze-substituting, and then preparing them for radioautography. The grains were not present in greater amount in the constricted areas where a higher concentration of neurofilaments and neurotubules are contained, indicating that these elements are not being continuously manufactured and spun out of the somas. This result, however, does not eliminate the possibility that these linearly organized elements are involved in transport (Ochs, 1967).

The importance of microtubules for the maintenance of form and transport in a wide variety of cell types was emphasized by Porter, 1966, and John and Joyce, 1969. The similarity of microtubules to neurotubles suggested that colchicine, which blocks cell division (Borisy and Taylor, 1967) may also disrupt the microtubules (neurotubules) of neurons. A blocking action of colchicine on axonal transport has since been reported by Dahlström (1968), Karlsson and Sjöstrand (1969), and by Kreutzberg (1969), though in some cases the concentrations of colchicine applied to peripheral nerve were high, allowing for the possibility of a non-specific effect. Schmitt (1968), considering that only particulates are fast transported, hypothesized that the particles have almost matching sites on them similar to sites on the neurofilaments. Then, as in the sliding filament theory of muscle, the particles roll down the fiber by making and breaking bonds at the matching sites. However, present concepts of the sliding filament mechanism in striated muscle appear to be more complex than those earlier envisioned, with a special ratchet action at the meromyosin ends of the myosin filaments (Huxley, 1969). Most difficult to encompass in the small particle schematization is our finding that soluble proteins, polypeptides, and free leucine—as well as small particulates—are carried down the axons by the fast transport mechanism (Ochs et al., 1967; Ochs et al., 1969; Ochs, 1969; Kidwai and Ochs, 1969; Ochs et al., 1969a).

An hypothesis which might account for the transport of all these diverse species is that the microtubules or neurofilaments have sites along their axial lengths along which a transporting protein leaving the soma moves as one member of a sliding system, as is the case in muscle contraction. All the varied species carried down by fast transport would be bound to this transport protein and in this way materials are moved distally in the axons. In accord with this concept actomyosin-like contractile proteins have been extracted from brain tissue (Berl and Puszkin, 1969).

A related problem is how mitochondria are transported down the nerve fiber. The amount of labeled activity we find in the mitochondrial fraction is not as relatively high as in the P fraction, and we may have to envision transport of mitochondria as a separate process.

However, some studies indicate that mitochondria may be transported down the axon from a somal site of synthesis at a fast rate. While Barondes (1966) had earlier found their appearance in the nerve terminals to be consistent with slow transport, some faster-appearing downflow of mitochondria was also noted. Weiss and Pillai (1965) reported the accumulation of mitochondria proximal to a dammed site, a finding earlier indicated by the appearance of the mitochondrial enzyme succinic dehydrogenase proximal to a nerve interruption (Friede, 1959). Of marked interest was the accumulation of mitochondria found in the short time of an hour in fibers just proximal to a nerve constriction by Kapeller and Mayor (1969). Such studies may indicate that a fast transport of mitochondria can also occur by means of the postulated transporting filament. The mitochondria are necessary to maintain oxidative metabolism through a local axonal supply of ATP and the energy required for fast transport as well as the energy for the ionic mechanisms underlying nerve excitability (Ochs et al., 1970; Ochs and Ranish, 1970b; Ochs and Hollingsworth, 1970).

One final implication of the dependence of the fast transport system on oxidative metabolism remains to be noted. With a nominal mean axon path length of 300 μ for intracortical neurons and a fast axoplasmic transport rate of 400/mm day, it would take only a minute or so for transmitter materials or modifiers to move from a somal site of synthesis to the synaptic terminals. Asphyxiation of the brain could, in addition to its other known effects, also block fast transport and in this way affect synaptic transmission and in turn higher functions.

Acknowledgments

This work was supported by National Science Foundation Grant GB 7234X, National Institutes of Health Grant RO1-NB 08706, and the Hartford Foundation.

References

Austin, L., and Morgan, I. G. (1957). J. Neurochem., 14:377.

Autilio, L. A., Appel, S. H., Pettis, B., and Gambeth, P.-L. (1968). Biochemistry, 7:2615.

Banks, P., Mangnall, D., and Mayor, D. J. (1969). Physiol., 200:745.

Barondes, S. (1966). J. Neurochem., 13:721. ——— and Samson, F., eds. Axoplasmic Transport. Neurosci. Res. Prog. Bull., 5.

Berl, S., and Puszkin, S. (1969). Internat. Soc. Neurochem., 2nd Meeting, Milan. Proc., p. 87.

Borisy, G. G., and Taylor, E. W. (1967). J. Cell. Biol., 34:525 and 535.
Bray, J. J., and Austin, L. J. (1968). J. Neurochem., 15:731. ——— and Austin, L. J. (1969). Brain Res., 12:230.
Burdwood, W. O. (1965). J. Cell. Biol., 27:115A.
Dahlström, A. (1968). Europ. J. Pharm., 5:111. ——— and Haggendal, J. (1966). Acta Physiol. Scand., 67:278.
Droz, B. (1969). Int. Rev. Cytol., 25:363. ——— and Barondes, S. H. (1969). Science, 165:1131.
Edström, A., Edström, J.-E., and Hokfelt, T. (1969). J. Neurochem., 16:53.
Friede, R. (1959). Exp. Neurol., 1:441. ——— (1966). Topographic Brain Chemistry. New York: Academic Press.
Gerard, R. W. (1932). Physiol. Rev., 12:469.
Grafstein, B. (1969). In: Adv. Biochem. Psychopharmacology, Costra, E., and Greengard, P., eds. 1:11.
Huxley, H. E. (1969). Science, 164:1356.
Iversen, L. L. (1967). The Uptake and Storage of Noradrenaline in Sympathetic Nerves. Cambridge: Cambridge University Press.
Jahn, T. L., and Bovee, E. C. (1969). Physiol. Rev., 49:793.
Johnson, J. (1970). Brain Res., 18:4270.
Kapeller, K., and Mayor, D. (1969). Proc. Roy. Soc. B., 172:39.
Karlsson, J.-O., and Sjöstrand, J. (1968). Brain Res., 11:431. ——— and Sjöstrand, J. (1969). Brain Res., 13:617.
Kerkut, G. A., Shapira, A., and Walker, R. J. (1967). Comp. Biochem. Physiol., 23:729.
Kidwai, A. M., and Ochs, S. (1969). J. Neurochem., 16:1105.
Koenig, E. (1967). J. Neurochem., 14:437.
Kreutzberg, G. (1969). Proc. Nat. Acad. Sci. U.S.A., 62:722.
Lajtha, A. (1964). Int. Neurobiol., 6:1.
Lasek, R. (1967). In: Axoplasmic transport, Barondes, S., ed. Neurosci. Res. Prog. Bull., 5:314 ——— (1967). In: Axoplasmic transport, Barondes, S., ed. Neurosci. Res. Prog. Bull., 5, No. 4. ——— (1968). Brain Res., 7:360.
Lehmann, J. E. (1937). Am. J. Physiol., 119:111.
Lehninger, A. L. (1965). The Mitochondrium. New York: W. A. Benjamin.
Lipmann, F. (1969). Science, 164:1024.
Livett, B. G., Geffen, L. B., and Austin, L. (1968). J. Neurochem., 15:931.
Maruhashi, J., and Wright, E. B. (1967). J. Neurophysiol., 30:434.
McEwen, B. S., and Grafstein, B. (1968). J. Cell Biol., 38:494.
Miani, N. (1963). J. Neurochem., 10:859. ——— Di Girolamo, A., and Di Girolamo, M. (1966). J. Neurochem., 13:755.
Miledi, R., and Slater, C. R. (1970). J. Physiol., 207:507.
Morgan, I. G., and Austin, L. (1969). J. Neurobiol., 1:155.
Ochs, S. (1966). In: Macromolecules and Behavior, Gaito, J., ed. New York: Appleton-Century-Crofts. ——— (1967). In: Axoplasmic transport, Barondes, S., and Samson, F., eds. Neurosci. Res. Prog. Bull., 5. ——— (1970). In: Protein Metabolism of the Nervous System, Lajtha, A., ed. New York: Plenum Press. ——— Dalrymple, D. E., and Richards, G. (1962). Exp. Neurol., 5:349–363. ——— and Johnson, J. (1969) fl. J. Neurochem., 16:945. ——— Johnson, J., and Ng, M.-H. (1967). J. Neurochem., 14:317. ——— and Ranish, N. (1970a). J. Neurobiol., 1:247. ——— and Ranish, N. (1970b). Science, 167:878. ——— Sabri, M. I., and Johnson, J. (1969a). Science, 163:686. ——— Sabri, M. I., and Ranish,

N. (1969b). Biophys. J., 9:A15. ——— Sabri, M. I., and Ranish, N. (1970a). J. Neurobiol., 1:329. ——— Sabri, M. I., Hollingsworth, D., and Helmer, E. (1970). Am. Neurochem. Soc., 1:59, and Fed. Proc. (1970a), 29:264.

Porter, K. R. (1966). In: Principles of Biomolecular Organization (CIBA Foundation Sympos.), Wolstenholme, G. E. W., and O'Connor, M., eds. Boston: Little, Brown & Co.

Singer, M., and Salpeter, M. M. (1968). J. Morphol., 120:281.

Sjöstrand, J. (1969). Exp. Brain Res., 8:105. ——— and Karlsson, J.-O. (1969). J. Neurochem., 16:833.

Schmitt, F. O. (1968). Neurosci. Res. Prog. Bull., 6:114.

Weiss, P., and Hiscoe, H. B. (1948). J. Exp. Zool., 107:315. ——— and Pillai, A. (1965). Proc. Nat. Acad. Sci. U.S.A., 54:48.

Wettstein, F. O., Noll, H., and Penman, S. (1964). Biochem. Biophys. Acta, 87:525.

Wright, E. (1946). Am. J. Physiol., 147:78. ——— (1947). Am. J. Physiol., 148:174.

Yarmolinsky, M. B., and de la Haba, G. L. (1959). Proc. Nat. Acad. Sci. U.S.A., 45:1721.

Chapter 10

A Molecular Basis for Regional Differentiation of the Excitable Membrane

EDWARD KOENIG

State University of New York (Buffalo)
Buffalo, New York

Irrespective of the degree of complexity, the point of departure for an analysis of functional activity in the nervous system lies with the neuronal plasmalemma. This is a fundamental axiom of neurophysiology. It is based upon the fact that function is a bioelectrical phenomenon which derives its origin from intrinsic properties and activities of the excitable membrane. It follows, therefore, that any attempt to link a biochemical parameter to function must ultimately show an effect on membrane properties and/or biological activity. It is self-evident that such relationships can be established only after greater understanding has been achieved of the molecular biology of the excitable membrane. The classical paucimolecular model of membrane structure and organization, while having served its purpose well as a formative construct, must now be regarded as being anachronistic and too simplistic, especially in light of recent chemical and spectroscopic analyses of various membrane systems. However, in addition to the need to formulate new concepts regarding molecular organization and dynamics, another area concerning the membrane that must receive attention is the origin of membrane macromolecules (i.e., specification and synthesis of membrane proteins).

The cell has traditionally been viewed as a compartmentalized structure, in which specialization of metabolic function is localized or sequestered. Consequently, there is often a presupposition that specialized metabolic activity may be restricted to one region or compartment of the cell to the tacit exclusion of other potential sites of activity. One such example is that of protein synthesis, which occurs principally in the cytoplasm. However, there is also evidence of protein-synthesizing activity in the nucleus (Reid and Cole, 1964; Burdman and Journey, 1969) as well as in the mitochondrion (see Nass, 1969). There is no reason a priori for not extending this activity to the plasmalemma of certain cell types, such as the neuron, as well. The most compelling argument for an intrinsic membrane protein-synthesizing machinery is that the neuron is functionally differentiated on a regional basis.

Regional Differentiation of Neuronal Plasmalemma

It is long-recognized that the neuronal plasma membrane is not uniform throughout its extent with respect to its functional properties. Thus, notwithstanding physical continuity of the membrane structure, there are topographically discrete regions or areas that exhibit properties differing from those of their contiguous surroundings. Examples include subsynaptic regions of the soma and dendrites that are responsive to excitatory and inhibitory transmitters. These circumscribed areas are capable of only graded, nonpropagated responses that are governed by activation of particular ionic "gating" mechanisms—i.e., principally sodium and potassium for excitatory synapses and potassium and chloride for inhibitory synapses (see Eccles, 1964). In some nerve cells there is evidence that regions of the dendrite and soma are capable of propagated action potentials in addition (Eccles, 1964; Purpura, 1967). In any case, the axon is certainly an example par excellence for having a "spike" membrane. Unlike the local, graded response, different gating mechanisms underlie the propagated spike—i.e., voltage-dependent and time-dependent sodium and potassium conductances which are referred to generally as activation and inactivation processes (Hodgkin, 1958). The peripheral terminals of sensory axons, on the other hand, exhibit sensitivities to specific forms of physical and chemical stimuli and generate receptor potentials which are not propagated but depolarize the preterminal spike membrane to the firing level to initiate an action potential.

The highly selective action of the neurotoxin, tetrodotoxin, which blocks sodium activation (Narahashi et al., 1967), and the Q_{10} for sodium activation (Hodgkin et al., 1952) indicate that a specific protein is responsible for the sodium gating underlying the spike mechanism. It is probable that the selective increase in sodium permeability entails a transient conformational change of the gating protein. Improved precision in the measurement of positive and negative heat generated during the action potential indicates the occurrence of entropy changes—e.g., alteration in the organized state of the membrane (Howarth et al., 1968). Recent use of fluorescent probes (Tasaki et al., 1969) lend additional support to the notion of a steric alteration's taking place during the excitatory process.

It seems likely that there may be other functional proteins in addition to the so-called gating proteins just considered which are distributed in a nonuniform manner throughout the membrane's physical extent. In any case, regional localization of a single type of biological activity requiring a specific protein is sufficient basis for postulating the existence of a topographic differentiation of the membrane. It follows, therefore, that the membrane is not everywhere identical with respect to protein composition. This poses some very important theoretical questions. How are membrane proteins specified? Where is the protein-synthesizing machinery localized? How is protein turnover controlled and regulated? What is the basis for regional differentiation and is it reversible?

Membrane Renewal: Growth Versus Local Turnover

The cytoplasmic ribosomal system (i.e., protein-synthesizing machinery) is designed to synthesize protein for export to other regions of the cell to satisfy local metabolic needs. With certain possible exceptions, such as enzymes that may be "packaged" at the Golgi complex (e.g., membrane-limiting storage granules, lysosomes, etc.), or proteins that are transported proximodistally in the axon (Droz, 1967), the intracellular movement of the bulk of newly synthesized protein is presumably undirected and a random process. As far as it is known presently, the neuronal plasma membrane in the adult nervous system is not undergoing constant proliferation or net growth. Two known exceptions to this assumption are the special case of the rod photoreceptor outer segment, and the abnormal circumstance of axonal regeneration following axotomy. Hence, local protein turnover rather than de novo membrane formation (i.e., growth) would seem to be the basis for membrane renewal of the excitable plasmalemma. The rod cell, on the other hand, is an example of a cell type in which renewal is accomplished by a continuous growth process. It appears that the cell's centralized machinery (i.e., cytoplasmic ribosomes) synthesizes membrane proteins that subsequently assemble and form de novo membrane at a site remotely situated from where they are synthesized (Young, 1967; Young and Droz, 1968). A cursory look at this specialized system would be instructive.

The visual cell is divided into anatomically distinct regions: an outer segment which is composed of a stacked, transverse array of membrane-limiting discs or sacs where the photopigment, rhodopsin, comprises a significant proportion of membrane protein (Wolkin, 1966); an inner segment, connected to the outer segment by a modified cilium, where the cell's metabolic machinery (e.g., ribosomes, mitochrondria) is localized for the most part; a nucleus, often separated from the inner segment in mammals by an interconnecting fiber; and an axon having its origin at the nuclear region. The very illuminating studies of Young (1967) and Young and Droz (1968) clearly showed by light and electronmicroscopic autoradiography that radioactive amino acids were initially incorporated into the inner segment of the cell, the site of protein synthesis. Labeled proteins were subsequently shuttled through the modified cilium into the outer segment, where they apparently aggregated to form de novo membrane. The outer segment membrane discs are formed normally from an invagination of the limiting plasma membrane; they later bud off from the plasmalemma a short distance from the basal region (Nilsson, 1964). The earliest appearance of silver grains in the outer segment occurred as a discrete band over the basal invaginations, which gradually became displaced with time toward the apical portion as new membrane was formed. In the rat, it takes approximately nine days for this displacement to be completed (Young, 1967). Thus the outer segment is seen to be in a constant state of growth in which new membrane is formed at the basal region from precursors synthesized in the inner segment, transformed subsequently into discs which are ultimately phagocytized when reaching the apical end by the continuous pigment epithelial cell (Young, 1967).

Here, then, is a membrane system that obviously does not have an intrinsic protein-synthesizing machinery. In addition, the rod cell outer segment is noted for being poorly endowed with respect to a number of enzyme systems tested (see Pearse, 1961). Its membrane ultrastructure shows distinctions that are unique and different from other membranes (Nilsson, 1964; Blasie et al., 1965). Furthermore, there is evidence that the inner and outer segments are in a highly metastable state (Koenig, 1967c; Koenig, 1971). That is, in the absence of electrolytes, inner and outer segments *selectively* undergo complete disruption in the presence of osmolar concentrations of certain neutral organic amphiphiles (i.e., electrically uncharged molecules having polar and nonpolar portions). All other cellular elements of the retina are spared disruption; of interest is the fact that the rod cell axon, to which the nucleus remains attached, is also spared. When disruption occurs there is a solvation of two membrane proteins (Koenig, unpublished data). Disruption (under such mild conditions) by chemically unreactive compounds signifies that the membrane is stabilized by weak, attractive forces and interactions (i.e., noncovalent). As will be seen below, the neuronal plasmalemma is a very highly stabilized structure and stands in contrast to the rod cell in this respect, and in other respects as well. In summary, the very specialized membrane of the rod cell, axon excluded, is relatively simple and stereotyped in its makeup and composition (Koenig, unpublished data)—one whose formation could easily occur from a spontaneous self-assembly of precursors under appropriate physicochemical conditions typical for a number of supramolecular assemblies or aggregates that have been studied (see Reed, 1967).

As in the case of membrane renewal by growth, involving undirected self-assembly of newly formed membrane subunits just considered, the requirements of membrane renewal by local turnover could conceivably be achieved by similar means; i.e., a centralized machinery could provide membrane proteins from a distance, which by random introsusception into preexisting membrane become integral with it. However, in the regionally differentiated membrane, such a means of renewal seems inadequate for no other reason than the regional differentiation necessitates a spatially discriminatory mechanism for directing introsusception of specific proteins in circumscribed, topographically discrete regions.

The Axon as an Heuristic Model for Membrane Protein Synthesis

Long before it was appreciated that Nissl or tigroid bodies were the site of protein synthesis in the nerve cell, Schaffer (1893) had noted that the axon was devoid of this basophilic substance. Indeed, the region of the cell from which the axon emerges is conspicuous for its paucity of basophilia. The so-called Nissl bodies were eventually shown by Palay and Palade (1955) to be composed of aggregates of ribosomes in association with regional complexes of endoplasmic reticulum. These workers further verified the absence of Nissl substance from axoplasm at the electronmicroscopic level. Thus, if the axon is capable of synthesizing axolemmal protein, the absence of visible ribosomes from axoplasm notwithstanding, this would be presumptive evidence for an intrinsic membrane protein-synthesizing machinery. In addition to the absence of an apparent axoplasmic protein-synthesizing machinery, there is also the advantage of a very large surface-to-volume ratio, especially in the mammalian axon. This means that the axolemma constitutes a very significant proportion of the total protein mass of the axon—e.g., 30 percent w/w for the average axon of the rabbit spinal accessory nerve (see below). These considerations provide a justification for regarding the axon as an excellent heuristic model for studying protein synthesis concerned with the excitable membrane.

There is now substantial evidence that the axon is capable of autochthonous protein synthesis. A recent review of the earlier work on the subject may be found in the *Handbook of Neurochemistry* (Koenig, 1969). Basically, the evidence derives from three principal modes of investigation: (1) the demonstration of a local restoration of the membrane-bound enzyme acetylcholinesterase (AChE) in peripheral nerves after irreversible inactivation by organophosphorylation (Koenig and Koelle, 1961; Clouet and Waelsch, 1961) and the block of local AChE restoration with inhibitors of protein synthesis (Koenig, 1965a; 1967a); (2) studies of in-vitro incorporation of radioactive amino acids into proteins of vertebrate axonal systems (Edström, 1966; Koenig, 1967b; Edström and Sjöstrand, 1969; Koenig, 1968b; 1970) and in isolated giant axons of the squid (Fischer and Litvak, 1967; Guiditta et al., 1968); and (3) studies of in-vitro incorporation of radioactive amino acids into proteins of purified synaptosome (i.e., axonal terminals) fractions (Austin and Morgan, 1967; Morgan and Austin, 1968; Autilio et al., 1968).

Two vertebrate axonal systems have been most extensively investigated thus far. Edström has employed the very large axon of the Mauthner neuron, which has its origin in the medulla and extends throughout the spinal cord of fish, while the present author has limited his studies principally to the axons of the intracranial portion of the spinal accessory nerve root of rabbit or cat. Although some distinctions between the two axonal systems have emerged, the results have been comparable. The findings may be summarized as follows. Amino acids are incorporated into axonal protein at a maximal rate for several hours before falling off significantly (Edström, 1966; Koenig, 1967b; Edström and Sjöstrand, 1969). Puromycin, which inhibits peptide elongation by causing a premature separation of the peptidyl tRNA from the ribosome, produces significant inhibition of amino acid incorporation (Edström, 1966; Koenig, 1968; Edström and Sjöstrand, 1969). Chloramphenicol produces no apparent effect (Koenig, 1967b; Edström and Sjöstrand, 1969). The significance of this latter observation lies with the reported selective inhibition of this antibiotic on mitochondrial protein synthesis, but not on microsomal (i.e., cytoplasmic protein-synthesizing system) protein synthesis (Kroon, 1965). Compatible with the view that axonal mitochondria do not contribute measurably to overall incorporation in the axon is seen by experiments in which the potent inhibitor, cycloheximide, or its close analogue, acetoxycycloheximide, were utilized. In contrast to chloramphenicol, these latter inhibitors have no apparent effect on mitochondria, but they do inhibit microsomal protein synthesis (Siegel and Sisler, 1965; Beattie, Basford and Koritz, 1967; Lamb et al., 1968). In the axon cycloheximide and acetoxycycloheximide inhibit amino acid incorporation significantly (Edström and Sjöstrand, 1969; Koenig, 1970).

It is gratifying that the results of Morgan and Austin (1968) and Autilio et al. (1968) on purified synaptosome fractions from rat cerebral cortex were in accord with those obtained on the isolated axonal systems. For example, in the studies by Morgan and Austin (1968), purified synaptosomes incorporated amino acids at a maximal rate for at least 30 min. Subfractionation of the axonal endings into membrane, vesicle, soluble, and mitochondrial fractions revealed that the highest specific radioactivity was found in the membrane and soluble fractions, with the mitochondrial fraction having slightly more than half of that observed for either of the former. Moreover, the inhibitory effect of cycloheximide was marked for membrane and soluble fractions, but insignificant for the mitochondrial fraction; the converse was true for chloramphenicol.

The experiments considered above demonstrate that the axon contains a protein-synthesizing machinery that is independent of that localized in axoplasmic mitochondria. Furthermore, the experiments by Morgan and Austin directly exclude the satellite cell (e.g., oligodendrocyte, Schwann cell) as an only source of in-vitro synthesized axonal protein. Nonetheless, there could be a contribution by the satellite cell under conditions in which the axon-satellite relationship is undisturbed. Pulse-chase experiments on rabbit nerve (Koenig, 1970), however, now make it appear unlikely that there is a translocation of protein from the Schwann cell to the axon, at least over a period of 17 hr in vitro.

Structural Stabilization of Axolemma: The Possibility of Protein Introsusception

The microanalytical method for protein determination (Koenig, 1968a), which is employed for analysis of myelin-free axons, utilizes concentrated phosphoric acid (i.e., 55 percent or more) and an elevated temperature of 110 to 115° C for 4 to 5 min to solubilize completely an axon clump (i.e., a sample 5 to 10 cm of denuded axon, cumulative length). The same conditions apply for solubilizing isolated nerve cells. The need for such drastic conditions in order to bring about complete solubilization indicates that the membrane is stabilized by covalent chemical bonds. This is not to imply that weak forces based on electrostatic attraction of polar and charged groups as well as hydrophobic interactions of nonpolar moieties do not ordinarily contribute to structural stabilization. In any case, these latter weak cohesive forces, especially hydrophobic interactions, can be disrupted readily by a solvating system made up of aqueous phenol with an organic base and acid (Koenig, 1970b).

Applying the phenol solvating system (PSS) to axonal clumps results in an extraction of almost 80 percent of the total protein mass of the axon. The insoluble residue constitutes a residuum of the axolemma (Koenig, 1970b). It appears likely that in addition to the extraction of axoplasmic protein by PSS, there may have been an extraction of as much as a third of the axolemmal protein as well. However, this estimate is based on theoretical considerations of the contribution made by axoplasmic protein to the total axonal protein mass (Koenig, 1970b), an estimate that is crude at best.

The PSS-insoluble residuum may be viewed as a polymeric protein fabric cross linked by interpeptide chain disulfide bonds. Selective disulfide bond reduction of axonal clumps render the axons completely soluble in

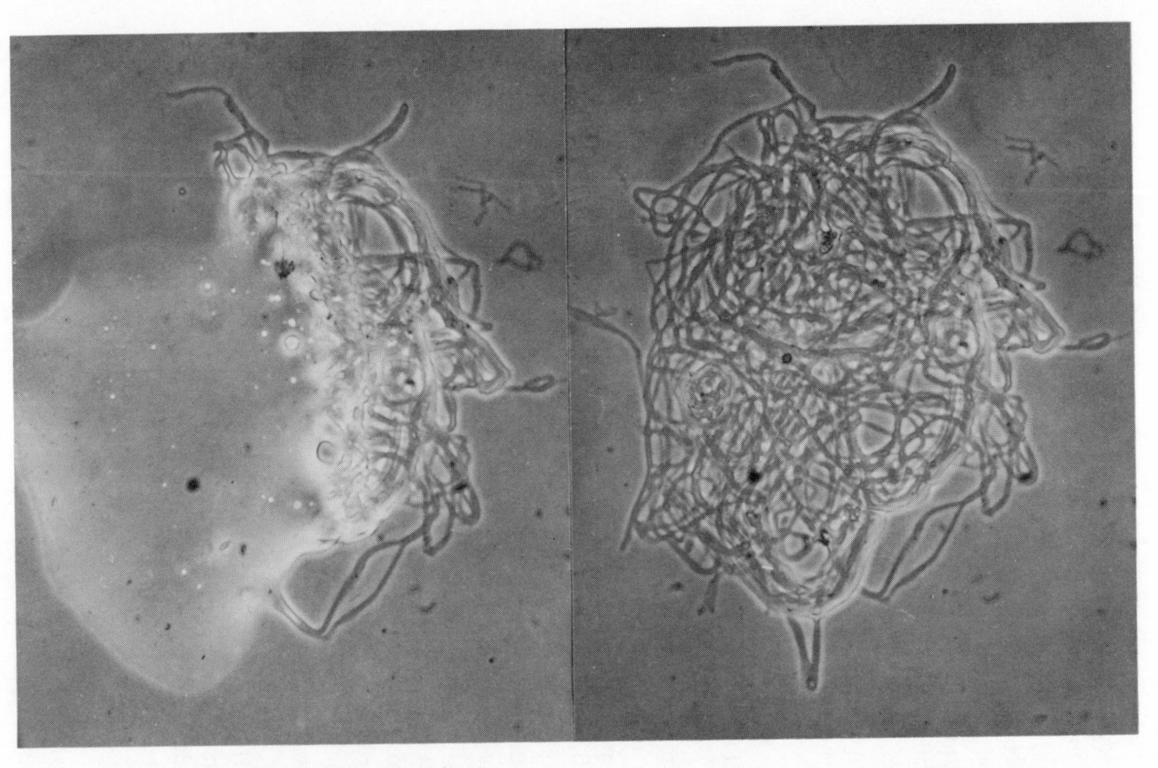

FIG. 1. (A) A phase-contrast photomicrograph of an axon clump (*ca.* 10 cm) in which axons had been taken from a rabbit spinal accessory nerve after it had been incubated in a medium (pH 8) containing dithiothreitol (disulfide bond reducing agent) for 1 hour at room temperature. (B) The same axon clump after the phenol solvating system had been applied to half of it, resulting in immediate solvation. (From Koenig. 1970. *In* Costa and Giacobini, eds. Biochemistry of Simple Neuronal Models. Courtesy of Raven Press.)

PSS (Fig. 1); the same finding holds for isolated nerve cells. Oxidation following reduction restores the PSS-insoluble residue. Alkylation of the free sulfhydryl groups, generated by reduction, blocks reoxidation and restoration of PSS insolubility. A summary of these experiments is depicted in Figure 2.

Distribution of radioactivity between the PSS-soluble and PSS-insoluble portions indicate that incorporation of amino acid had occurred in both fractions. Both fractions exhibited a similar specific radioactivity (Koenig, 1970b). In this respect, there is agreement with the results of Morgan and Austin (1968) on synaptosome membrane and soluble fractions (see above). However, at present there is no way of discerning whether the soluble protein constitutes a class of axoplasmic protein or whether it may have been released or extracted from the axolemmal membrane. Irrespectively, the incorporation of amino acid into protein that is in covalent linkage in the membrane structure suggests that an introsusception of newly synthesized protein may have taken place, reflecting the next step after synthesis in the process of membrane renewal (i.e., local protein turnover).

RNA IN THE AXON

Axonal RNA was first demonstrated in the Mauthner axon of goldfish by Edström et al. (1962), who used ultramicroanalytical procedures developed by Edström (1964). This study removed the single major objection to the proposition that there could be a local autochthonous protein synthesis in the axon based on indirect evidence of local acetylcholinesterase synthesis (Koenig and Koelle, 1961). RNA was subsequently demonstrated in other axonal systems, including crustacea (Grampp and Edström, 1963), mammals (Koenig, 1965b), and molluscs (Lasek, 1970). Review of this subject (Koenig, 1969) may be consulted for further details regarding quantitative and qualitative aspects of axonal RNA on a comparative basis.

The localization of RNA in the axon is of considerable theoretical interest. Inhibition of protein synthesis by puromycin or cycloheximide (see above) indicates that a robosomal-type mechanism is probably operative. The absence of visable ribosomes in axoplasm leads to the inference that a machinery, functionally equivalent to that of a ribosomal system, may form an integral part of the membrane. Two indirect lines of evidence point to the possibility of an axolemmal RNA. First, virtual solubilization of all organized axonal structure is required for complete extraction. Second, digestion of axon clumps by ribonuclease (RNAase) or 1 N HCl at 37° releases a fraction of RNA only very slowly and over a

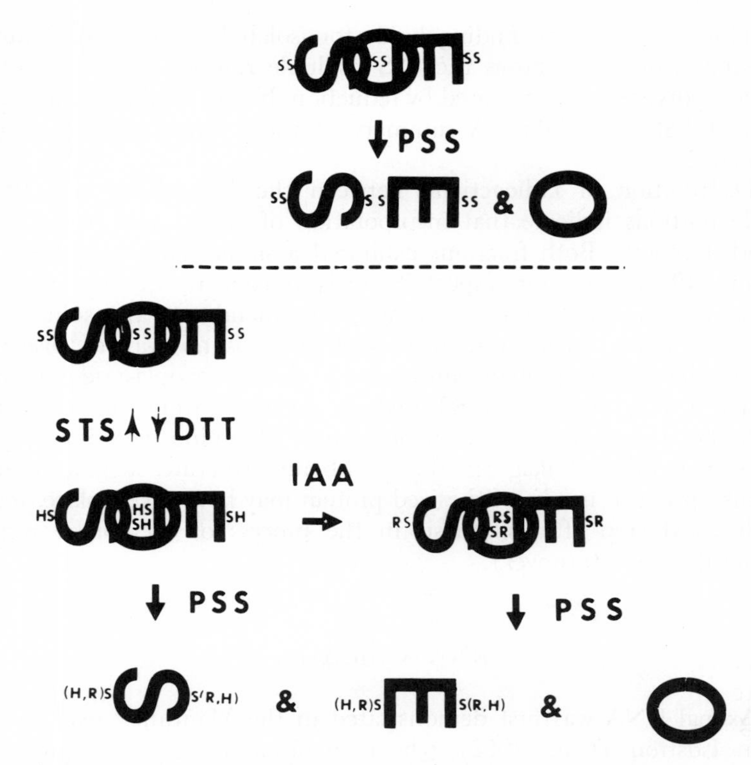

FIG. 2. A schematization showing a fragment of the polymeric protein fabric comprising the membrane and the action of the phenol solvating system (PSS). (A) The PSS extracts about one-third of the membrane proteins that are presumably not in disulfide linkage (–SS–) (*i.e.*, those proteins stabilized by noncovalent forces and interactions). (B) The reduction (HS–) by dithiothreitol (DTT) which can then lead to complete dissolution (*i.e.*, disaggregation of membrane proteins) by PSS, or be reversed by oxidation with sodium thiosulfate (STS) to restore the PSS-insoluble residuum; the reversal can be blocked by alkylation (RS–) with iodoacetamide (IAA).

long period of time (Koenig, 1968b); this indicates the likelihood of its being sequestered and/or highly stabilized. Furthermore, in their study on synaptosomes, Austin and Morgan (1967) observed substantial amounts of RNA associated with membrane fractions. These considerations are grounds for supporting the idea of the membrane's being a locus for protein synthesis.

The origin and metabolic stability of axonal RNA are equally important theoretical questions. Other than the mitochondrion, which is now well established as a site of extranuclear RNA synthesis (see Nass, 1969), the nucleus is still regarded as the primary, if not the only, source of cellular RNA. Therefore, it would be expected, a priori, that (1) axonal

RNA is synthesized in the nerve cell nucleus, and (2) axonal RNA is metabolically stable (i.e., not rapidly degraded), owing to the distance that the axon is removed from the nucleus. However, RNA becomes labeled in the axon following incubation in-vitro with radioactive precursors (Koenig, 1967b). The incorporation was inhibited by actinomycin D. Digestion with deoxyribonuclease (RNAase), after exhaustive digestion with RNAase, released almost 20 percent of the total radioactivity incorporated (Koenig, 1970a). This DNAase-releasable radioactivity seemed to represent labeled nascent polyribonucleotide fragments that may have still been bound to DNA, and could be taken as evidence for an intraaxonal DNA template (1968b). Therefore, it did not seem unreasonable that axoplasmic mitochondria might be responsible for the labeled RNA in the axon. However, other experiments showed that most of the radioactively labeled RNA was extramitochondrial in distribution (Koenig, 1970). Three possibilities, none of which were mutually exclusive, were evident: (1) that an extramitochondrial DNA template was present in the axon; (2) that mitochondria synthesized RNA for export to the axoplasm; and/or (3) there was an extraaxonal source of RNA (i.e., Schwann cell nucleus).

Edström et al. (1969) also observed incorporation of radioactive RNA precursors in vitro in the Mauthner fibers (i.e., axon with adherent myelin) from goldfish and carp. The extracted RNA contained sufficient radioactivity to permit sedimentation analysis in linear sucrose gradients, having added purified, unlabeled RNA from *E. coli* as carrier. Fibers left in situ in the spinal cord during incubation and later dissected out before extracting RNA separately from axon and myelin sheath yielded sedimentation values in three regions, viz., 4 S, 16 S, and 28 to 30 S. Of interest was the fact that incubation of isolated Mauthner fibers with myelin sheath yielded *only* an S value in the 4 S region. Although the dissection procedure could have disturbed the synthesizing machinery, an alternative interpretation offered by these workers was that the higher-molecular-weight RNA in axons incubated in situ in the cord may have its origin in the satellite cell nucleus (i.e., glia).

Preliminary results from pulse-chase experiments in the rabbit spinal accessory nerve (Koenig, 1970) were also interpreted as indicating a possible translocation of polyribonucleotide from Schwann cell nucleus. This explanation was invoked to account for a delayed enhanced labeling of axonal RNA that appeared during the postincubation period in a "chase" medium (i.e., the medium contained unlabeled precursors in 100-fold excess and actinomycin D to block further RNA synthesis). The interpretation of these experiments was tentative only, pending verification of the preliminary findings.

An autoradiographic study on the uptake of ^{3}H-uridine in peripheral nerve of the amphibian, *Triturus*, by Singer and Green (1968) also lent support to the idea that labeled RNA is transported into myelin sheath and axon from the Schwann cell nucleus, although transport of precursor which is locally incorporated could also not be excluded. Additional observations on freshly excised peripheral nerves as noted by Singer and Bryant (1969) indicated that pulsatile movements within myelin sheaths occur to a variable degree, a finding that seems to provide a mechanism for translocation of substances from Schwann cell to axon. Singer and Bryant (1969) inferred that metabolic precursors and even macromolecules reach the myelin sheath and subjacent axon via the spirals of Schwann cell cytoplasm that exist at the Schmidt-Lantermann clefts and nodes of Ranvier.

The evidence clearly shows that there is a synthesis of neural RNA that is not dependent on nuclear RNA of the nerve cell. Whether this RNA could be mRNA cannot be answered with certainty. Actinomycin D, which blocks DNA-dependent RNA synthesis, has been found to have a variable effect upon incorporation of amino acids in vitro, depending on the particular axonal system examined. Edström noted a highly significant inhibition of amino acid uptake in the Mauthner axon of the carp fish after 1-hr incubation (1967). Later, however, he found only a slight inhibition of amino acid uptake in the goldfish Mauthner axon (Edström and Sjöstrand, 1969). In the mammalian axon, on the other hand, no inhibition by actinomycin D was detected (Koenig, 1967b, 1968b). This was true for short-term experiments which tested for short-lived mRNA (i.e., degraded shortly after synthesis), as well as for long-term experiments in which preincubation with actinomycin lasted for 18 hr before transfer to radioleucine incubation. Lack of an effect cannot be attributed to a diffusion barrier to actinomycin D because RNA synthesis is inhibited under the same incubation conditions. Thus, assuming that the action of actinomycin was specifically to block DNA transcription in the Mauthner axon, one must conclude that protein synthesis in this system is dependent to a variable extent upon metabolically labile mRNA. In the mammalian motor axon, protein synthesis would appear to be dependent on stable messenger.

Although actinomycin D does not affect amino acid incorporation in mammalian axon, there is, nevertheless, evidence that synthesis of specific axonal proteins may be affected by it in a selective fashion. The manner in which synthesis is affected is *not* in the direction of inhibition, but in the direction of stimulation. The case in point is acetylcholinesterase.

AChE is associated with membranous structures of the neuron, including the axolemma (de Lorenzo et al., 1969). Previous studies (Koenig, 1965b, 1967b) provided evidence that this enzyme is synthesized in the axon. The experiments with actinomycin D were carried out on the peripheral hypoglossal nerve of the cat. Actinomycin is quite water-insoluble, so that its action can be limited locally by the technique of topical application of dry powder to the epineurium of the nerve trunk, covering a segment of 1 to 2 cm. AChE was assayed two days later in the treated segment, a segment of the same nerve centrally removed from the treated region, and assayed also in two corresponding segments of the untreated, contralateral nerve. Enzymic activities were compared in corresponding segments of treated and untreated nerves.

Two days after application, there was a 40 percent *increase* in AChE activity of the *treated* segment. The segment of the treated nerve outside the applied area showed no significant difference from that of the contralateral control nerve. The increase was induced in animals that had a normal enzyme complement as well as in animals that had most of the enzyme irreversibly inactivated by organophosphorylation preceding actinomycin application. This signified that the relative increase in enzymic activity did not stem from a reduction in rate of enzyme degradation. Furthermore, puromycin or 5-fluororotic acid, each of which inhibits AChE synthesis in the nerve (Koenig, 1965a, 1967a), blocked the increase in AChE produced by actinomycin D. This seemed to indicate that the circumscribed increase in enzymic activity probably reflected an increase in de novo enzyme synthesis.

Actinomycin D is known to block DNA-dependent RNA synthesis (Reich, 1964). Hence, it inhibits protein synthesis indirectly, i.e., if the proteins are specified by short-lived mRNA. A priori, actinomycin would be expected to cause either an inhibitory effect or no effect. A stimulatory effect is an apparent paradox unless it is interpreted as a "release" phenomenon; i.e., mRNA specifying a particular protein is stable, but cannot be translated owing to repression by metabolically labile DNA dependent product. According to this line of reasoning, it would seem that AChE-mRNA is stable; however, the full potential for AChE synthesis is not normally realized in the axon because its synthesis is held in check at the stage of translation by a labile DNA-dependent product. Similar "paradoxical" increases of certain inducible enzyme systems of the liver are also produced by actinomycin D (see Tomkins et al., 1969), indicating that AChE is not an isolated example.

The differences in results brought out by the two experimental approaches (i.e., in-vitro and in situ) used to evaluate the effects of actinomycin D on the mammalian axonal machinery deserve further comment. It is important to distinguish between a negative result that arises from failing metabolic machinery and a negative result that may stem from an inadequate sensitivity of detection. The excised nerve incubated in-vitro is a deteriorating metabolic system. Using amino acid incorporation as a criterion of viability, one can say that the rate of deterioration increases very rapidly after several hours. Therefore, the lack of an effect by actinomycin D under in-vitro conditions probably has no significance with respect to stability of axonal mRNA beyond several hours. In this context, too, it should be pointed out that attempts to reproduce the in-situ stimulation of AChE by actinomycin D in-vitro met with failure. In-vitro incubation of excised hypoglossal nerves for two days with and without actinomycin did not yield any difference in specific AChE activity between treated and untreated nerves (Koenig, 1967a). Thus, the use of specific enzymic activity can be regarded as a convenient amplification device for providing a sensitive quantitative measure of a specific protein. On the other hand, overall incorporation of amino acid into protein may be too insensitive as a technique to reveal possible labile mRNA that specifies a very small number of unique proteins, or for that matter to uncover a subtle control mechanism that may be operating for a very few unique proteins (e.g., repression of translation).

Hypothesis for Regional Differentiation of Membrane

It is difficult to draw any conclusions about the origin of the mRNA specifying membrane proteins, especially since there is evidence for local RNA synthesis. In the Mauthner axon, there is also evidence that the locally synthesized RNA has a messenger function (Edström, 1967; Edström et al., 1969)—but not necessarily for membrane proteins. In his thesis summary, Edström (1970) reported results that were negative with respect to transport of axonal RNA from the Mauthner perikaryon. If these observations are confirmed in other axonal systems, then it would mean that membrane protein mRNA is: (1) synthesized by a local template (i.e., localized in the axon and/or satellite cell); (2) synthesized in the nerve cell nucleus only during initial growth process (e.g., neurogenesis, growth, regeneration); or (3) a combination of both these possibilities. The second proposition has the merit of providing a *raison d'être* for some of the profound structural changes induced in the perikaryon following axotomy. The chromotolysis, swelling and eccentric displacement of nucleus, early signs attendent to axonal regeneration, reflect an obvious reorganization of the cell's metabolic machinery (i.e., a regression perhaps to a less differentiated state not unlike that of a neuroblast). Part of the reorganization may involve synthesis of classes of RNA that become incorporated into the newly formed membrane structure as part of the local protein-synthesizing machinery. This would mean, however, a metabolic stability for membrane RNA that is very unusual or atypical for macromolecules of a biological system.

In any case, I should like to suggest two models as alternative working hypotheses that may serve to explain mechanisms underlying topographical or regional differentiation of the excitable membrane. They are

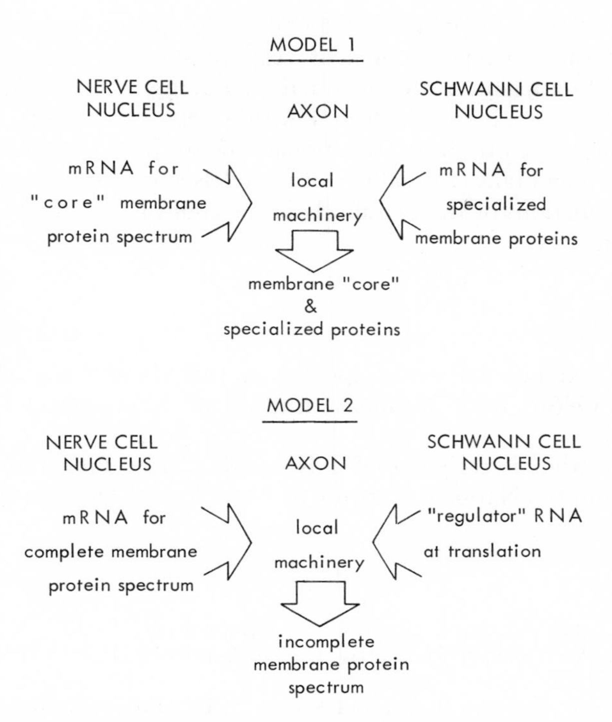

FIG. 3. Two alternative models that are not mutually exclusive to account for regional differentiation of the axon. The models are not intended to be restricted to the axon only, but can be generalized to include all excitable membranes by substituting "EXCITABLE MEMBRANE" for "AXON" and "SATELLITE" for "SCHWANN" in the schemata.

depicted in Figure 3. The basic assumption is that all or most of the mRNA specifying membrane proteins is synthesized in the nerve cell nucleus. In Model 1, the mRNA specifies a "core" spectrum of proteins that comprises all proteins common throughout the physical extent of the membrane. Local differentiation would be induced by superposition of locally synthesized mRNA, specifying regionally specific membrane proteins; synthesis would occur in extraneuronal elements (e.g., glia, Schwann cells, perineural epithelial cells). Model 2 proposes that *all* membrane proteins are specified by mRNA synthesized in the nerve cell nucleus. Local differentiation would then be induced by a local DNA-dependent regulatory control, exercised by extraneuronal elements in which there would be a selective repression at translation; this would result in a regional omission of specific membrane proteins. The stimulation of AChE by actinomycin D would best be accounted for by Model 2. A conclusion that can be drawn from either model is that regional differentiation of the membrane is reversible. It predicts that under appropriate circumstances, such as altering the cell type or element (e.g., presynaptic axon) coming into apposition to a region of the neuronal membrane, it should be possible to transduct the membrane, such that within certain inherent limitations it can be transformed from one functional type to that of another (e.g., spike to graded membrane). Finally, both models are not mutually exclusive (i.e., combinations of the two mechanisms could be operative at the same or different sites).

Acknowledgments

The experimental work reported in this chapter was supported by Grant NSO4656 from National Institute of Neurological Diseases and Stroke.

The author is recipient of a Research Career Program Award (NB 14254) from the National Institute of Neurological Diseases and Stroke.

References

Austin, L., and Morgan, I. G. (1967). J. Neurochem., 14:377.

Autilio, L. A., Appel, S. H., Pettis, P., and Gambetti, P. L. (1968). Biochemistry, 7:2615.

Beattie, D. S., Basford, R. E., and Koritz, S. B. (1967). J. Biol. Chem., 242: 4584.

Blasie, J. K., Dewey, M. M., Blaurock, A. E., and Worthington, C. R. (1965). J. Mol. Biol., 14:143.

Burdman, J. A., and Journey, L. J. (1969). J. Neurochem., 16:493.

Clouet, D., and Waelsch, H. (1961). J. Neurochem., 8:201.

de Lorenzo, A. J. D., Dettbarn, W. D., and Brzin, M. (1969). J. Ultrastruct. Res., 28:27.

Droz, B. (1967). J. Microscopie, 6:201.

Eccles, J. C. (1964). The Physiology of Synapses. New York: Academic Press.

Edström, A. (1966). J. Neurochem., 13:315. ——— (1967). J. Neurochem., 14:239. ——— (1970). Acta Physiol. Scand. (in press). ——— Edström, J.-E., and Hökfelt, T. (1969). J. Neurochem., 16:53. ——— and Sjöstrand, J. (1969). J. Neurochem., 16:67. ——— (1964). In: Methods of Cell Physiology, Prescott, D., ed. New York: Academic Press. ——— Eichner, D., and Edström, J.-E. (1962). Biochim. Biophys. Acta, 61:178.

Fischer, S., and Litvak, S. (1967). J. Cell Physiol., 70:69.

Grampp, W., and Edström, J.-E. (1963). J. Neurochem., 10:725.

Giuditta, A., Dettbarn, W. D., and Brzin, M. (1968). Proc. Nat. Acad. Sci., U.S.A., 59:1284.

Hodgkin, A. L. (1958). Proc. Roy. Soc., Ser. B., 148:1. ——— Huxley, A. F., and Katz, B. (1952). J. Physiol., 116:424.

Howarth, J., Keynes, R., and Ritchie, J. (1968). J. Physiol., 194:745.

Koenig, E. (1965a). J. Neurochem., 12:343. ——— (1965b). J. Neurochem., 12:357. ——— (1967a). J. Neurochem., 14:429. ——— (1967b). J. Neurochem., 14:437. ——— (1967c). J. Cell Biol., 34:265. ——— (1968a). J. Cell Biol., 38:562. ——— (1968b). In: Macromolecules and the Function of the Neuron, Lodin, Z., and Rose, S. P. R., eds. Amsterdam: Excerpta Medica. ——— (1969). In: Handbook of Neurochemistry, Vol. 2, Lajtha, A., ed. New York: Plenum Press: (1970). In: Biochemistry of Simple Neuronal Models, Costa, E., and Giacobini, E., eds. New York: Raven Press. ——— (1970a). In: Protein Metabolism of the Nervous System, Lajtha, A., ed., New York: Plenum Press. ——— and Koelle, G. B. (1961). J. Neurochem., 8:169.

Kroon, M. (1965). Biochim. Biophys. Acta, 108:275.

Lamb, A. J., Clark-Walker, G. D., and Linnane, A. W. (1968). Biochim. Biophys. Acta, 161:415.

Lasek, R. J. (1970). J. Neurochem., 17:103.

Morgan, I. R., and Austin, L. (1968). J. Neurochem., 15:41.

Narahashi, T., Anderson, N. C., and Moore, J. W. (1967). J. Gen. Physiol., 50:1413.

Nass, M. M. K. (1969). Science, 165:25.

Nilsson, S. E. G. (1964). J. Ultrastruct. Res., 11:581.

Palay, S. L., and Palade, G. E. (1955). J. Biophys. Biochem. Cytol., 1:69.

Pearse, A. G. E. (1961). In: The Structure of the Eye, Smelser, K., ed. New York: Academic Press.

Purpura, D. P. (1967). In: The Neurosciences, Quarton, G. C., Melnechuk, T., and Schmitt, F. O., eds. New York: Rockefeller Univ. Press.

Reed, L. J. (1967). In: The Neurosciences, Quarton, G. C., Melnechuk, T., and Schmitt, F. O., eds. New York: Rockefeller Univ. Press.

Reich, E. (1964). Science, 143:684.

Reid, B. R., and Cole, R. D. (1964). Proc. Nat. Acad. Sci. U.S.A., 51:1044.

Schaffer, K. (1893). Neurol. Centralbl. (Leipzig), 12:849.

Siegel, M. R., and Sisler, H. D. (1965). Biochim. Biophys. Acta, 103:558.

Singer, M., and Bryant, S. V. (1969). Nature (London), 221:1148. ——— and Green, M. R. (1968). J. Morph., 124:321.

Tasaki, I., Carnay, L., Sandlin, R., and Watnabe, A. (1969). Science, 163:683.

Tompkins, G. M., Gelehrter, T. D., Martin, D. G. D., Samuels, H. H., and Thompson, E. B. (1969). Science, 1474.

Wolkin, J. J. (1966). Vision. Springfield, Ill: Thomas.

Young, R. W. (1967). J. Cell. Biol. 33:61. ——— and Droz, B. (1968). J. Cell Biol., 39:169.

Chapter 11

Macromolecular Synthesis in Synapses

STANLEY H. APPEL

Duke University, Durham, North Carolina

The synapse has recently been recognized as the critical and even rate-limiting factor in intercellular communication and information processing within the nervous system. An understanding of synaptic organization and synaptic biochemical properties may, therefore, hold the key to understanding the mechanisms of short- and long-term information storage as well as the key to understanding those factors which modulate behavior. Only with the advent of electronmicroscopy and sophisticated electrophysiologic techniques has it been possible to characterize these vital junctions and demonstrate their normal variations, both morphologically and electrophysiologically.

From a biochemical point of view, studies of isolated preparations have been difficult to interpret because of the heterogeneity of brain slices and homogenates containing both neuronal and glial components. However, techniques have recently become available to isolate enriched populations of synapses from mammalian cortical tissues and to analyze their constituent protein, lipid and hormonal contents (De Robertis et al., 1963; Whittaker et al., 1964). These synaptosomes retain many of the morphologic characteristics of presynaptic terminals and are relatively free from axonal, nerve cell body, glial, and mitochondrial contamination (Autilio et al., 1968).

The usefulness of this preparation becomes evident when one attempts to analyze the factors in the synapse which are potentially rate-limiting in intercellular communication and information processing. We can indicate essentially three modifiable sites connected with the synapse: presynaptic sites, intersynaptic cleft, and postsynaptic sites. It is not necessary to postulate that information processing changes the actual number of synapses present within the nervous system. There is sufficient plasticity within established synapses to change the efficiency of any of the postulated rate-limiting reactions and thereby enhance the efficiency of overall intercellular communication. Presynaptic factors which could influence efficiency of such communication include the nature of the membrane, its structure and function, the coupling between membrane transmitter synthesis and energy processes, the interchangeability of transmitter pools, the resting membrane potential, and the ionic conductances and fluxes of the presynaptic terminals. In addition, the nature of the intercellular matrix, its width, and relative resistance to the passage of neurotransmitters represent extra cellular factors which are modifiable. The modifiable sites on the postsynaptic side include the state of the receptor, the coupling between receptor conformation changes and membrane changes resulting in depolarization, the actual structure of the membrane, the resistance of the cytoplasm, and the width of the dendritic arborization, as well as other factors. In essence, potential synaptic modifiability is either related to neurotransmitter changes, to membrane changes, or to a combination of both neurotransmitter and membrane changes.

Early investigations of synaptosomes were directed toward studies of transmitter substances contained therein, their synthesis and degradation, and the molecular mechanisms involved in their storage and release. Such studies are reviewed by Marchbanks and Whittaker (1969). Within synaptosomes acetylcholine is present in at least two compartments, a labile and a stable bound form (Marchbanks and Whittaker, 1969). The labile compartment and its acetylcholine content can be rapidly exchanged with ^{14}C-acetylcholine or can be synthesized either from glycolytic or choline radioactive precursors. The stable compartment containing acetylcholine is not so readily exchanged. The enzyme synthesizing acetylcholine—namely, choline acetylase—has also been demonstrated to be present within synapses. The factors effecting the release of acetylcholine from synapses or synaptic vesicles have not been completely elucidated at the present time.

Similar investigations on catecholamine synthesis, degradation, and turnover within synapses have been performed primarily in peripheral tissues. More recently, studies of catecholamine metabolism of central synapses have been fruitfully investigated, and our understanding of control mechanisms and modulating drugs has been extended. Similarly, studies have been performed on other putative transmitters such as dopamine, 5-hydroxytryptamine, GABA, and the enzymes involved in the synthesis and degradation of these transmitters (Marchbanks and Whittaker, 1969).

Other studies have been concerned with synaptic macromolecules and energy metabolism. The continuous activity of the synapse with respect to membrane depolarization and repolarization, and transmitter synthesis, release, and degradation necessitate considerable expenditure of energy. It is not immediately obvious how such demands are met. The nerve endings frequently are located at vast distances from their cell bodies, and the question must be raised whether they meet their metabolic requirements by local synthesis or by transport from the cell body. Several experiments have confirmed the classic observation of Weiss and Hiscoe (1948) that constituents synthesized in neuronal perikarya may migrate down the axon (Droz and LeBlond, 1963; Barondes, 1964; Lassek, 1967; McEwen and Grafstein, 1968). (See Chapter 9, present volume.) However, there is data suggesting that proteins may be synthesized within axons (Edström, 1966; Koenig, 1967; see also Chapter 10, present volume). Furthermore, at the time we initiated our studies, it was possible to infer that protein synthesis occurred in synapses from the observation that brain slices incorporated radioactive amino acids into synaptosomal proteins without any time lag (Austin and Morgan, 1967).

Experiments from our laboratory have been concerned with membrane and molecular metabolism in isolated synaptosomes. In brief, the data have directed attention to the independent metabolism of synapses in vitro and the extremely tight link between membrane activities and macromolecular metabolism. In our studies nerve endings isolated from rat brain cortex incorporated amino acid and glucosamine into membrane glycoproteins in vitro. Such synthesis was modulated by ionic constituents and endogenous energy availability (Autilio et al., 1968). Under the same conditions, potassium was actively transported into synaptosomes (Escueta and Appel, 1969). Investigations of amino acid uptake, sodium-potassium-ATPase, and oxygen uptake demonstrated tight coupling of membrane function, ionic flux, and amino acid incorporation into membrane protein (Appel et al., 1969; Festoff et al., 1970). They indicated that in-vivo synaptic needs may be met by local synthesis as well as by axonal transport. The specific modulation by ions and the role of the membrane in coupling ionic flux and macromolecular metabolism suggest that synapses may store information in the potentiality of the plasma membrane to translocate ions. The studies to be reported below represent a summary of our experiments attempting to document the synthetic capacities of synapses in vitro and the specific ionic alterations which may affect such processes.

Experimental Work

In-vitro Incorporation of ^{14}C-leucine into Protein

When rat cortical synaptosomes isolated by discontinuous Ficoll gradients are incubated with ^{14}C-leucine, radioactivity is incorporated linearly into lipid-free hot trichloroacetic acid precipitates for 20 min and continued for approximately 50 min (Autilio et al., 1968). Preincubation

TABLE 1. Effect of Energy Sources or Inhibitors on Synaptosome Protein Synthesis

	Control (%)
Complete System	100
+ATP (10^{-4} M)	100
+ADP (10^{-4} M)	112
+AMP (10^{-4} M)	115
+Glucose (10^{-4} M)	112
+Dinitrophenol (10^{-4} M)	20
+KCN (10^{-3} M)	48
+Oligomycin (10^{-4} mg/ml)	40
+Mg (5×10^{-3} M)	95
+Ca (5×10^{-3} M)	90
+α-Ketoglutarate (5×10^{-3} M)	100
+Glutamate (10^{-3} M)	78
+Succinate (5×10^{-3} M)	87

of the synaptosomes for 10 min in the complete medium reduced the leucine incorporation by 20 percent. Almost complete activity was lost after 6 hr at 4° C.

Of interest was the fact that the addition of adenine nucleotides (AMP, ADP, or ATP) did not significantly affect the incorporation observed. GTP similarly had minimal effect on incorporation. As indicated in Table 1, glucose, magnesium, and calcium had minimal effects, as did α-ketoglutarate, glutamate, and succinate. However, dinitrophenol, potassium cyanide, and oligomycin had quite striking inhibitory effects. A number of other inhibitors were employed, as indicated in Table 2. Ribonuclease was found to have minimal effects at concentrations known to inhibit brain ribosomal and microsomal systems. Chloramphenicol, which is known to inhibit mitochondrial but not ribosomal or microsomal protein

TABLE 2. Inhibitors of Synaptosomal Protein Synthesis

	Amount	Control (%)
Complete System		100
+Ribonuclease	10 μg	85
+Chloramphenicol	50 μg	75
+Puromycin	1 μg	75
	10 μg	25
	50 μg	12
+Cycloheximide	1 μg	36
	10 μg	28
	50 μg	12
+Ouabain	5×10^{-4} M	50
	1×10^{-3} M	40

synthesis (Clarke-Walker and Linnane, 1966), inhibited only 25 percent. This inhibition of 25 percent may, therefore, indicate an upper limit to the contribution by mitochondria to the protein synthesis observed. Puromycin, which has been demonstrated to inhibit both ribosomal (Campbell et al., 1966) and mitochondrial (Wheeldon and Lehinger, 1966) systems, was found to be a potent inhibitor in our system. In addition, cycloheximide, which is thought to inhibit ribosomal and microsomal—but not mitochondrial—protein synthesis, also was most effective.

Some of these studies are relevant to the consideration of whether contaminating bacteria, microsomes, or extrasynaptosomal mitochondria may contribute to the protein synthesis observed. The bacteria may be excluded by the following observations: (1) the limited inhibition by chloramphenicol (25 percent); (2) the short period of linear protein synthesis compared with the exponential rate of synthesis in bacteria; (3) the loss of synthesizing capacity in synaptosomes kept for 6 hr at 4° C and its retention with bacteria under the same circumstances; (4) the effectiveness of cycloheximide as an inhibitor, despite its negligible effect on bacterial protein synthesis; and (5) the limited presence of bacteria noted on direct culturing of the fractions. The lack of an inhibitory effect of ribonuclease and the failure of ATP to enhance protein synthesis appears to rule out a significant contribution of extrasynaptosomal ribosomes or microsomes. The inhibition of ouabain, the lack of effect of exogenous substrate, and the rate occurrence of free mitochondria as revealed by electronmicroscopy suggest that intrasynaptosomal, but not extrasynaptosomal, mitochondria contribute to the protein synthesis.

Ionic Activation of Protein Synthesis

The most striking aspect of these studies is the stimulation noted in the presence of sodium and potassium. Either sodium or potassium stimulated incorporation, although sodium was always more effective than potassium (Fig. 1). In initial studies, 100 mM sodium in the absence of potassium was found to be optimal. However, the addition of 10 mM potassium in the presence of 100 mM sodium gave an even greater enhancement. In order to exclude osmotic effects, studies were performed at a cation concentration of 110 mM. A peak incorporation of amino acid into protein was noted at 60 to 100 mM sodium and 10 to 50 mM potassium (Fig. 2). The critical factor appeared to be the ratio of sodium to potassium, with 5:1 representing the optimal ratio. Under these same circumstances, 5×10^{-4} M ouabain was found to inhibit amino acid incorporation into protein by approximately 50 percent.

These results suggested the possible participation of a sodium-potassium-activated ATPase. With further studies the enzyme was found to be

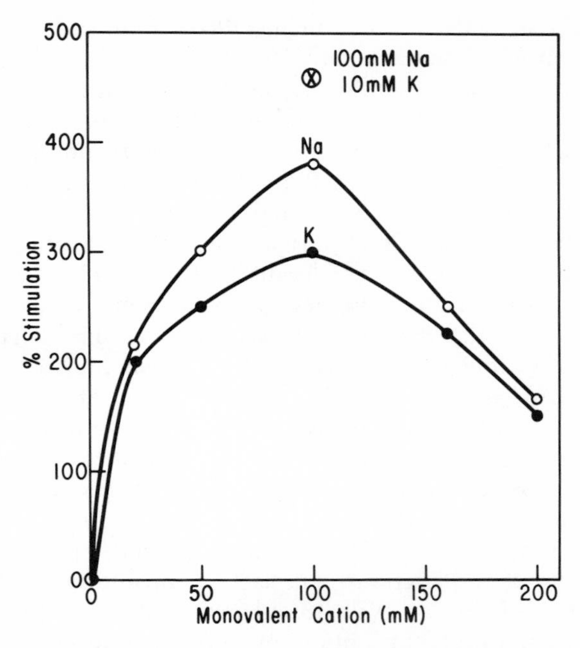

FIG. 1. Effect of ions on synaptosomal protein synthesis. Incubations were performed in the presence of sodium or potassium alone at the indicated concentrations or a combination of sodium (100 mM) and potassium (10 mM). The ionic activation is expressed as percent stimulation over the control which contained no ions and 0.1 M sucrose. (From Autilio et al. 1968. *Biochemistry,* 7:2615.)

highly active in our synaptosome preparations (Appel et al., 1969). In the presence of 50 mM sodium, 10 mM potassium, and 7.5 mM magnesium, the enzyme specific activity was 70 μmoles inorganic phosphate per milligram synaptosomal protein per 30 min. Half maximal velocity was obtained with 0.4 to 0.6 mM potassium (at 50 mM sodium), with 5 mM sodium (at 10 mM potassium), and 8.5×10^{-5} M ATP concentrations. When the pattern of activation of sodium-potassium-ATPase was performed with a total of 120 mM ions, a similar peak was observed at 60 to 100 mM sodium and 20 to 60 mM potassium, as had been noted with the amino acid incorporation into protein. In addition, those substances such as ouabain, which inhibited sodium-potassium-ATPase activity, also inhibited amino acid incorporation into protein.

Potassium Transport

The studies described above suggested a link between synaptic membrane events and macromolecular metabolism which might be mediated by some direct or indirect consequence of the ATPase activity. Since the

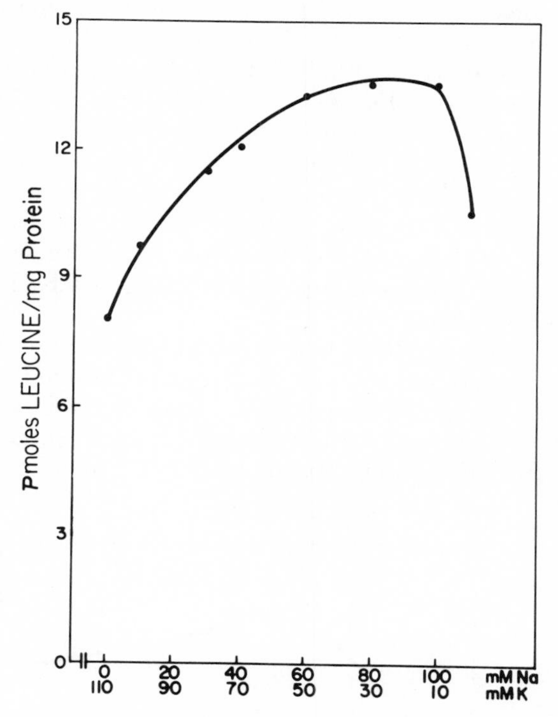

FIG. 2. Ionic activation of protein synthesis. Incubations were performed at 37° C for 20 min in 1 ml containing 0.033 M Tris-HCl (pH 7.6), 0.1 M sucrose, varying concentrations of ions, 1 μC of ^{14}C-leucine, and 0.7 mg of synaptosomal protein. (From Appel et al. 1969. *J. Biol. Chem.*, 244:3166.)

potassium content might be so regulated, experiments were designed to test this hypothesis directly (Escueta and Appel, 1969). Immediately after isolation, synaptosomes contained an average of 0.092 μmoles potassium per milligram protein. At 23° with 100 mM sodium and 10 mM potassium in the external medium, the synaptosomal potassium content increased to 0.25 μmoles potassium per milligram protein after 6 min. The optimal potassium accumulation occurred with 50 mM sodium present in the medium. At this level of sodium, potassium concentrations in the media greater than 10 mM produced minimal increase in synaptosome potassium accumulation.

As demonstrated in the studies of amino acid incorporation into protein, neither ATP, ADP, glucose, increased oxygen concentration, nor the substrates α-ketoglutarate, succinate, fumarate or glutamate influenced potassium accumulation. Ouabain was an effective inhibitor of potassium accumulation only when both sodium and potassium were present in the medium. In sodium-free media, ouabain had essentially no effect on the accumulation noted. The optimal effect of ouabain inhibition occurred at 50 mM sodium and 10 mM potassium, which were the ionic concentrations at which ouabain had been the most effective inhibitor of amino acid incorporation into protein (Fig. 3). In the presence of 50 mM sodium, 10 mM potassium, and 10^{-4} M ouabain, the total potassium content of the synaptosomes declined by 30 to 75 percent. Both 2,4 dinitrophenol (10^{-4} M) and potassium cyanide (10^{-3} M) inhibited potassium accumulation as extensively as ouabain did (Fig. 3). Studies performed with ^{42}K gave results essentially similar to those noted with cold potassium; this supported the conclusion that net potassium uptake was being observed rather than merely exchange diffusion.

Total synaptosomal volume was measured with the use of radioactive antipyrene and inulin, and the potassium content per milligram protein was expressed as synaptosomal potassium concentration. With 50 mM sodium and 10 mM potassium, there was an increase of intrasynaptosomal potassium concentration from approximately 0.075 M following isolation

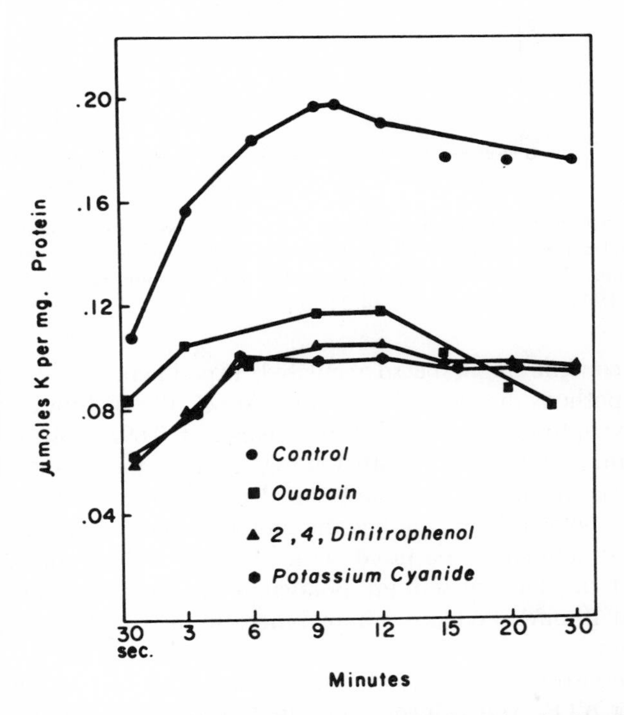

FIG. 3. Potassium accumulation in synaptosomes. Following incubation for the times indicated in 100 mM sodium and 10 mM potassium with ouabain (10^{-4} M), 2, 4 DNP (10^{-4} M), or KCN (10^{-4} M), the synapses were processed for intrasynaptosomal potassium. (From Escueta and Appel. 1969. *Biochemistry,* 8:725.)

to 0.104 M, representing a significant accumulation of potassium against a concentration gradient. These experiments, therefore, demonstrated that rat brain synaptosomes may actively transport potassium against a concentration gradient in vitro. They further suggested that the optimal parameters for sodium-potassium-ATPase activity and amino acid incorporation into protein parallel those noted with the active transport of potassium. Similarly, those substances which inhibited potassium transport similarly inhibited amino acid incorporation into protein and sodium-potassium-ATPase activity.

Oxygen Uptake

When oxygen uptake was examined in the synaptosomal preparations, a similar sodium and potassium activation and ouabain inhibition were noted (Appel et al., 1969). In the absence of ions, the basal uptake of oxygen was approximately 0.12 μl of oxygen per minute per milligram synaptosomal protein. With the addition of 10 mM potassium, no significant change in oxygen uptake was noted; however, with the further addition of 50 mM sodium, there was a 48 percent enhancement of oxygen uptake. The addition of 10^{-4} M ouabain inhibited 38 percent. Potassium cyanide produced 80 percent inhibition of oxygen uptake.

Amino Acid Transport

The parallel behavior of potassium transport, sodium-potassium-ATPase activity, and amino acid incorporation into protein indicated that certain metabolic consequences of membrane activity affected synaptosomal macromolecular metabolism (Appel et al., 1969). One possibility was that amino acid uptake was rate-limiting in this system, and, therefore, activation or inhibition of amino acid uptake would be reflected in activation or inhibition of radioactive amino acid incorporation into protein. In these studies the uptake of ^{14}C-leucine into synaptosomes was extremely rapid and linearly dependent upon the concentration of synaptosomal protein. Optimal uptake occurred at 20 to 40 mM sodium in the absence of potassium. The apparent K_m for leucine uptake was 3.0 mM in the absence of sodium and 0.7 mM in the presence of 50 mM sodium (Fig. 4). Sodium, therefore, appeared to enhance the affinity of the amino acid for components of the transport process. Potassium, on the other hand, appeared to inhibit this uptake. In the presence of 40 mM sodium, approximately 45 percent inhibition was noted by addition of 10 mM potassium, compared to incubation with sodium alone. An increase in potassium to 40 mM only slightly increased the inhibition. Ouabain had no effect in the absence of sodium and potassium. In the presence of 100 mM sodium and no potassium, ouabain produced 16 percent inhibition, whereas in the presence of 100 mM sodium and 10 mM potassium, ouabain produced 48 percent inhibition. Thus, although sodium gave the maximal stimulation, in the absence of potassium ouabain was a poor inhibitor. Only when sodium and potassium were present was ouabain an effective inhibitor. Maximal inhibition by ouabain was, therefore, dependent upon conditions optimal for sodium-potassium-ATPase function.

The mechanism for leucine accumulation demonstrated may thus be similar to the mechanism of sugar transport noted in intestinal tissue by Crane (1965), in which facilitated transport of amino acid and sodium may occur. The concentration gradient of amino acid would be achieved by pumping out the sodium. The ouabain inhibition of amino acid transport might then be explained by its inhibition of active extrusion of sodium, which would indirectly limit amino acid uptake.

Although sodium-potassium-ATPase activity was found to be indirectly related to amino acid uptake, no definitive parallelism was noted between amino acid transport studies and those previously described for protein synthesis, potassium transport, and sodium-potassium-ATPase activity.

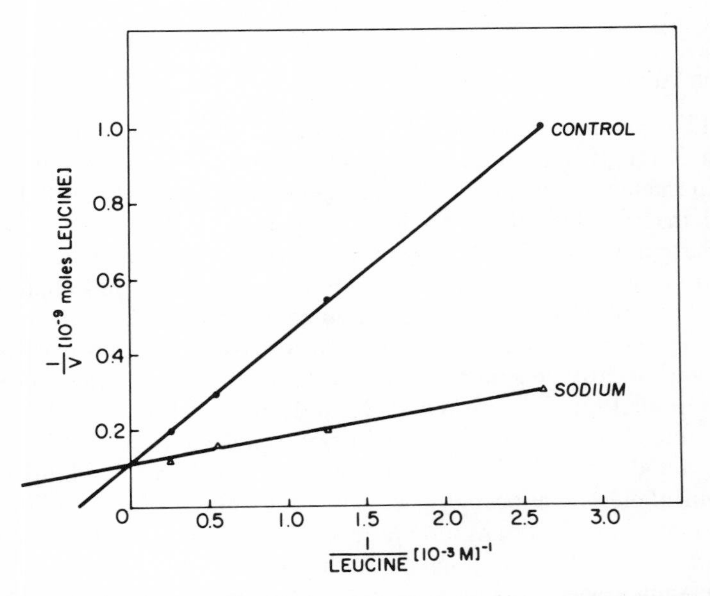

FIG. 4. Kinetics of leucine uptake. Five minute points were selected for ^{14}C-leucine uptake at varying concentrations of leucine and in the presence or absence of 50 mM sodium. The data are plotted as the reciprocals according to the method of Lineweaver and Burk. (From Appel at el. 1969. *J. Biol. Chem.*, 244:3166.)

Distribution of Radioactive Proteins

The distribution of radioactivity in various synaptosomal fractions was examined next. Following osmotic shock and centrifugation in a discontinuous sucrose gradient, the protein bound radioactivity was distributed in supernatant (22 percent), mitochondrial (25 percent), and membrane particulate fractions (53 percent). Of particular interest was the fact that the highest specific radioactivity was found in the membrane fraction (0.8 M) that also yielded the highest sodium-potassium-activated ATPase specific activity, and which was thereby identified as the synaptic plasma membrane. At this point, it was not clear how many different membrane proteins were being synthesized, although it was hoped that this number was small so that they could be individually characterized. When the membranes were solubilized in 1 percent SDS and electrophoresed in polyacrylamide gels, at least 20 protein bands could be identified, as noted in Figure 5. Radioactivity was associated with a minimum of 5 rapidly migrating bands. The pattern was extremely similar to that described with liver endoplasmic reticulum by Kiehn and Holland (1968) and HEP cells by Spear and Roizman (1968). When the total radioactivity incorporated into synaptosomes in vitro was increased or decreased, the level of all protein bands increased or decreased *pari passu*. Experiments are presently continuing in our laboratory to characterize the individual protein components of the membrane and to determine whether any neuronal activity may be associated with changes of single or a limited number of membrane components.

Synthesis of Glycoproteins

The localization of the radioactivity within the membrane prompted us to explore whether some of the recently synthesized proteins were in fact carbohydrate-containing macromolecules—namely, glycoproteins (Festoff et al., 1970). When synaptosomes were incubated with ^{14}C-glucosamine, radioactivity was found predominantly in the particulate fraction. In these membrane fractions, radioactive carbohydrates were bound in covalent linkage requiring acid hydrolysis or enzymatic digestion for release. Incorporation was linear from 10 to 45 min. Over half the particulate radioactivity was extracted with chloroform:methanol and partitioned into the organic phase. Of the radioactivity 20 percent was incorporated into the protein residue as glucosamine, which was identified by paper chromatography, and more definitively with the use of the amino acid analyzer. Radioactivity was also identified in NANA of the protein residue by paper chromatography, and by enzyme suggestion with NANAldolase and identification of the resulting mannosamine derivative with the amino acid

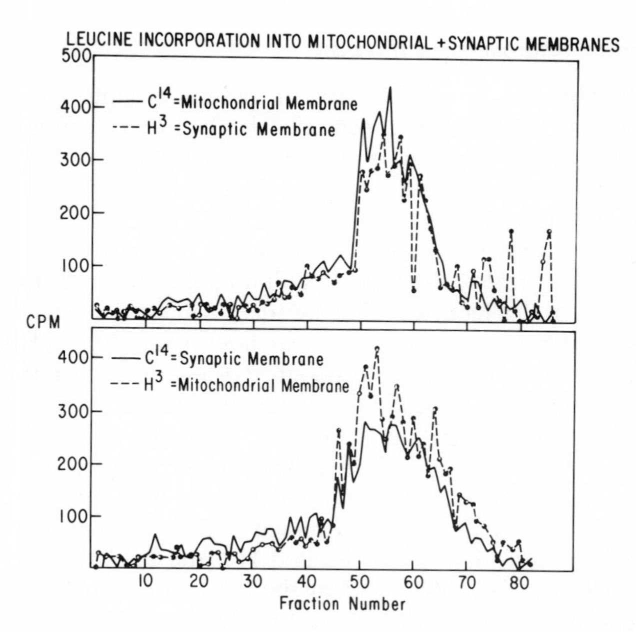

FIG. 5. **Radioactive protein profile of synaptic plasma and mitochondrial membranes. Following incubations with ^{14}C-leucine, synaptic plasma and mitochondrial membranes were separated on discontinuous sucrose gradients, dialyzed, and solubilized in SDS, β-mercaptoethanol, and urea. The solubilized membranes electrophoresed in 5 percent acrylamide-SDS gels at 3 mA/tube for 18 hours at pH 7.2. The gels were washed in 5 percent TCA, crushed and fractions collected for radioactive counting. The top of the gel is on the left. Albumin in the same system appears at tubes 40 to 44, with the radioactivity in tubes 45 to 75 being smaller molecular weight.**

analyzer. Labeled glucuronic acid was identified either by hyaluronidase digestion or by acid hydrolysis followed by chromatography and electrophoresis. The wide range of metabolites labeled was related to the presence of glucosamine-6-phosphate deaminase which would shunt the radioactivity extensively into the glycolytic pathway.

The carbohydrate incorporation was stimulated by ions previously demonstrated to enhance amino acid incorporation and was distributed in macromolecules of all subsynaptosomal fractions. Sialic acid and glucosamine were found to be significant components of the synaptic plasma membrane, and it was not surprising that the highest percentage of radioactivity was found in this membrane. When synaptosomes were incubated with ^{14}C-glucosamine and ^{3}H-leucine, both radioactivities were identified in tryptic digests of the membrane following electrophoresis in polyacrylamide gels (Fig. 6), or following combined chromatography and electrophoresis. By peptide mapping, ninhydrin-positive residues could be isolated which contained both ^{14}C-glucosamine and ^{3}H-leucine, thereby strongly suggesting the local synthesis of glycopeptides in synaptosomes (Fig. 7).

Den and Kaufman (personal communication) have demonstrated the presence of enzymes which transfer sialic acid and galactose to glycoprotein and glycolipids in chick embryo brain fractions. The present experiments provide evidence that rat cortex synaptosomes contain the necessary metabolic machinery not only for the final step of sialic acid and glucosamine incorporation into macromolecules, but also for the preceding steps necessary to convert free ^{14}C-glucosamine to glycoprotein-bound glucosamine and sialic acid.

Discussion

Employing in-vivo injections of ^{14}C-glucosamine, Barondes and Dutton (1969) concluded that polypeptide acceptors are transported to nerve endings and carbohydrates added at that point. In our experiments, following incubation, tryptic digestion, and subsequent chromatography in

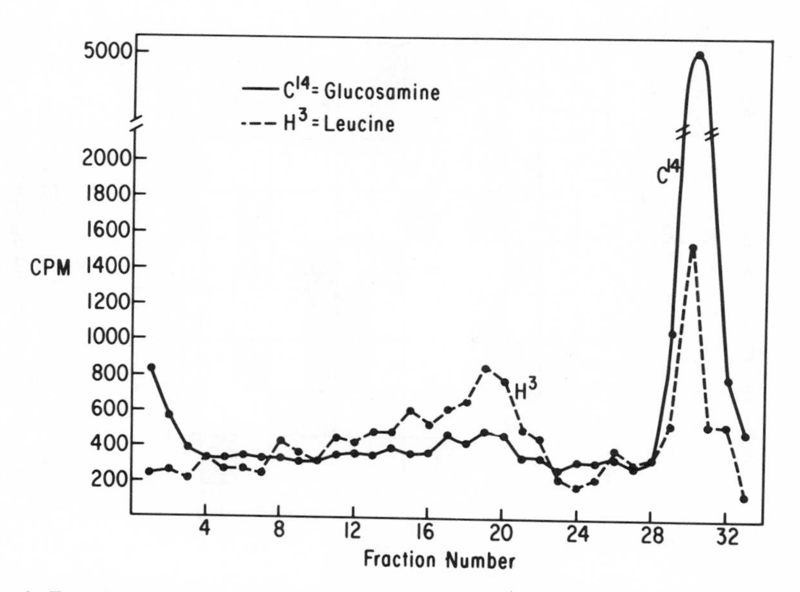

FIG. 6. Tryptic digestion of synaptic membranes. Synaptic membranes isolated following incubation with ^{14}C-glucosamine and ^{3}H-leucine were digested in trypsin and electrophoresed in 10 percent acrylamide bisacrylamide gels at 5 mA/tube for 1 hour at pH 9.2. The anode is at the right. No SDS is present in this preparation. (From Festoff et al. 1970. Unpublished data.)

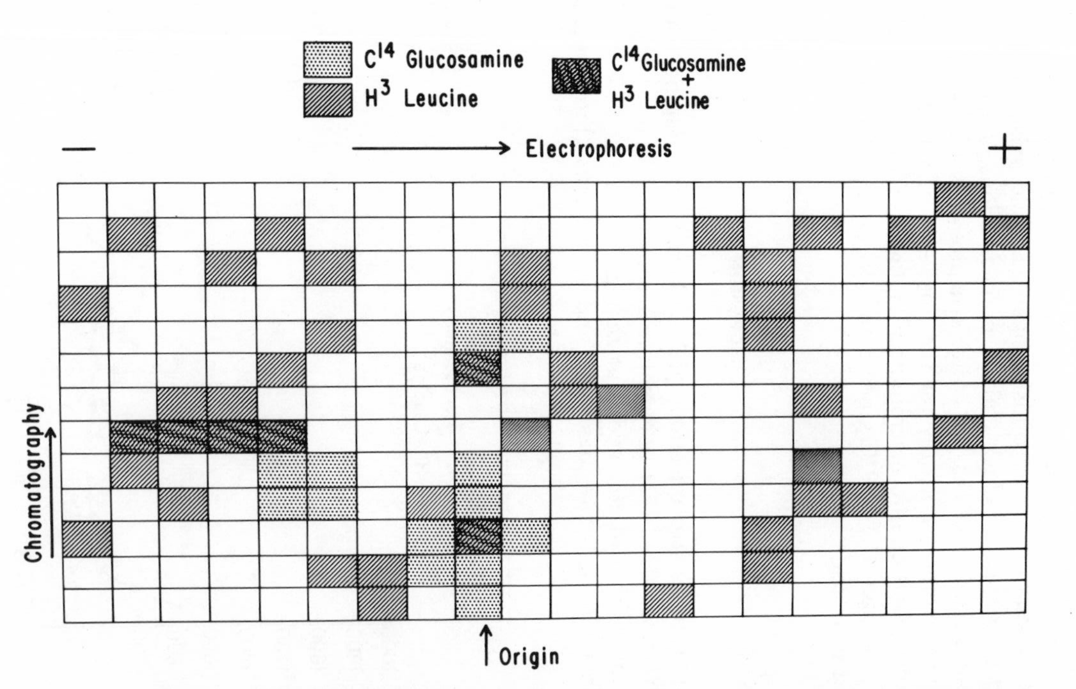

FIG. 7. Mapping of radioactive membrane peptides and glycopeptides. Tryptic digests as in Figure 6 were chromatographed and then electrophoresed as indicated. The paper was then cut into 260 squares and radioactivity determined. (From Festoff et al. 1970. Unpublished data.)

electrophoresis, glycopeptides could be isolated which contain both radioactivities. We must then conclude that both amino acid and carbohydrate incorporation into membrane glycoproteins may take place in synaptosomes, and that protein acceptors destined for membranes may be synthesized locally as well as possibly being transported to nerve endings.

The highest concentrations of sialic acid and glucosamine were found to reside in the synaptic plasma membrane fraction. These same fractions contained the largest percentage of particulate radioactivity hexosamine and sialic acid. These carbohydrates could be readily released from the membrane by tryptic digestion and isolated in glycopeptides. It is of interest that sialic acid appears to be an important component of the synaptic membrane, as suggested by the electronmicroscopic observations of Bondareff and Sjöstrand (1969) and the experiments of Sellinger et al. (1969), in which neuraminidase treatment of synaptic membrane fractions alters the migration of membrane particles in an electrical field. Bloom and Aghajanian have recently demonstrated that ethanolic phosphotungstic acid staining material is located at synaptic terminals. This material may well represent carbohydrate containing macromolecules and be of special significance to synaptic function.

Studies of amino acid and carbohydrate incorporation into membrane macromolecules support the concept that the synaptic region is capable of an independent metabolism and need not rely on axonal transport for all macromolecular constituents required. Therefore, it becomes most critical to assess whether any of the dynamic metabolic parameters observed in vitro possess the potential for regulation of synaptic function. The most critical fact is that our experiments demonstrated a very tight coupling of membrane function in macromolecular synthesis. Both amino acid and carbohydrate incorporation into protein were strongly affected by the ionic constituents of the medium. The composition which was most stimulatory was identical to that which stimulated sodium-potassium-ATPase activity. Different combinations of sodium and potassium produced similar effects on sodium-potassium-ATPase activity and on macromolecular synthesis. Ouabain inhibited ATPase activity, potassium accumulation, and macromolecular synthesis. Furthermore, several agents which inhibited the sodium-potassium-ATPase also inhibited macromolecular synthesis.

The critical question is how ATPase function in protein synthesis may be related. Experiments by Whittam and Blond (1964) with the brain homogenates suggest that membrane ATPase in active transport may not only represent the major component of energy utilization but may also act as a pacemaker for oxygen uptake and energy production. They suggested ADP as a potential link between membrane function and energy production. The levels of ADP might increase the result of ATPase activity and would enhance oxygen uptake in ATP synthesis.

The presence of increased intracellular potassium is another explanation of how ATPase function may be coupled to macromolecular synthesis. Lubin (1967) demonstrated that in animal and bacterial cells the level of potassium appears to influence the weight of macromolecular synthesis. However, his data did not distinguish between the direct effect to potassium levels of macromolecular synthesis and the indirect effects to mitochondrial function, glycolysis, or ATP metabolism. In our own studies, a direct proportionality was demonstrated between ouabain-sensitive potassium accumulation and ouabain-sensitive amino acid incorporation into protein or carbohydrate into macromolecules. These correlations do not distinguish between direct or indirect effects of potassium. However, they do suggest that in-vitro alterations in ATPase function may lead to alterations in potassium accumulation and in turn to alterations in membrane structure.

Figure 8 presents a model which best integrate the series of experiments reported above. Although only the radioactive amino acid incorporation is depicted, a similar scheme could be devised for carbohydrates. The radioactive compound is noted to pass the membrane barrier in association with sodium. The sodium in turn is pumped out with the help of the sodium-potassium-ATPase. In conjunction with the extrusion of sodium, potassium is pumped into the synaptosomal particle. Also in this process, ATP is broken down to ADP. The combination of the increased potassium and increased ADP stimulates mitochondrial uptake of oxygen and in turn results in restoration of high-energy phosphate for the synapse. Potassium may have direct and indirect effects on glycolysis and mitochondrial function as well as on the formation of the peptide bond. On the other hand, neither the effects of potassium nor the levels of ADP generated by the ATP broken down in the pumping process necessarily accounts for the stimulation of protein synthesis observed. Parker and Hoffman (1967) demonstrated that a major rate-determining step for the metabolism of glucose to lactate by human red blood cells was the conversion of 1,3-DPG + ADP to 3-PG + ATP catalyzed by phosphoglycerate kinase. The enzyme was found to be membrane-bound. It was proposed that this enzyme was the point at which the active transport mechanism can influence the glycolytic rate of the cell. The suggestion was made that the specific interaction between the cation pump and phosphoglycerate kinase occurs on or in the cell membrane, and that compartmentalization of ADP within the membrane may offer a means whereby the activity of the sodium pump could act as a pacemaker for high-energy phosphate production. Thus increased activity of the sodium pump would be associated with enhanced glycolysis and increased ATP synthesis from membrane-bound ADP. A similar scheme might be proposed for the synapse where intracellular potassium, membrane ADP, as well as glycolytic intermediates, might represent the link between membrane events and macromolecular synthesis. The active transport of sodium and potassium would, therefore, be associated with an enhancement of high-energy phosphate levels in the synapse.

Although the specific mechanism by which synaptic membrane activity may be coupled to macromolecular synthesis is difficult to specify precisely, the existence of the tight coupling of membrane activity with macromolecular synthesis offers an intriguing explanation for information storage within the nervous system. A simple hypothesis may be offered that sodium and potassium influence the rate of membrane protein and glycoprotein synthesis, and that newly synthesized glycoproteins may have significant effects upon the subsequent flux of sodium and potassium. Such effects on ion flux would have far-reaching consequences upon transmitter output, receptor interaction, and receptor-depolarization coupling. In brief, at a new steady state a level of activity previously incapable of

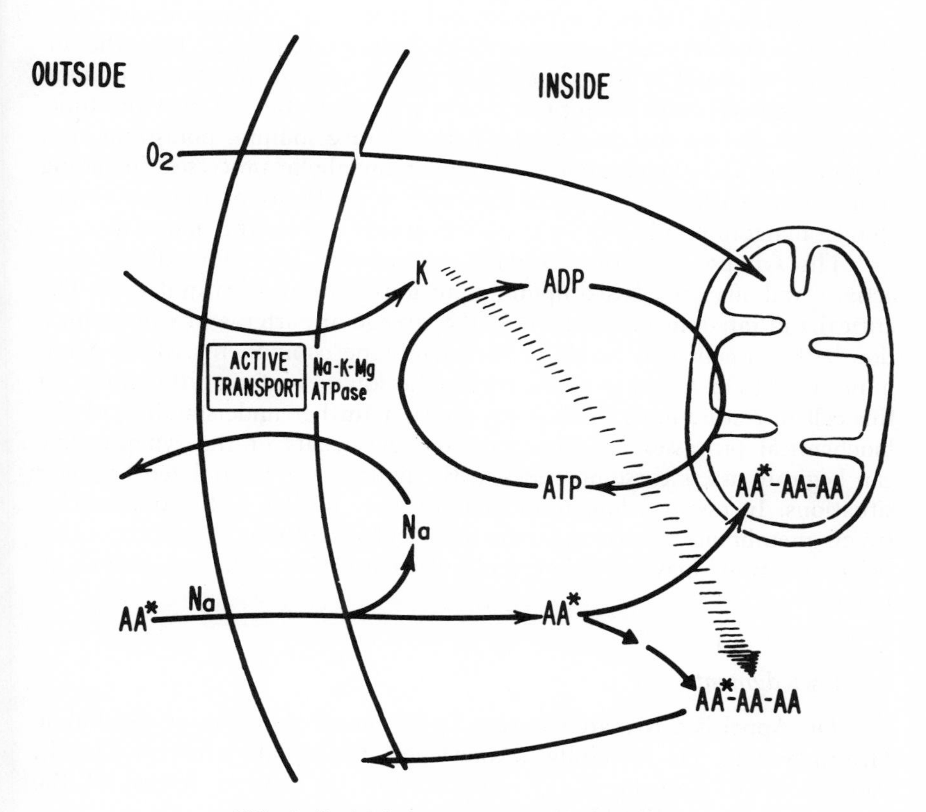

FIG. 8. Model for synaptic protein synthesis.

stimulating and sustaining macromolecular synthesis might now be able to do so because the ion translocation process and membrane-linked energy processes would have been previously activated. Such a scheme is based upon the assumption that information storage in the nervous system is associated with a change in membrane constituents and, more specifically, with a change in the macromolecular constituents of the synaptic membrane. Such a hypothesis has never been directly validated. On the other hand, the modulating effect of ions and neurohormones on the synthesis of carbohydrate constituents of the synapse membrane in vitro make such a proposal reasonable. Short-term information storage could be associated with synaptic membrane conformation changes, whereas long-term storage could be associated with covalent structural alterations in the synapse membrane, specifically in carbohydrates or amino acids on the external surface of the cell which would influence the ionic environment and probability of monovalent and divalent cation interaction and subsequent transport. The three basic elements of this theory are: (1) that information is stored at selected synapses and that such storage results in a changed probability of communication between cells; (2) that the information is stored in macromolecules of the synapse plasma membrane whose synthesis results from prior synaptic activity; and (3) that the function of the storage macromolecules is to enhance sodium, potassium, and calcium flux and thereby enhance cellular metabolic processes, including transmitter synthesis and release, as well as synthesis of the membrane storage macromolecules whose presence already defines the new state.

The synapse may thus contain the capacity for self-modulation of activity and efficiency based upon its previous patterns of stimulation. The critical and rate-limiting element in the storage of patterns of communication between cells may reside in the membrane components which determine the rate of response to environmental ionic events and the ability of this cell to restore ionic imbalances. With a further understanding of the biochemical processes underlying physiologic events of the synapse, one can begin to assess these processes in the light of certain isolated behavior situations. In essence, long-term memory may involve tight coupling between membrane events and macromolecular synthesis. Similarly, other behavioral traits may have their molecular correlates at the synaptic membrane.

Acknowledgments

Dr. Appel is a Research Career Development Awardee of the Public Health Service. He gratefully acknowledges the collaboration of Drs. L. Autilio, A. V. Escueta, B. W. Festoff, and E. Day, in several of the studies reviewed in this Chapter.

This research was supported by Grant NB-07872-03, U.S. Public Health Service, and by The Robert McManus Memorial Grant 558-B-3 from the National Multiple Sclerosis Society.

References

Appel, S. H., Autilio, L., Festoff, B. W., and Escueta, A. V. (1969). J. Biol. Chem., 244:3166.
Austin, L., and Morgan, I. G. (1967). J. Neurochem., 14:377.
Autilio, L., Appel, S. H., Pettis, P., and Gambetti, P. L. (1968). Biochem., 7:2615.
Barondes, S., and Dutton, G. R. (1969). J. Neurobiol., 1:99. ——— (1964). Science, 146:779.
Bloom, F. E., and Aghajanian, G. K. (1968). J. Ultrastruct. Res., 22:361.
Bondareff, W., and Sjöstrand, J. (1969). Exp. Neurol., 24:450.
Campbell, L., Mahler, H., Moore, W. J., and Tewari, W. (1966). Biochem., 5:1124.
Clarke-Walker, G. D., and Linnane, A. W. (1966). Biochem. Biophys. Res. Commun., 25:8.
Crane, R. K. (1965). Fed. Proc., 24:1000.
De Robertis, E., Rodriguez, D. L. A., Salganicoff, L., Pellegrino, D. I. A., and Zieher, L. H. (1963). J. Neurochem., 10:225.
Droz, B., and LeBlond, C. P. (1963). J. Comp. Neurol., 121:325.
Edström, A. (1966). J. Neurochem., 13:315.
Escueta, A. V., and Appel, S. H. (1969). Biochemistry, 8:725.
Festoff, B. W., Appel, S. H., and Day, E. (1970). Unpublished data.
Kiehn, E. D., and Holland, J. J. (1968). Proc. Nat. Acad. Sci. U.S.A., 61: 1370.
Koenig, E. (1967). J. Neurochem., 14:437.
Lassek, R. J. (1967). Neurosci. Res. Program Bull., 5:314.
Lubin, M. (1967). Nature (London), 213:451.
McEwen, B. S., and Grafstein, B. (1968). J. Cell. Biol., 38:494.
Marchbanks, R. M., and Whittaker, V. P. (1969). In: Biological Basis of Medicine, Bittar, E. E., and Bittar, N., eds., Vol. 5. New York: Academic Press.
Parker, J. C., and Hoffman, J. F. (1967). J. Gen. Physiol., 50:893.
Sellinger, O. Z., Borens, R. N., and Nordrum, L. M. (1969). Biochim. Biophys. Acta, 173:185.
Spear, P. G., and Roizman, B. (1968). Virology, 36:545.
Weiss, P., and Hiscoe, H. B. (1948). J. Exp. Zool., 107:315.
Wheeldon, L. W., and Lehninger, A. L. (1966). Biochem., 5:3533.
Whittaker, V. P., Michaelson, I. A., and Kirkland, R. J. A. (1964). Biochem. J., 90:293.
Whittam, R., and Blond, D. M. (1964). Biochem. J., 92:147.

Chapter 12

Chemical and Anatomical Plasticity of Brain: Replications and Extensions, 1970

MARK R. ROSENZWEIG, EDWARD L. BENNETT,
AND MARIAN C. DIAMOND

University of California (Berkeley)
Berkeley, California

The search for changes in the brain resulting from experience dates back at least two centuries. Only in the last few years, however, has it become clear that environmental variables, including learning and memory, can produce neurochemical and anatomical changes in the brain. In the present volume many of the chapters discuss experiments in which variables in the environment are shown to cause measurable changes in the synthesis and/or composition of neurochemicals.

This chapter will summarize our group's findings to date on anatomical and chemical changes induced in the rodent brain by differential experience, and it will include much unpublished work. Significant cerebral effects have been produced in long-term experiments. Control experiments have served to rule out many factors (such as stress or accelerated maturation) that have been suggested as alternatives to learning in accounting for effects. The observed anatomical and chemical brain effects will be considered in the *Discussion* to show their possible importance for mechanisms of learning and memory.

The earliest research that we have been able to find on this topic was done in the 1780s by a young Italian anatomist, Michele Vincenzo Malacarne (1819). His experimental design anticipated in important respects the one that we have employed in our own research. As subjects, Malacarne chose two littermate dogs and pairs of birds (each pair coming from the same clutch of eggs). In each pair of animals, he gave intensive training to one and no training to the other. After a few years he sacrificed the animals and compared the brains of the trained and untrained members of each pair. Apparently Malacarne found the results to be encouraging, but they were not convincing to others, and his research along these lines seems not to have been continued. Among many later investigators who speculated about growth of brain induced by training we may mention the phrenologist, Spurzheim, the neurologist-anthropologist, Broca, and the physiologist, Donaldson.

In spite of sporadic research dating back nearly two centuries in which differences in brain anatomy of animals differing in past experience were sought, little positive evidence could be produced until quite recently. At a 1963 conference *The Anatomy of Memory* (Kimble, 1965) the participants were pessimistic that detectable and meaningful anatomical differences could be found. This pessimism seemed unwarranted to us, because we had already demonstrated changes in chemistry and then in weight of rat brain as a function of differential experience (Krech et al., 1960; Rosenzweig et al., 1962). ~~Even though earlier work of others had been inconclusive or negative, we believed that there were good reasons why our use of prolonged and varied experience should produce effects (Rosenzweig et al., 1962)~~. Even though earlier work of others had been inconclusive or negative, we believed that there were good reasons why our use of prolonged and varied experience should produce effects (Rosenzweig, 1966; 1970). We were also encouraged by reports of alterations in neurochemistry during learning that Hydén had published by this time (1962), some of which were presented at the 1963 Conference. Our own findings on biochemical and anatomical effects were reported briefly by Krech at the 1963 Conference. This report elicited questions concerning our techniques, the reproducibility of the effects we reported, factors responsible for the changes, and interpretations of their possible significance for learning and memory.

In 1964 we published in *Science* a fuller summary of our findings, along with a discussion of many of the questions that have frequently been raised concerning them (Bennett et al., 1964). This has been the most general, and most often cited, and most available account of our work, and it has reached a wider audience than have later summaries of our research (Rosenzweig, 1966 and 1968; Bennett, et al., 1970). In the present chapter we will therefore summarize how the work has progressed since 1964 in terms of corroborating or modifying results already obtained and extending findings into several further areas; finally we will discuss current interpretations.

Current Types of Research on Cerebral Effects of Environment and Training

One way of comparing and contrasting our research with those of others in this volume is in terms of the duration of typical experiments. Studies in which the effects of environment and training on brain chemistry and anatomy have been investigated can be divided into three groups of durations: (a) those studies in which the period from start of training to removal of brain tissue lasted only from minutes to a few hours; (b) those in which a few training trials per day over a few days were given; and (c) those with differential experience which lasted for weeks. The short-term experiments have recently been undertaken by a number of investigators—Glassman, Wilson and associates (Adair et al., 1968; Kahan et al., 1970), Bowman and Stroebel (1969), Machlus and Gaito (1968, 1969), Gold et al. (1969), and Beach et al. (1969). These experiments are discussed in depth in Chapters 3 and 4 of this volume.

The intermediate-term approach by Hydén has a somewhat longer history. In addition, the unique approach of the Swedish group has concentrated on presumably active sites of learning, and, within such sites, on separate cell types. Initially, these experiments focused on composition and amount of RNA and protein. Only recently has this group looked for altered rates of synthesis in nerve cells utilizing radioactive tracers. These experiments are discussed in more detail in other chapters of this volume.

The experiments to be discussed in this chapter are representative of the third class: long-term treatments after which relatively modest changes in the neurochemistry and neuroanatomy of the brain and in the behavior of the animal have been sought. This approach of using long-term experience can be seen as a continuation of the earliest experimental method to be brought to the task of detecting changes in the brain associated with memory, i.e., that of Malacarne, which was described earlier. Furthermore, if some of the short-term changes described above reflect synthesis of RNA and protein in short-term learning, presumably there are also persisting chemical and structural alterations that maintain long-term memory.

In addition, recent reviews by Glassman (1969), by Rose (1969), and by Bennett and Rosenzweig (1970) have dealt rather completely with the literature on neurochemical changes through 1969. Anatomical changes in the brain associated with differential experience have been reviewed by Rosenzweig and Leiman (1968).

Another type of research on effects of environment on brain chemistry should be mentioned briefly here, although it will not be taken up further in this chapter. These are the long-term studies of Garattini and collaborators (Giacalone et al., 1968, Grattini et al., 1969) in Italy, and of Welch and Welch (1969, 1970) in the United States; they are aimed at elucidating brain mechanisms of aggressiveness or emotional behavior in biochemical terms.

Cerebral Effects of Experience

Now let us turn to our research, in much of which animals have been subjected to differential environments: standard colony (SC), enriched condition (EC), or impoverished condition (IC). We will list first some of our main recent findings, then review the major effects reported in our 1964 paper, next show how these have stood up under replication, and then take up new findings that have stemmed from the earlier work and that have helped to give it greater meaning and generality. In a later section we will take up tests of alternative hypotheses to account for the cerebral effects of experience. This chapter will be confined to reviewing cerebral effects of experience and will not review attempts to relate them to behavioral effects; we have recently discussed possible relations between cerebral and behavioral effects elsewhere (Bennett et al., 1970; Rosenzweig, 1970).

A Preview of Recent Findings

By way of "advance notice," some of our recent findings are listed below, along with the pages on which they will be discussed:

1. In addition to the EC-IC differences in cortical thickness reported previously, histological analyses have also revealed changes in glial/neuronal ratio, and in cross-section area of neuronal cell bodies and nuclei.

2. Cerebral changes with experience are easier to produce than had been supposed because a 30-day period of differential experience is sufficient to induce full-fledged differences in some cerebral measures, and only 2 hr per day over 30 days need be spent in the enriched conditions to produce the effects.

3. Different brain measures show changes developing at different rates as animals remain in the experimental environments.

4. EC-IC brain effects can be produced not only in weanlings and young adults, as previously reported, but also in year-old rats and in sucklings.

5. Cholineacetylase activity appears to follow the pattern of acetylcholinesterase differences between EC and IC. Acetylcholine concentration in whole brain does not differ significantly between SC and IC.

6. In EC, as compared to IC, and in light-raised, as compared to dark-raised, there is significantly lesser DNA per unit of weight of occipital and somesthetic cortex, and RNA/DNA ratios are greater.

7. The relative effects on weights of brain sections are identical whether wet weights (fresh tissue) or dry weights are taken, so fluid content of the brain is not a differentiating factor.

8. Drugs can modulate the cerebral effects of experience, with an excitant heightening the changes and a depressant reducing them.

9. The effects on brain weights and brain chemistry previously reported for the rat have, in large part, been found in mice and gerbils, so there is considerable generality across species of rodents.

10. Although the cerebral effects are easier to produce than previously supposed (see 2 above), certain treatments are ineffective. These include:
(a.) stress of either electrical shock or tumbling.
(b.) visual or auditory stimulation from outside the cage.
(c.) being handled and placed daily in an apparatus other than the home cage.

11. Typical EC-IC effects can be produced in hypophysectomized rats, thus indicating that the effects are not produced by altered functioning of any of several endocrine organs.

12. Effects of enriched experience can be produced when single rats are put in EC cages, provided their responsiveness is heightened by an excitant drug or by receiving the exposure during the dark part of the diurnal cycle.

Procedural Outline

Basic Procedures

The procedures involved in giving animals differential experience and in seeking to detect cerebral effects have been described in our reports in scientific journals, and many of these procedures are available in greater detail in dittoed form. Basic procedures will be summarized briefly here, but readers desiring further specifications should consult the references below or should address us. Variants employed in particular cases will be taken up later in the context of specific experiments.

Behavioral Procedures

Unless otherwise stated, all subjects were male rats of the Berkeley S_1 strain, bred and born in the psychology department's colony. The three main environmental conditions are standard colony (SC), enriched condition (EC), and impoverished condition (IC). In SC, three animals of the same sex were housed in a colony cage (32 by 20 by 20 cm) made of wire bars and with shavings on the floor. In EC, to enrich experience above the SC baseline, animals are housed in a same-sex group of 10 to 12 in a large cage (70 by 70 by 46 cm), and the cage is provided with stimulus objects ("toys") from a standard pool. Each day there are six or more toys in the EC cage; some may be in new positions from the previous day and some are newly taken from the pool. Rats in an EC cage were shown in Figure 1 of the 1964 *Science* paper; the pool of objects was shown in Figure 1 of Rosenzweig and Bennett (1969). Each day the EC animals are placed for 30 min in the field of a Hebb-Williams apparatus (75 by 75 cm) where the pattern of barriers is changed daily. During this time the selection and arrangement of objects in the EC cage are changed. Through 1965 the animals also received one or two trials a day in various standardized mazes for sugar pellet rewards, but we found that this procedure could be omitted without affecting the cerebral results. It is for this reason that we now use the designation EC (enriched condition) instead of the previous ECT (environmental complexity and training). Usually the EC and SC cages are in the same room.

In IC, to reduce experience below the SC baseline, animals are housed in individual cages similar in size to SC cages. Furthermore, in all of the earlier and many current experiments, the IC cages have solid side walls so that the animals cannot see each other, and the IC cages are placed in a separate, quiet, and dimly illuminated room. In these special IC cages, the floors are made of wire bars and pans of shavings below the cages can be changed without touching the animals. Later experiments have indicated that effects of isolation can be obtained even when single animals are housed in the regular colony cages on the same racks as SC animals. That is, the cerebral effects are found although the IC rats have visual contact with other animals.

All animals—SC, EC, and IC—have laboratory pellets and water available ad lib. All animals are weighed regularly, usually at weekly intervals.

In most experiments, littermates are assigned semirandomly among the number of groups required. Assignment is semirandom in that, if necessary, exchanges are made to insure that the distributions of body weights among groups will be similar. Runts and animals whose weight departs far from other littermates are excluded. The groups so chosen are then assigned at random to the experimental conditions.

Dissection of Brain Samples

Animals are brought to sacrifice under code numbers that do not reveal to the analysts the behavioral condition of any subject; littermates are sacrificed in immediate succession, the EC rat being first in some pairs and the IC first in others. When weights of brain samples are to be taken and chemical analyses are to be done, the brain is divided rapidly by dissection, usually into these six samples: (1) sample of occipital cortex;

(2) sample of somesthetic cortex; (3) remaining dorsal cortex; (4) ventral cortex and adjacent tissues including corpus callosum, hippocampus, and amygdala; (5) cerebellum and medulla; and (6) the rest of the subcortex, including the olfactory bulbs. Sections 5 and 6 together will be referred to as "rest of brain" or "subcortex." The method of dissection was described and diagrammed in the 1964 *Science* paper (p. 612). The samples are weighed to 0.1 mg and then stored at $-20°C$ until chemical analysis.

Acetylcholinesterase and cholinesterase. Through 1962, brain acetylcholinesterase (AChE) was determined by measuring the rate of hydrolysis of acetylcholine at constant pH (the "pH stat" method Rosenzweig et al., 1958). Since that date a spectrophotometric assay has been employed in which acetylthiocholine is the substrate. Since AcSCh is less specific than AcCh with respect to hydrolysis by cholinesterase (ChE), we have used promethezine to inhibit ChE activity in this essay. For analysis of ChE, butyrylthiocholine is the substrate and AChE is inhibited with BW284C51 (1,5-bis [4-allyldimethylammoniumphenyl]-pentan-3-one diiodide). Analyses are made in duplicate; two AChE values usually agree within two percent and ChE values, within three percent.

Acetylcholine. As acetylcholine (ACh) is a very labile constituent of brain, the brain was frozen in situ to eliminate or minimize changes in the "normal" ACh content during sacrifice and brain dissection. The rat was placed in a small restraining cage and completely immersed in liquid O_2. The frozen brain (excluding cerebellum, medulla, and pons) was removed, weighed rapidly, crushed and extracted with an acetic acid-ethanol mixture. The ACh in the extracts was determined by a bioassay procedure employing the rectus abdominus muscle of the frog *R. pipiens*, sensitized with neostigmine bromide. The isometric contractions were measured by a transducer. Duplicate assays of the extracts generally agreed within 5 percent (Bennett et al., 1960; Crossland, 1961).

Choline acetyltransferase. Our procedure for the determination of cholineacetyltransferase (ChAt) has combined and adapted features of the methods described by Fonnum (1966) and McCaman and Hunt (1965). Since our procedure has not yet been presented elsewhere, we will outline the principles here; as with our other procedures, a detailed description is available from the authors.

$$(1)\ \text{ATP} + \text{acetate} + \text{CoA} \xrightleftharpoons{\text{acetyl-CoA synthetase, } Mg^{++}} \text{acetyl-CoA} + \text{AMP} + \text{PP}$$

$$(2)\ \text{Acetyl-CoA} + \text{choline} \xrightleftharpoons{\text{choline acetyltransferase, } Mg^{++}} \text{acetylcholine} + \text{CoA}$$

Under appropriate conditions, the rate of acetyl-CoA synthesis is sufficiently high to be nonlimiting so that the rate of the second reaction is determined by the ChAt activity of the brain sample used.

Insert side heading, CHEMICAL PROCEDURES, after line 7.

Line 14. Replace "essay" with "assay."

Before reaction number 1 insert the following sentence: "The successive synthetic reactions involved in the synthesis of ACh from acetate and choline may be formulated as follows:"

Brain samples were dissected, weighed, and frozen in the same manner as in our AChE procedure. The stability of ChAt in frozen brain was not established for periods longer than one month. Less than 10 percent loss of activity was noted in this time, and most analyses were made within two weeks of sacrifice. The samples were homogenized at concentrations of 30 to 40 mg/ml in 0.5 percent Triton X-100 to activate or solubilize the ChAt. This resulted in 30 percent more activity than was obtained with frozen brain homogenized in water alone, and the method was more convenient than the conventional ether treatment. Furthermore, our use of the ether method gave only about 30 percent increase in activity rather than the fourfold increase reported by Fonnum.

The incubation mixture described by Fonnum to carry out reaction (1) was modified principally by the substitution of potassium salts for sodium salts. A high Na^{+}-K^{+} ratio is inhibitory for the acetyl-CoA synthetase system. Mg^{++}, an activator of the kinase system, was also included.

The reaction mixture was preincubated for 30 min to form an excess of acetyl-CoA before initiating reaction (2). An equal volume of brain homogenate was then added; incubation was continued for 60 or 120 min. At the end of this time, the reaction was stopped by the addition of a trichloroacetic acid-ACh solution. The ACh was subsequently precipitated as the reineckate from an aliquot of the supernatant, washed thoroughly, and finally dissolved in acetonitrile for scintillation counting to determine the radioactivity of the ACh formed from the added acetate-^{14}C. In spite of the color of the choline reineckate, relatively large amounts can be counted by appropriate modification of discriminator settings.

The detailed procedure as finally developed and used in these analyses is a highly reproducible, relatively convenient method for the determination of ChAt in a large number of brain samples. Typically, duplicate analyses of each brain section were made; the average difference between duplicate assays was 2.5 percent. Less than 10 percent of the duplicate assays differed by more than 5 percent, while nearly 20 percent checked within 1 percent. The absolute values of ChAt activity found in rat brain were higher than had been previously reported. We found activities of 8.3 μM ACh/hr/g for total cortex and 13.9 for the subcortical brain excluding medulla and cerebellum (see Table 12).

RNA and DNA. As the excellent and comprehensive review of Munro and Fleck (1966) makes abundantly clear, there is no single procedure that is optimal for the quantitative analysis of nucleic acids from all organs. Each tissue presents its own particular problems for the determinations of RNA and DNA. Logan et al. (1952), Zamenhof et al. (1964), and Santen and Agranoff (1963) have pointed out specific difficulties in the quantitative determination of nucleic acid in brain. Of the procedures described at the time we investigated the effects of environment on RNA and DNA (1964–1965), the UV spectrophotometric method described by Santen and Agranoff appeared to be the most appropriate. However, this procedure suffers from the necessity of correcting empirically for ultraviolet absorbing breakdown products which interfere with the determination.

Since our total procedure differs in several respects from that of other investigators and since we have not previously described it, we will give it here in some detail. After separating DNA from RNA by the Schmidt-Thannhauser procedure, we enzymatically converted the purines to uric acid as suggested by Heppel et al. (1957) and determined the decrease in absorbance at 262 nm and the increase at 292 nm. The successive reactions were:

$$\text{(1) adenine} \xrightarrow{\text{adenase}} \text{hypoxanthine} \xrightarrow{\text{xanthine oxidase}} \text{uric acid}$$

$$\text{(2) guanine} \xrightarrow{\text{guanase}} \text{xanthine} \xrightarrow{\text{xanthine oxidase}} \text{uric acid}$$

Brain samples for nucleic acid analyses were dissected, frozen, and stored in the same manner as for our AChE procedure. We did not specifically investigate the effects of method of sacrifice or length of storage on the amount of nucleic acids recovered. May and Grenell (1959) reported that the apparent RNA content was less in brain which had been frozen as compared to brain which was homogenized immediately in trichloroacetic acid (TCA). On the other hand, Santen and Agranoff did not find significant differences between the RNA content of fresh and frozen brain, but did report a difference in the DNA content. In our experiments, rats from all conditions were sacrificed, stored, and analyzed in concert.

At the time of analysis, tissue samples were homogenized at known concentrations of approximately 10 mg/ml in cold 0.1 M NaH_2PO_4-Na_2HPO_4 buffer, pH 8.0. The subsequent procedure followed quite closely the conventional Schmidt-Thannhauser procedure for the elimination of nucleotides and the subsequent separation of RNA from DNA. The tissue was precipitated from a 5.0-ml aliquot of the homogenate by the addition of one-fifth volume of cold 50 percent TCA; after centrifugation, the precipitate was washed twice with 5 ml cold 10 percent TCA, once with 4 ml 95 percent EtOH, once with 3 m CHC_3-CH_3OH-H_{20} (38:27:3), and once with 4 ml ether. The RNA in the tissue precipitate was hydrolyzed in 500 μl of 1 N KOH for 1½ hr at 37° C, the DNA and protein were subsequently precipitated by the addition of 250 μl of 18 percent HCO_4, and the supernatant containing the nucleotides from the RNA was transferred to a 2-ml volumetric flask. The precipitate was washed twice with 300 μl

213 Line 34. Replace "3 m CHC_3-CH_3OH-H_{20}" with "3 ml $CHCl_3$-CH_3OH-H_{20}."

Line 37. Replace "HCO_4" with $HClO_4$."

214 Line 1. Replace "1 N HCO_4" with "1 N $HClO_4$."

of 1 N HCO_4, and the washes were combined with the initial extract. The purine nucleotides were hydrolyzed to free adenine and guanine by heating 1½ hr at 100° C in a heating block. DNA was extracted from the tissue residue by heating at 80 to 90° with 600 μl of 1 N HCl for 30 min and the residue was reextracted with 300 μl of 1 N HCl (heated 10 min at 80°). To ensure complete conversion to free purines, the combined DNA extracts in 2 ml volumetric flasks were heated at 95 to 100° C for 1 hr.

Both the RNA and DNA samples were stored in acid solution until immediately prior to analyses, at which time the solutions were neutralized with 4 M KOH and diluted to 2.0 ml with a Tris-PO_4 buffer, pH 7.35 prepared by mixing equal volumes of 0.4 Tris-Tris HCl buffer, pH 8.15 and 0.4 M NaH_2PO_4-$NaHP0_4$ buffer, pH 6.6.

To determine the amount of purine bases in the RNA and DNA hydrolysates, aliquots were transferred to microcuvettes, diluted to 800 μl with 0.2 M Tris- 0.2 M phosphate buffer, pH 7.3, warmed to 37° C, and the absorbance determined at 262 and 292 nm with a Gilford model 2000 automatic spectrophotometer equipped with a digital absorbance meter. The desired initial absorbance was about 2.0. Adenine aminohydrolase (adenase) was added to convert adenine to hypozanthine resulting in a decrease in absorbance at 262 nm. The adenase was prepared from *Azotobacter vinelandii* by $(NH_4)_2SO_4$ fractionation followed by further purification on a Biogel P200 column. It is essential that this enzyme be free of guanase activity. In addition, this enzyme, as well as the other enzymes used in the procedure, should be as pure as possible to minimize the enzyme blanks caused by their additions. Typically, the adenase preparation contributed a blank absorbance of about 0.02 at 262 and 292 nm and completely deaminated 60 mμ moles adenine in about 10 to 15 min.

After completion of the deamination, xanthine:oxygen oxidoreductase (xanthine oxidase) was added to convert hypoxanthine to uric acid, resulting in further decrease in absorbance at 262 nm and a large increase in absorbance at 292 nm. Xanthine oxidase, partially purified by ammonium sulfate frationation, was obtained from California Biochemical Corporation with a stated activity of 0.5 enzyme units (EU) per milligram at 25° C. In actual practice, we found the activity to be about 1.5 EU/Mg at 37° C. It was necessary to use about 10 μg/assay and this amount contributed approximately 0.03 absorbance. The reaction was completed in approximately 15 min.

After completion of these two reactions, guanine was converted to uric acid by the addition of guanine aminohydrolase (guanase from California Biochemical Corporation). This enzyme from rabbit liver typically had a specific activity of 0.07 EU/mg protein. For each assay 40 μg of this enzyme was used. This amount contributed a blank absorbance of approximately 0.04. The conversion of guanine to uric acid was completed in approximately 20 min.

After making appropriate corrections for the enzyme blanks and for volume changes due to the addition of the enzymes, the amounts of adenine and guanine were calculated by using the following molar values for differences between initial and final extinction ($A_{final}—A_{initial}$):

		262 *nm*	292 *nm*
adenine	→ uric acid	−10,300	+12,000
hypoxanthine	→ uric acid	− 4,300	+12,200
guanine	→ uric acid	− 4,500	+ 9,200

With use of known samples of adenine, guanine, hypoxanthine, and xanthine, as well as mixtures of adenine and guanine, values within 5 percent of calculated were obtained. We used as a measure of RNA and DNA the average values determined by both the adenine to uric acid and hypoxanthine to uric acid conversion plus the value determined by the guanine to uric acid conversion, i.e., $\frac{\text{adenine} + \text{hypoxanthine}}{2} + \text{guanine}$.

Changes in absorbance at both 262 and 292 nm were weighted equally. These results were expressed in terms of mμmoles/g wet weight sample.

Since to our knowledge this is the first time this method has been applied to the analysis of crude RNA and DNA, it may be well to compare our values with some of those obtained by other methods—with particular reference to rat brain. Our analyses were done on tissue from 138 rats in five experiments; occipital and somesthetic cortex was analyzed in each animal, and the other brain section were analyzed from 42 rats. Values are given for the young adult EC rat. Let us consider first DNA; in doing so we will take up the A:G ratio, then A+ G concentration (our primary measure), and, based on this, the weight of DNA.

Our analyses yielded an average value of 1.39 for the ratio of A to G in DNA which is in excellent agreement with the literature. A recent compilation of the A:G ratio for DNA of numerous rat tissues give ratios ranging from 1.16 to 1.49, and the average of these 15 values is 1.34 (Shapiro, 1968). For brain, Jacob et al. (1966) have reported a ratio of 1.35, while Yajima (1966) reports a value of 1.40. The A + G content for the occipital area of the EC rats was 1.22 mμmoles/g (the content of the subcortical brain (including cerebellum and pons) was 2.56 mμmoles/g and whole brain was 1.92 mμmoles/g. Other cortical areas had approximately 90 percent of the DNA concentration of the occipital area. Assuming that

215-216 A/G and A:G are used interchangeably but should, of course, be consistent.

A = T and G = C, a "tetranucleotide" weight for DNA of 1,235 can be calculated. The DNA content of the cortex was 665 μg/g the subcortex, 1,575 μg/g; while total brain had a content of 1,188 μg/g. For comparison, a few selected values from the literature can be quoted. Six values for total brain ranging from 850 μg/g to 2,000 μg/g have been quoted by Rappoport et al. (1969); Santen and Agranoff (1963) report a value of 2,200 μg/g; Bondy (1966) reports 1,450 μg/g. Zamenhof et al. (1964) report a value of 700 μg/g for a brain sample which omitted the cerebellum and medulla, areas which are high in DNA, and May and Grenell (1959) found about 700 μg/g for rat cortex—which is close to our value.

For brain RNA, we found the A:G ratio to be 0.67. The base ratio of ribosomal RNA (approximately 75 percent of total brain RNA) is given as 0.57 by both Jacob et al. (1966) and Mahler et al. (1966). A slightly higher value of 0.61 would be estimated when the contribution of tRNA and mRNA is considered. Thus our value would appear to be perhaps 10 percent higher than expected for the A/G ratio. The A + G content of the brain areas sampled differed only slightly, the occipital area had the highest value of the areas analyzed, 2.44 mμmoles/g, and the content of total brain was 2.16 mμmoles/g. Assuming that A = U and G = C (an approximate assumption), a "tetranucleotide" weight of 1,288 can be calculated. By utilizing this figure, the RNA content per gram ranges from a high of 1,572 μg/g for the occipital cortex to a low of 1,346 μg/g for the remaining subcortex. Values ranging from 1,100 to 3,700 μg/g are quoted in the review by Rappoport et al. (1969); Santen and Agranoff (1963) report a value of 1,580 μg/g; Bondy (1966) gives 180 μg/g; while Dellweg et al. (1968) report 2,000 μg/g for ribosomal plus tRNA. Grenell and May found the cerebral cortex to contain about 1,550 μg RNA/g.

Line 25. Replace "180" with "1,800."

The ratio of RNA:DNA (μg/g) was 2.1 ± 0.1 for the four cortical areas, 0.82 for the subcortex, and 1.1 for the whole brain. May and Grenell (1959) found 2.1 for cerebral cortex, while Jacob and Mandell (1966) report 1.77. Values for the RNA:DNA ratio of whole brain range from a low of 0.7 (Santen and Agranoff) to a high of 2.67 (Leslie, 1955), but the majority of the values quoted are in the range of 1.2 to 1.4. It is obvious from this cursory review of the literature that more work needs to be done to compare and standardize the methods utilized for the determination of RNA and DNA in the brain.

Histological Procedures

For light-microscopic examination of the depth or thickness of the cortex, both frozen and celloidin sections were prepared. The celloidin sections were also used for differential neuron-glial counts (Diamond et al.,

1966), and for cross-sectional area measurements of the perikarya and their nuclei (Diamond, 1967).

Depth measurements were taken on ten transverse sections of the rat brain, which were outlined with the use of a microslide projector (magnification ×22.5). On the drawings, lines 2 mm apart were extended up from the corpus callosum to the dorsal surface of Layer II and were measured with a millimeter ruler. The lines were grouped for comparisons into four segments, B, C, D, and E, depending upon the width of the section. (See Fig. 4 for locations of the sections and the segments.)

For the cell counts, overlapping photomicrographs were taken of occipital cortex in Section 9, Segment B. Each composite photomicrograph included the cortex immediately lateral to the elevation of the corpus callosum and it extended from the pial surface through Layer VI. For a sample composite picture, see Fig. 2 in Diamond et al. (1966). On these pictures, neurons and glia were identified and counted independently by two technicians. Differences in their counts were then discussed, leading to further agreement on criteria.

On the same sections photographed for cell counts, perikarya and their nuclei were measured with a planimeter on outlines traced from microfilm projections. The cortex was divided into upper, middle, and lower thirds, since the specific layers are not clearly defined in the rat occipital cortex.

Results of 25–105 Day Experiments

Early Findings

In 1964 we reported a number of ways in which brains of littermate rats differed after the animals had been put into ECT or IC at weaning (about 25 days of age) and kept there for approximately 80 days. Let us review briefly the main findings and then note how these have been confirmed and extended by later work. Seven such experiments had been run with male rats of the S_1 strain from 1960 through July 1963, employing a total of 77 littermate pairs. The ECT brains, in comparison with IC, showed:

1. Greater weight of total cerebral cortex (4.6 percent, $p < .001$).
2. Slightly lesser weight of the rest of the brain (−1.2 percent, $p < .05$).
3. The largest cortical weight differences in the occipital (then called "visual") region (6.4 percent, $p < .001$) and smallest in the somesthetic region (2.7 percent, $p < .05$).

4. About 9 percent less body weight, so that increased cortical weight did not reflect body weight.

5. Increased total activity of AChE in cortex (2.7 percent, $p < .01$) and in the rest of the brain (2.1 percent, $p < .001$). Since the increase of total AChE in cortex was less than the increase in weight, AChE:weight was significantly lower in the cortex of EC than of IC; for total cortex the difference in AChE:weight amounted to −1.8 percent ($p < .01$). In the rest of the brain, AChE:weight was greater for EC than for IC (3.4 percent, $p < .001$).

6. Increased total activity of ChE in cortex (8.9 percent, $p < .001$), but a slight decrease in rest of brain. Since the increase of total ChE in cortex exceeded the increase in weight, ChE:weight in cortex was also greater in EC (4.8 percent, $p < .001$).

7. Increases in cortex of total protein and of total hexokinase activity that closely parallel the increase in cortical weight.

8. Increased thickness of cortex in the occipital ("visual") region (6.2 percent, $p < .001$) and in the somesthetic region (3.8 percent, $p < .01$). This was based on 20 littermate S_1 pairs for the occipital area and 18 pairs for the somesthetic area.

Three of the first seven S_1 experiments had also included an SC group, and this permitted us in the 1964 report to discriminate between effects of enrichment and impoverishment, measured from the colony baseline. A figure in that paper (Fig. 5, 1964, p. 617) showed weight and total AChE values for both cortex and rest of brain of ECT and IC rats in relation to SC. "The bulk of the effects on cerebral weight and [total] acetylcholinesterase activity is due to enriching rather than to restricting the experience of our colony animals" (1964, p. 615). We later found that for total ChE, on the contrary, the main difference occurred between IC and SC (Rosenzweig, 1968).

We also noted in 1964 that five lines of rats beside the S_1s had been run in ECT-IC experiments and had shown similar effects in brain weights and AChE activity.

Replications

Further EC and IC (and sometimes SC) groups were subsequently run, not principally to replicate these findings but chiefly because other variant conditions were being tested and we wanted to compare them as directly as possible with previous conditions. Since there is some variability among replications, we prefer, when possible, to make comparisons among groups run simultaneously. Consequently, by December 1969, we had obtained weight and chemical results on 16 experiments with S_1 rats run in EC and IC from 25 to 105 days of age (175 littermate pairs); there have been two more EC-SC-IC experiments, making five of these (58 sets of S_1 triplets). It should be noted that the actual experimental durations have varied on both sides of 80 days because of exigencies of calendar and laboratory schedule; durations have ranged from 73 to 82 days. On most measures the newer results have not changed substantially the values that we reported in 1964, but in one respect there has been a change that is worth reporting.

A direct comparison of the results available through July 1963 with the subsequent results (December 1963 through 1969) is given in Table 1 in terms of EC-IC percentage differences. Overall values for all 16 experiments, 1960–1969, are also given. It will be seen that the results for tissue weights are closely similar for the two sets of experiments. The rank order of magnitudes of effects among the six brain samples are identical, with occipital cortex showing the largest effects, remaining dorsal cortex next, and negative differences for rest of brain in both the earlier and the later experiments. In each cortical region except occipital cortex, however, the EC-IC difference is smaller in the replication experiments than in the original series.

The EC-IC enzymatic measures do not show as much stability over time as do the brain weight effects. In acetylcholinesterase activity per unit of tissue weight (AChE:wt), the more recent experiments yield consistently larger negative differences in the cortex than did the earlier series. Thus for total cortex the 1963–1969 experiments show a difference of −2.9 percent ($p < .001$) as against −1.8 percent ($p < .01$) for 1960–1963. Furthermore, the increase in the rest of the brain is smaller in the later series (1.3 percent, $p < .05$) than in the earlier experiments (3.4 percent $p < .001$). The cortical-to-subcortical ratio of AChE:weight does show quite stable effects, since the EC-IC difference amounted to −5.1 percent ($p < .001$) earlier and is −4.2 percent ($p < .001$) in the replication series.

In total activity of AChE, the more recent experiments fail to show the significant EC-IC differences that characterized the earlier results. Although three of the four cortical areas yield small positive effects in total AChE, only in the remaining dorsal cortex is this significant, and the 0.6 percent difference for total cortex falls short of statistical significance. The results above in the table for weight and AChE:weight reveal why total AChE no longer gives significant differences: the percentage gains in tissue weight in the replication experiments are scarcely larger than the percentage decreases in AChE:weight. While the overall results for total AChE are significant, it is clear that we have not been able to replicate fully this aspect of our 1964 report.

It should be noted that the AChE analyses for five of the seven 25 to 105-day experiments included in the 1964 report were done by the "pH

220 Table 1. Heading of second column on right should be Cortex/Rest rather than Cortex/Rest.

TABLE 1. Comparison of EC(T)-IC Percentage Differences in Seven Initial Experiments (1960-63) and Nine Replication Experiments (1963-69), Based on S_1 Rats Run from 25 to 105 Days of Age

	N (pairs)	Cortex: Occipital	Cortex: Somesthetic	Cortex: Dorsal	Cortex: Ventral	Cortex: Total	Rest of Brain	Total Brain	Cortex Rest	Terminal Body Wt.
Tissue Weight										
1960-63	77	6.4***	2.7*	5.2***	4.0**	4.6***	−1.2*	1.2*	5.9***	−8.8***
1963-69	98	6.4***	1.4	4.8***	2.1*	3.6***	−1.0*	0.9*	4.7***	−8.3***
Overall	175	6.4***	1.9**	5.0***	3.0***	4.0***	−1.1**	1.0**	5.2***	−8.5***
AChE/Weight										
1960-63	73	−3.0***	−0.7	−1.8**	−1.5	−1.8**	3.4***	1.1*	−5.1***	–
1963-69	98	−4.6***	−2.6***	−3.1***	−2.0**	−2.9***	1.3*	−0.5	−4.2***	–
Overall	171	−3.9***	−1.8***	−2.5***	−1.8**	−2.4***	2.2***	0.2	−4.6***	–
Total AChE										
1960-63	73	3.2**	2.1	3.1**	2.6	2.7**	2.1***	2.2***	0.6	–
1963-69	98	1.5	−1.2	1.6*	0.1	0.6	0.3	0.4	0.2	–
Overall	171	2.2**	0.1	2.2**	1.2	1.5**	1.0**	1.1**	0.4	–
ChE/Weight										
1962-63	34	7.3*	8.6***	4.6*	3.9*	4.8***	0.6	1.0	4.5***	–
1963-69	98	3.2***	1.4	2.6***	4.1***	3.2***	1.2**	1.2**	1.9**	–
Overall	132	4.2***	3.0***	3.1***	4.0***	3.6***	1.1**	1.1***	2.6***	–
Total ChE										
1962-63	34	11.9**	10.3***	10.3***	6.4*	8.9***	−1.0	1.7**	10.0***	–
1963-69	98	9.7***	2.7*	7.5***	6.4***	6.9***	0.1	2.1***	6.6***	–
Overall	132	10.2***	4.4***	8.2***	6.4***	7.3***	−0.1	2.0***	7.5***	–
ChE/AChE										
1962-63	34	9.8**	7.7***	5.7*	3.5	5.4***	−3.2**	−0.8	9.0***	–
1963-69	98	8.3***	4.1***	5.9***	6.1***	6.2***	−0.2	1.6***	6.4***	–
Overall	132	8.7***	4.9***	5.8***	5.5***	6.0***	−1.0*	1.0**	7.0***	–

*$p < .05$; **$p < .01$; ***$p < .001$

stat" method in the absence of any inhibitor of ChE. For two reasons, we do not believe that the difference in assay procedures is responsible for the discrepancy in total AChE results before and after 1963. First, in three experiments, both methods of assay were used with essentially identical results. Second, careful studies and evaluation of the interference of ChE in the assay of AChE indicate that ChE in the earlier analyses did not contribute importantly to the AChE values.

The less specific enzyme ChE had been analyzed in only three EC-IC experiments at the time of the 1964 report was written. These three experiments had yielded relatively large effects in cortical ChE activity; in fact, the very first of these experiments gave larger cortical effects than we have ever seen subsequently, so that perhaps the results of this initial experiment should be discounted. The 1963–1969 replication series has continued to show highly significant positive EC-IC differences in ChE:weight. In the later experiments the positive effects appear not only in the cortex but also in the rest of the brain, so the EC-IC effect in the cortical/subcortical ratio of ChE:weight is considerably smaller in the recent experiments (1.9 percent, $p < .01$) than in the earlier series (4.5 percent $p < .001$).

Total ChE activity continues to yield significant EC-IC differences, since both tissue weight and ChE:wt do so. In total cortex the EC-IC affect amounts to 6.9 percent ($p < .001$) in the replication series, as against 8.9 percent ($p < .001$) in the original series.

Finally, the ratio of activity of the two enzymes (ChE:AChE) merits attention, since this is a purely chemical measure in which tissue weight does not enter. On this measure the two series of experiments show closely similar EC-IC effects in all respects but one: whereas the earlier experiments gave a significant effect for rest of brain (−3.2 percent, $p < .01$), the later experiments show no EC-IC difference outside the cortex.

The nine experiments conducted since we wrote the *Science* paper thus confirm that differential experience significantly alters both brain weights and activities of AChE and ChE. All of the major effects reported earlier have been replicated except that the increase in total activity of AChE did not reach significance in the later experiments.

The replication series attests to stability of effects across space as well as over time, since the first five ECT-IC experiments were conducted in the psychology laboratories in the Life Sciences Building, whereas the 11 subsequent EC(T)-IC experiments were conducted in our new quarters in Tolman Hall (since Fall 1962). In Tolman Hall the IC animals can be isolated more stringently than in the Life Sciences Building, although, as will be discussed later, we do not believe that this has any major influence on the brain measures we take.

222 Replace "Firchmin" with "Ferchmin."

Let us now inspect the results in greater detail, taking up first brain weights, then chemical measures, and then histological measures.

Brain weight effects. The largest mean EC-IC effect occurs in the occipital region (6.4 percent, $p < .001$), and it is significantly larger than the percentage effects in somesthetic or ventral cortex (Rosenzweig et al., 1969). Since this tissue sample is small (and perhaps also because its boundaries are defined rather arbitrarily) there is considerable variability in the magnitude of the effect in the occipital sample. Nevertheless, only one of the 16 experiments failed to yield a positive EC-IC difference in weight of the occipital cortical sample, as can be seen in Figure 1, which presents brain weight effects experiment by experiment. Twelve of the 16 experiments yielded significant effects, as shown by asterisks in the figure. The nearby somesthetic sample yields a significantly smaller overall EC-IC effect (1.9 percent) but one which is nevertheless clearly above the chance level ($p < .01$). Four of the 16 experiments failed to show a positive EC-IC effect in this region, and none was by itself significant. In weight of total cortex, the EC animal exceeded its IC littermate in 140 of 175 pairs—80 percent of the cases—and 13 of the 16 experiments yielded significant results on this measure.

In the rest of the brain, on the contrary, EC weighed 1.1 percent less than IC, and this also was a consistent finding ($p < .01$). We will soon see that this loss of subcortical weight by the EC rats may be related to the fact that their terminal body weight is significantly lower than that of the ICs. In total brain weight, EC exceeded IC by only 1.0 percent ($p < .01$), since the positive change in cortical weight was partly offset by the loss in subcortical weight. The fact that this small difference is significant statistically is due to its reliability in a long series of experiments; it is not a result to be sought in only one or a few experiments. Similarly, measures of brain length and width showed nonsignificant EC-IC effects of about 1 percent in magnitude (Rosenzweig and Bennett, 1969), and intracranial measures showed no EC-IC effect (Diamond et al., 1965). On the other hand, the ratio of cortical to subcortical weight shows a particularly stable effect -5.2 percent ($p < .001$). EC exceeded IC on this measure in 155 of the 175 pairs (89 percent of all cases).

Attempts to find EC-IC effects on measures of whole brain have had a mixed history. Geller et al. (1965) reported EC to exceed IC in weight of whole brain, and recently Firchmin et al. (1970) found differences of 6.3 percent and 2.9 percent in two experiments ($p < .001$ in each case). Altman et al. (1968) reported a significant increase in the length but not the width of the cerebral hemispheres with prolonged EC or training. Bennett and Rosenzweig (1969) found only nonsignificant 1 percent increases in both length and width in a 30-day EC-IC experiment. Walsh et al (1971) observed a significant increase in length (2.5

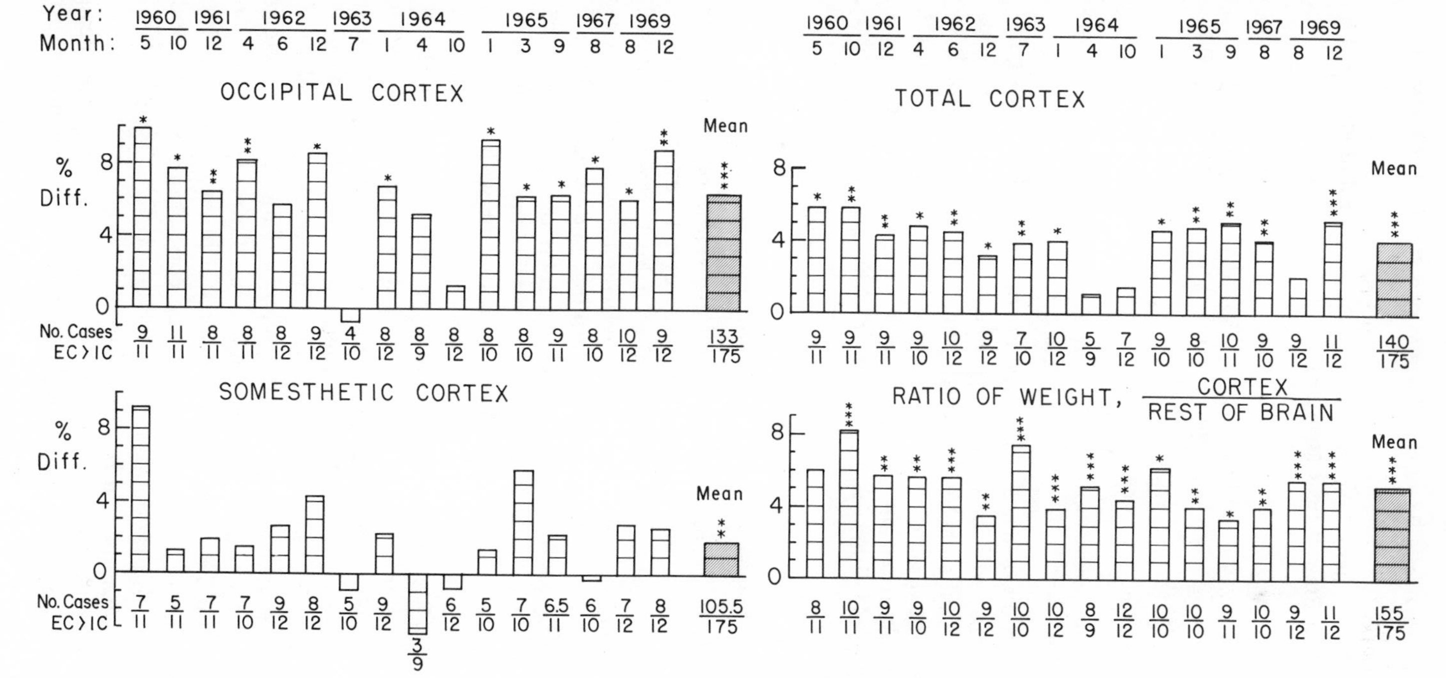

FIG. 1. Percentage EC-IC brain weight differences in 16 25- to 105-day EC-IC experiments with Berkeley S_1 strain male rats. For each experiment, the figure gives the month and year, and for each of the four measures it shows the percentage difference, the significance of the effect and the number of littermate pairs in which the EC value exceeded the IC value. $^{*}p < .05$, $^{**}p < .01$, $^{***}p < .001$.

224 Table 2. Right-hand column heading should be TBW vs. Total Cortex / Rest of Brain rather than TBW vs. Total Cortex. Rest of Brain

TABLE 2. Within-Experiment Correlations between Terminal Body Weights and Certain Brain Weight Measures*

	TBW vs. Total Cortex	TBW vs. Rest of Brain	TBW vs. Total Brain	TBW vs. Total Cortex Rest of Brain
EC(T)	.60	.65	.61	.01
IC	.70	.69	.68	.13

**Nine S_1 experiments, run from 25 to 105 days of age. Mean correlations were obtained by using the r to z transformation.*

percent, $p < .001$) in an 80-day EC-IC experiment, but only a nonsignificant increase in length (1.3 percent) in a 30-day experiment; they thus replicated both the results of Altman et al. and of Rosenzweig and Bennett. It should be noted that Walsh et al. failed to obtain a significant difference in total brain weight in these experiments of either duration. Finally, Diamond et al. (1965) found no effect of 80-day EC-IC on intracranial dimensions.

Influences of Body Weight on Brain Weights

To understand the brain weight effects fully, body weight must be taken into account. We will see that the differential environments influence brain weights not only directly but also, at least in young animals, through effects on body weight. The right-hand column of Table 1 shows that EC rats finished the experiments with terminal body weights 8.5 percent less than IC rats ($p < .001$), even though total brain weights were greater in EC than in IC. Although brain weight is known to vary with body weight, it does not do so in a linear manner, but within a limited range a linear approximation can be used. Since a given change in body weight entails a much smaller relative change in brain weight, it is not proper to use ratios of brain weight to body weight. The relations between brain weights and body weights in our experiments can be seen both within and among experiments. Within each experiment, brain weights correlate positively with body weights. Mean values of these correlations for 9 S_1 25- to 105-day experiments are presented in Table 2. The correlations range between .60 and .71 and are highly significant ($p < .001$) between body weight and all of the direct measures of brain weight. For the derived measure, total cortex/rest of brain, the correlations are low and nonsignificant. Thus the ratio measure is not affected by body weight, and this is one of the reasons that it is so stable.

Among experiments, the relation can be seen by plotting the mean EC-IC percentage difference in brain weight for a given experiment against the mean EC-IC percentage differences in body weight for that experiment; each experiment thus furnishes a single point. Two such plots are given in Figure 2, one for the total cortical weight effect versus the body weight effect, and the other for the EC-IC effect on weight of the rest of the brain versus body weight. It will be seen in the figure that the greater the negative EC-IC difference in body weight was in a given experiment, the less did the EC gain in total cortical weight tend to be. The correlation in this case was .55 ($p < .10$). Similarly, the EC-IC effect in weight of the rest of the brain tended to assume larger negative values with the greater differences in body weight; here the correlation was .50 (nonsignificant, presumably chiefly because of the small *N*, since each experiment yields only a single point in this analysis). With an overall EC-IC percentage body weight difference of −8.5 percent for the 16 experiments, the brain weight differences were +4.0 percent for total cortex and −1.1 percent for subcortex. Here the values on the regression lines for −8.5 percent change in body weight confirm the values of Table 1. If there had been no difference in body weight, then the figure reveals (reading the values of

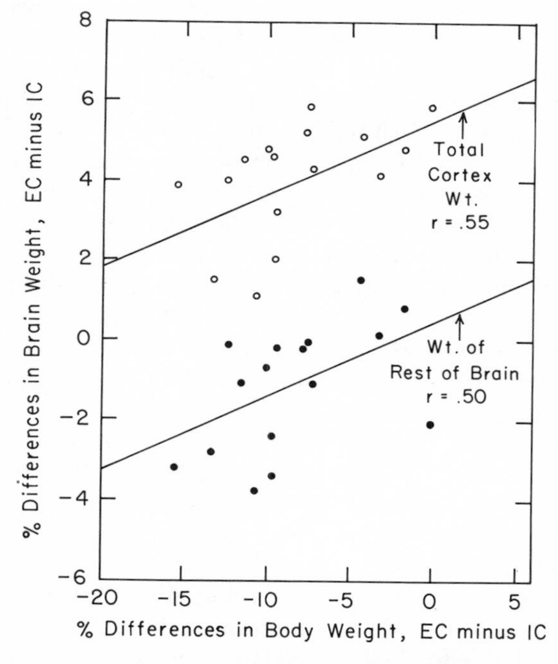

FIG. 2. **Relations between mean percentage EC-IC differences in brain weights and in terminal body weights. Each point represents values from one of 16 experiments with S_1 rats in EC or IC from 25 to 105 days of age. The straight lines represent the regression of total cortex weight and of weight of rest of brain on terminal body weight.**

the regression lines for 0 percent change in body weight) that cortical weight would have been 5.5 percent greater in EC than in IC (instead of 4.0 percent), and that EC-IC subcortical difference would have been 0.5 percent (instead of −1.1 percent).

It appears that the experimental environments affect brain development both by affording differential experience and by affecting body weights. The EC rats tend to have heavier brains, especially in the cortex, because of the enriched experience, but they also develop lighter bodies which in turn cause lighter brains. For every 10 percent difference in body weight, under these conditions, there is a 1.8 percent difference in brain weights.

Still broader consideration of the changes in body weights and their influences on brain weights will be possible after we have seen the ways in which EC-IC effects vary with starting age and duration of experiments.

Effects Measured from Standard Colony Group as Baseline

In trying to understand how enrichment or impoverishment affect the brain it would be useful to have a baseline condition against which to measure the effects of the experimental treatments. We have, since our first experiments in this area, used the standard colony condition as the only available baseline. Admittedly, it has its drawbacks, since the colony does not represent the natural conditions for which rodents evolved. Nevertheless, this is the reference group that most anatomists, biochemists, and psychologists have used in a great variety of studies. We have discussed elsewhere (Rosenzweig, 1970) the problems of baseline and our hopes to cope with them through experiments on feral animals, so we will not go into the underlying questions here. Rather, we will turn to comparisons of the separate effects of enrichment and impoverishment, as measured from the SC baseline. This will be done first for absolute brain weight measures, then for weights corrected for covariance on body weights, and then for the biochemical assays.

An SC group was included in five 25- to 105-day experiments with S_1 rats. Since the numbers of cases are smaller than in the 16 EC-IC experiments, the results are somewhat less secure, but the opportunity to measure separately the effects of enrichment and impoverishment provides a countervailing advantage.

The use of the SC group as a baseline must be qualified because of the special effects of the SC condition on body weight and the unknown consequences of this for brain weights. All three experimental conditions lead to differences in growth of body weight, as can be seen in Figure 3, which presents mean body weight data for five 25- to 105-day EC(T)-SC-IC experiments. By the end of the first experimental week, the IC rats

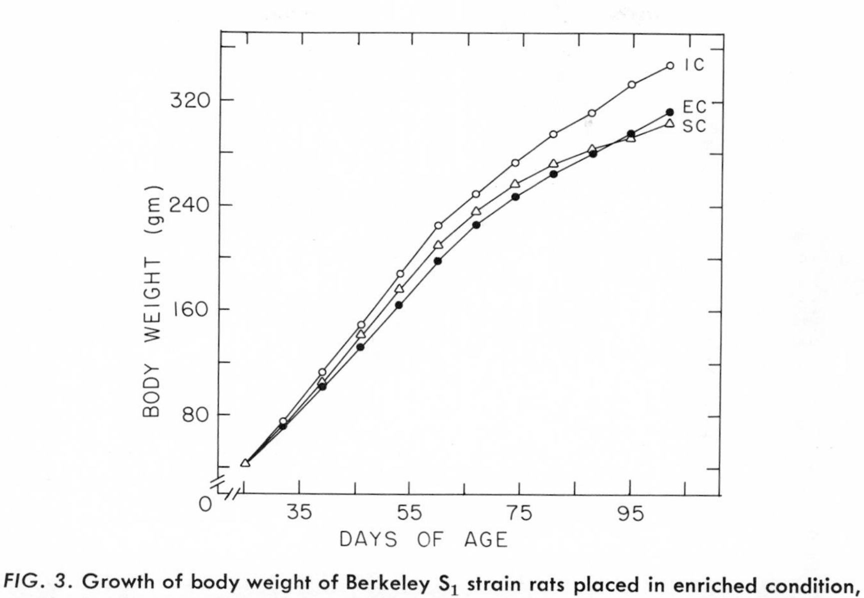

FIG. 3. **Growth of body weight of Berkeley S_1 strain rats placed in enriched condition, standard colony condition, or impoverished condition at weaning (25 days of age). The values are means of five experiments run from 1962 to 1969, and *N* is 58 per group. Since weights were not taken at exactly comparable days in all experiments, some values for individual experiments were obtained by linear interpolation between adjacent values.**

227 Line 8. Replace "Firchmin" with "Ferchmin."

weigh about 6 percent more than the other two groups. By the end of the third week, the SC group assumes an intermediate position between IC and EC. For about six weeks after weaning the weight gain of all groups is rapid and practically linear at a rate of over 30 g per week. Thereafter the growth rate decreases, with the IC and EC curves remaining roughly parallel, IC 20 or more grams above EC. The consistently greater weight of the IC group may occur because these animals are less active than their EC and SC littermates. Firchmin et al. (1970) have also found IC rats of their albino line to develop significantly greater body weights than EC littermates. Mayer has shown that both inactive rats (1964) and men (1956) eat more than is required by their energy expenditure and therefore gain weight in comparison with more active individuals. The SC animals after about the sixth experimental week show a still less rapid growth rate than the other two groups, so that their curve intersects the EC line and then falls below it. The SC rats thus end as the lightest of the three groups, and we have found this to be true in general when we have put or kept rats in SC beyond the age of about 60 days. It may be that our colony cages are too small for optimum growth of a group of three rats weighing 250 g or more. This may be true even

228 Table 3. Heading of second column on right should be Cortex/Rest rather than Cortex Rest.

TABLE 3. Brain Weights (mg) and Body Weights (gm) of Rats in EC, SC, or IC from 25 to 105 Days of Age*

	Cortex					Rest of Brain	Total Brain	Cortex Rest	Terminal Body Weight
	Occipital	Somesthetic	Rem. Dorsal	Ventral	Total				
SC Mean	70.5	54.2	283	284	691	938	1629	0.738	303
EC vs. SC									
% diff.[a]	2.7	1.9	3.2	2.6	2.8	0.5	1.5	2.3	2.4
p	<.05	<.10	<.01	<.05	<.001	N.S.	<.05	<.001	—
No., EC>SC[b]	37/58	34/58	40/58	38/58	42/58	30/58	36/58	39/58	35/58
IC vs. SC									
% diff.[c]	−3.0	−0.4	−1.7	0.4	−0.9	2.1	0.8	−3.0	14.4
p	<.05	N.S.	<.10	N.S.	N.S.	<.01	N.S.	<.001	<.001
No., IC>SC[d]	18/58	30/58	20/58	32/58	25/58	37/58	35/58	8/58	52/58

Based on 58 littermate sets of male S_1 rats.
[a] *100 × (EC-SC)/SC.*
[b] *Number of cases in which the value of the EC member of a pair exceeded the SC value.*
[c] *100 × (IC-SC)/SC.*
[d] *Number of cases in which the value of the IC member of a pair exceeded the SC value.*

though the cage size allows 230 cm^2 of surface per rat and thus exceeds the minimum of 180 cm^2 per rat recommended by the *Guide for Laboratory Animal Facilities and Care* of the Public Health Service (1968). The precise effects of these differences in body weight on brain weight cannot be stated because, although we know that brain weights are influenced by body weights, we do not know how rapidly or closely brain follows body in this regard. Bearing this qualification in mind, let us then consider how EC and IC brain measures vary in relation to those of SC littermates.

Brain weights. Brain weights and body weight results for the five EC(T)-SC-IC experiments are presented in Table 3. Here EC shows greater cortical weights than SC in each region—2.1 percent greater for total cortex ($p < .001$); IC shows lesser cortical weights than SC in each area except ventral cortex, and the IC-SC difference for total cortex is −0.9 percent (N.S.). Except for the occipital region, where the effects are virtually equal in magnitude, the EC-SC cortical differences are larger than the IC-SC differences. Thus the overall EC-IC effects in cortical weights appear to be due more to enrichment than to impoverishment of experience. In the rest of the brain, on the contrary, EC scarcely differs from SC in weight, whereas IC is 2.1 percent ($p < .01$) greater than SC. It is mainly because of this difference in subcortical weights that the percentage effect in the cortical/subcortical weight ratio is greater for IC-SC (−3.0 percent, $p < .001$) than for EC-SC (2.0 percent, $p < .001$).

The influence of body weights should now be taken into account, especially because in these five experiments the EC-IC differences in body weight were among the largest in the series of 16 S_1 experiments. The mean EC-IC difference in the EC-SC-IC experiments was 14.4 percent ($p < .001$). With the relatively large SC-IC body weight difference, the EC-IC cortical weight effect is somewhat smaller in these experiments than in the 16 experiments overall (3.0 percent versus 4.0 percent) and the decrease in weight of the rest of the brain was greater (−2.0 percent versus −1.1 percent). When the influence of body weight differences was removed by covariance analyses, we obtained the brain weight values shown in Table 4. The EC-SC differences in brain weights are scarcely altered from those of Table 3, since EC and SC body weights are closely similar, but the IC-SC cortical differences are now enhanced to the point where the negative IC-SC effects are larger than the positive EC-SC effects. The greater subcortical weight previously seen for IC disappears with the covariance analysis, so that total brain weight of IC falls below that of SC. Independent of body weight effects, it now appears that enrichment above the SC baseline is slightly less effective in altering brain weights than is environmental impoverishment. Or perhaps the firmest conclusion is that both the enrichment and the impoverishment effects in brain weights are statistically significant and that they are comparable in magnitudes. As we

229
Line 26. The following material was omitted after the word "was": "−10.5 percent ($p < .001$) and the mean IC-SC difference,".
Line 27. Replace "SC" with "EC."

230 Table 4. Right-hand column heading should be $\frac{\text{Cortex}}{\text{Rest}}$ rather than $\frac{\text{Cortex}}{\text{Rest}}$.

TABLE 4. Percentage Differences in Brain Weights Adjusted for Covariance of Brain Weights on Body Weights*

	Cortex							
	Occip-ital	Somes-thetic	Rem. Dorsal	Ventral	Total	Rest of Brain	Total Brain	Cortex Rest
EC vs. SC	2.7	1.5	2.6	1.8	2.1	−0.5	0.6	2.7
IC vs. SC	−6.3	−3.6	−5.0	−5.3	−5.4	−2.8	−3.4	−3.6

**Based on the same animals as Table 3—58 littermate sets of male S_1 rats.*

shall see later, for adult rats (105 to 185- and 290 to 380-day experiments) there is little complication of body weight differences, and the effects of enrichment are as large or larger than those of impoverishment.

AChE and ChE. Chemical results from EC-SC-IC experiments run from 25 to 105 days of age with S_1 rats are given in Table 5. (For the measures of enzymatic activity per unit of tissue weight, no correction need be made for group differences in body weight, since these measures are not correlated with body weight). The table reveals that the main chemical differences occurred between IC and SC, with only negligible differences between EC and SC. In the 1964 paper we had stated that "the bulk of the effects on cerebral weight and acetylcholinesterase activity is due to enriching rather than to restricting the experience of our colony animals" (p. 615) but it must be noted that the conclusion was with regard to *total* AChE activity, not to AChE activity per unit of tissue weight. Now the 1964 statement must be modified even with regard to total activity of AChE in the cortex, since we have seen that the last nine experiments did not yield a significant EC-IC difference in total AChE activity of cortex. Thus it appears that whereas enrichment from the colony baseline causes an increase in cortical weight, it is only impoverishment that modifies the enzymatic values. This conclusion holds for 25- to 105-day experiments but, as we will see shortly, the relative roles of enrichment and impoverishment on brain chemistry are altered for experiments performed with rats run as young adults (from 105 to 185 days of age).

Histological Effects in 25- to 105-Day EC-IC Experiments

The measures of cortical depth reported in 1964 were based on readings made in a small sector of the dorsal cortex, as was illustrated in Fig. 2 of that paper (1964, p. 612). The tissue sections of those and later ex-

TABLE 5. Brain Chemistry of Rats in EC, SC, or IC from 25 to 105 Days of Age*

	Cortex							
	Occip-ital	Somes-thetic	Rem. Dorsal	Ventral	Total	Rest of Brain	Total Brain	Cortex Rest
AChE/Weight								
SC Mean	5.64	6.80	7.02	10.62	8.33	18.34	14.09	0.455
EC vs. SC								
% diff.[a]	−1.9	2.3	0.3	0.2	0.3	1.0	0.4	−0.7
p	<.05	N.S.	N.S.	N.S.	N.S.	N.S.	N.S.	N.S.
No., EC>SC[b]	25/58	38/58	31/58	32/58	33/58	37.5/58	32.5/58	29/58
IC vs. SC								
% diff.[c]	2.3	3.4	3.1	0.9	2.2	−2.0	−0.4	4.2
p	<.05	<.001	<.001	N.S.	<.01	<.001	N.S.	<.001
No., IC>SC[d]	36/58	41/58	41.5/58	32/58	36/58	21/58	27/58	47/58
ChE/Weight								
SC Mean	0.327	0.340	0.307	0.291	0.305	0.540	0.440	0.566
EC vs. SC								
% diff.[a]	0.7	0.8	−0.2	−0.5	−0.2	0.4	0.0	−0.4
p	N.S.	N.S.	N.S.	N.S.	<.05	N.S.	N.S.	N.S.
No. EC>SC[b]	31.5/58	31.5/58	29/58	27/58	28.5/58	33.5/58	30/58	22/58
IC vs. SC								
% diff.[c]	−4.5	−4.9	−4.1	−4.1	−4.2	−0.8	−1.4	−3.5
p	<.01	<.001	<.001	<.001	<.001	N.S.	=.01	<.001
No. IC>SC[d]	17/58	12.5/58	20/58	18/58	13.5/58	24.5/58	19.5/58	16/58
10^2 × ChE/AChE								
SC Mean	5.82	5.01	4.38	2.74	3.67	2.95	3.12	1.246
EC vs. SC								
% diff.[a]	2.8	−1.3	−0.4	−0.9	−0.4	−0.6	−0.5	0.2
p	N.S.	N.S.	N.S.	N.S.	N.S.	N.S.	N.S.	N.S.
No., EC>SC[b]	34/58	24/58	26/58	28/58	21/58	24.5/58	25/58	27/58
IC vs. SC								
% diff.[c]	−6.6	−7.8	−6.8	−4.8	−6.2	1.2	−0.9	−7.2
p	<.001	<.001	<.001	<.001	<.001	N.S.	N.S.	<.001
No., IC>SC[d]	15/58	10/58	11/58	17/58	9/58	36/58	23/58	5/58

*Based on 58 littermate sets of male S_1 rats.

[a] 100 × (EC-SC/SC.

[b] Number of cases in which the value of the EC member of a pair exceeded the SC value. Numbers ending in .5 indicate tied values.

[c] 100 × (IC-SC)/SC.

[d] Number of cases in which the value of the IC member of a pair exceeded the SC value. Numbers ending in .5 indicate tied values.

231 Table 5. Right-hand column heading should be Cortex/Rest rather than Cortex Rest.

231 The units of measurement for AChE/wt and ChE/wt were omitted. They should be as follows:

"AChE activity is expressed in units of nanomoles acetylthiocholine hydrolyzed/min/mg."

"ChE activity is expressed in units of nanomoles butyrylthiocholine hydrolyzed/min/mg."

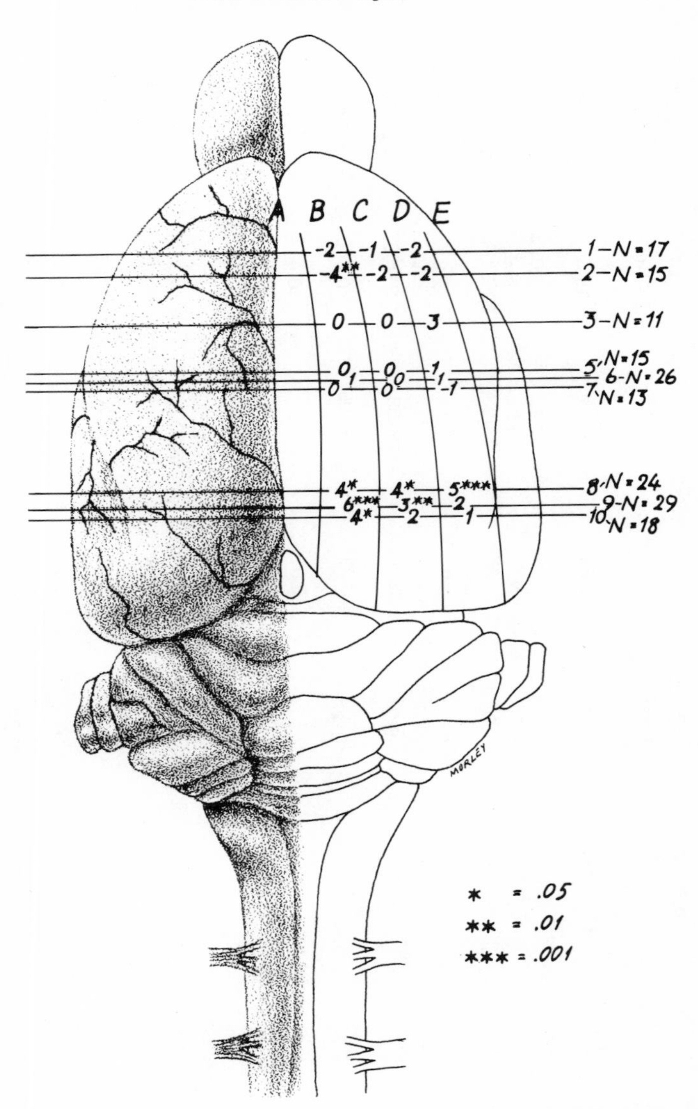

FIG. 4. Dorsal view of rat brain showing positions of transverse sections taken for cortical depth measurements. The numbers on the brain represent percentage differences in cortical depths between EC and IC. The measures were taken in both left and right hemispheres and were averaged to give the values shown. (Section 4 was discontinued after original experiments.) Note that all of the differences in sections 1 and 2 are negative; *i.e.*, the EC depth was less than the IC depth. (From Diamond et al. In preparation.)

periments were subsequently analyzed over a greater extent of the cortex, and the results are presented in Figure 4.

It is evident from the data of the figure that the greatest EC-IC differences in cortical thickness occur in sections from the occipital cortex, especially in the section taken with the posterior commissure as the subcortical landmark (Section 9). Table 6 was compiled to examine the consistency of results found over the years 1964 to 1968 in Segment B of Section 9. Here we see that the EC-IC difference is quite consistent in this region of the brain. Results of Walsh et al. (1969) have corroborated the greater depth of occipital cortex in EC than in IC.

From the cell counts we found no significant differences in the number of neurons between EC and IC rats in 17 littermate pairs. However, the numbers of cortical glia differed significantly by 14 percent ($p < .01$), with EC greater than IC in 12 out of 17 pairs (Diamond et al., 1966). Walsh et al. (1969), making cell counts in the hippocampus, have found a greater number of glia in EC than in IC rats. While Altman initially reported that rats in an enriched environment showed glial proliferation in the cortical radiation and corpus callosum (Altman and Das, 1964), he has now expressed reservations as to whether these cells are actually glia and refers to them as undifferentiated migrating elements (1967).

EC-IC differences were also noted in measurements of cross-sectional areas of perikarya and nuclei. The EC perikarya differed from the IC by 18 percent ($p < .001$) in the upper third of the cortex; by 7 percent ($p < .05$) in the middle third, and by 12 percent ($p < .01$) in the lower third. The nuclear measures for these same cells showed very similar differences to those found in the perikarya —20 percent ($p < .01$) for the upper third, 8 percent ($p < .01$) for the middle third, and 10 percent ($p < .05$) for the lower third (Diamond et al., 1967). Thus the values from both perikarya and nuclei confirm the hypothesis that the nerve cell

TABLE 6. Percentage Differences in Depth of Cortex from Rats in EC-IC from 25 to 105 Days of Age*

Exp.	Date	% diff.	No., EC>IC	p
1	1964	4.8	7/11	<.01
2	1964	8.2	9/9	<.001
3	1966	9.4	3/3	N.S.
4	1966	8.3	9/9	<.001
5	1968	2.3	6/9	N.S.
6	1968	7.0	11/11	<.001
		6.3	45/52	

**Section 9, Segment B, in six S_1 experiments.*

and its nucleus change size in response to environmental manipulation. Effects of environment on cell size in preweaning rats will be discussed in a later section.

EC-IC Experiments with Varied Durations and Starting Ages

Since the 1964 paper, we have varied both the durations of the EC-IC period and the starting age systematically in a series of experiments. We had already reported in the 1964 paper that rats exposed to the differential environments as adults (from 105 to 185 days of age) showed changes of the brain just as do animals started in the experiments as weanlings. The further experiments in this series revealed several new findings that could not have been gained by using the 25- to 105-day schedule exclusively.

Brain and Body Weights

Let us start with those experiments that concern brain weights and body weights. Table 7 presents weight results for the most frequently used experimental durations: 30, 80, and 160 days, and starting ages: 25, 60, 105, and 290 days of age.

EC-IC Effects in Adult Rats. Table 7 demonstrates that brain weight differences are induced in experiments begun after the age of weaning, even though body weight differences are not. The 1964 paper reported, on the basis of two 105- to 185-day experiments, "that the adult (rat) brain shows increases in cortical weight and total acetylcholinesterase activity as readily as the young brain" as a consequence of enriched experience (p. 616). The plasticity of the adult brain has been further borne out by recent experiments conducted by Walter H. Riege in our laboratories (1971). He has studied S_1 rats kept under colony conditions until almost a year old and then assigned to EC, SC, or IC. Some of his results are given in the bottom of Table 7 for the 290-day starting age. (The starting ages of Riege's experiments actually ranged from 270 to 310, and the longer duration was 90 rather than 80 days.) It will be seen that the percentage brain weight effects for the 90-day-duration experiments begun at 290 days of age are very similar to those of the 80-day-duration experiments begun at 105 days of age, and even closer to the percentages for the 30-day experiment begun at 60 days of age. These results make it clear that plasticity in terms of brain weights persists into adulthood.

Brain Weight Effects in 30-day Experiments. In 30-day experiments starting at 25 days of age, the percentage effects in occipital cortex and total cortex are seen in Table 7 to be larger than the corresponding

235 Table 7. Restore line omitted from 25-day group:

Total brain 2.8*** 1.0** -0.9

TABLE 7. EC-IC Percentage Differences in Terminal Body Weight and Brain Weights in S_1 Experiments of Different Starting Ages and Different Durations

Ages (Days)	Measures	Durations (Days)		
		30	80	160
25	N (pairs)	135	175	22
	Body weight	−11.2***	−8.5***	−6.8*
	Occip. cortex	10.4***	6.4**	11.4***
	Total cortex	5.9***	4.0***	2.5**
	Rest of brain	0.5	−1.1**	−3.3**
	Cortex/rest	5.3***	5.2***	6.1***
60	N (pairs)	87	—	—
	Body weight	−1.0		
	Occip. cortex	7.8***		
	Total cortex	5.9***		
	Rest of brain	2.1***		
	Total brain	3.7***		
	Cortex/rest	3.7***		
105	N (pairs)	—	45	—
	Body weight		−0.1	
	Occip. Cortex		10.9***	
	Total cortex		5.4***	
	Rest of brain		1.7*	
	Total brain		3.2***	
	Cortex/rest		3.7***	
290	N (pairs)	21	23	
	Body weight	−4.3	4.0	
	Occip. cortex	4.8***	9.6***	
	Total cortex	2.7**	5.8***	
	Rest of brain	0.2	2.1*	
	Total brain	1.2	3.6**	
	Cortex/rest	2.5**	3.6**	

$^*p<.05$, $^{**}p<.01$, $^{***}p<.001$.

25- to 105-day effects. "Rest of brain" showed a slight positive effect, rather than the negative effect characteristic of the 25- to 105-day experiments, so the ratio of total cortex to rest of brain was about the same for 25 to 55 days as for longer experiments.

When 30-day experiments were started at later ages (60 to 290 days), the EC-IC effects for occipital cortex and total cortex were smaller than in the experiments started at weaning. This can be seen by reading down the left column of the table. On the other hand, rest of brain shows greater EC-IC increases at the later ages, and we will shortly relate this to the lack of significant body weight effects among the older rats. It is clear that 30 days of differential experience is sufficient, at any of the three starting ages, to produce significant EC-IC brain weight effects.

EC-IC BODY WEIGHT EFFECTS ON BRAIN WEIGHTS. Table 7 reveals that sizable EC-IC body weight differences occur only in experiments begun at 25 days of age (weaning). Thus the complicating influences of body weight differences on EC-IC brain weight effects can be avoided by starting the experiments when the rats are 60 days of age or older.

Even though there is not a negative EC-IC effect on weight of rest of brain in the 25 to 55-day experiments, the relation between body weight differences and brain weight effects is even stronger than in the 25 to 105-day experiments; this can be seen by comparing Figure 5 with Figure 2. In the 30-day as in the 80-day experiments, the greater the negative EC-IC difference in body weight, the less the positive EC-IC difference in cortical weight. Both the slopes and the absolute values of brain effects are greater for the 30-day than for the 80-day experiments, and both the correlations of Figure 5 are significant at beyond the .001 level.

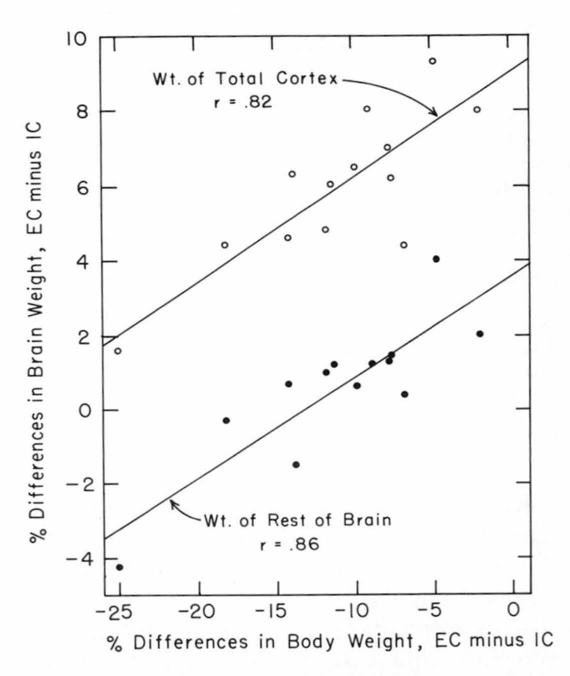

FIG. 5. **Relations between mean percentage EC-IC differences in brain weights and in terminal body weights. Each point represents values from one of 13 experiments with S_1 rats in EC or IC from 25 to 55 days of age. The straight lines represent the regression of total cortex weight and of weight of rest of brain on terminal body weight. When these results are compared with those for the 25- to 105-day period (See Figure 2), it is seen that the 25- to 55-day experiments show both steeper slopes and higher intercepts.**

Returning to Table 7, it should be noted that the percentage values for body weight differences may be deceptive unless it is realized that they are based on increasing absolute weights. Thus the percentage difference in body weights decreases from −11.2 to −8.5 to −6.8 percent as the experiments are prolonged from 30 to 80 to 160 days. But the absolute EC-IC weight difference does not decrease among these experiments; it goes from −22.7 g (EC, 179.3; IC, 202.0) to −28.6 (EC, 307.6; IC, 336.2) to −26.9 (EC, 371.0; IC, 397.9). In the case of brain weights, the variation in baseline with age is of lesser importance, since total brain weight increases by only about one-tenth from 55 to 185 days of age, whereas body weight doubles during this period.

TRANSITORY ASPECTS OF CERTAIN WEIGHT EFFECTS. Some of the brain weight differences pass through a maximum after around 30 days of differential experience and then decrease, as we have reported elsewhere (Rosenzweig, 1968; Bennett et al., 1970). Table 7 in the preceding section showed a larger EC-IC effect in weight of total cortex in the 30-day than in the 80-day experiments, and a larger effect in the 80-day than in the 160-day experiments. It should be noted, however, that the negative EC-IC difference in subcortical brain is larger in the longer experiments, and the cortical/subcortical ratio shows closely comparable effects for all these durations. To investigate this aspect further, we ran some experiments in which subgroups were sacrificed after successively longer differential experience—15, 30, 45 or 60 days—in experiments begun at 60 days of age. Some results of these experiments were presented in Bennett et al. (1970). The results for 15-day EC-IC were clearly smaller than the 30-day effects for weight of both total cortex and rest of brain, but for all four durations the cortical/subcortical ratio yielded a difference of about 3.5 percent. Still shorter EC-IC periods are now being studied.

Effects on AChE and ChE

The slower rate of change in ChE than in other measures became evident as we ran 30-day EC-IC experiments and found little effect in ChE/weight, despite clear effects in brain weights and AChE. Table 8 gives chemical results from such 30-day experiments. We needed to be certain, however, that the failure to obtain clear ChE effects in the shorter experiments was not due to any difficulty with our assay procedures, since we had largely given up 80-day experiments after 1966 when the shorter experiments proved to yield clear effects in brain weights. We therefore ran two experiments in 1969, each including three durations of treatment—25 to 55 days, 70 to 100 days and 25 to 100 days. This enabled us to compare directly the results of treatment among groups run for different

TABLE 8. EC-IC Percentage Differences in 30-Day Experiments with S_1 Rats, 1966-68

	Weight		AChE/wt.		ChE/wt.		ChE/AChE	
	Total Cortex	Rest of Brain	Total Cortex	Rest of Brain	Total Cortex	Rest of Brain	Total Cortex	Rest of Brain
25-55 days, 61 pairs								
% diff.[a]	6.6	0.7	−3.7	1.7	1.9	1.0	5.9	−0.6
p	<.001	N.S.	<.001	<.001	<.001	<.05	<.001	N.S.
No., EC>IC[b]	60/61	38/61	12/61	40/61	42.5/61	38/61	50/61	30/61
60-90 days, 51 pairs								
% diff.[a]	5.9	2.2	−3.1	0.3	−0.2	0.1	3.0	−0.2
p	<.001	<.001	<.001	N.S.	N.S.	N.S.	<.001	N.S.
No., EC>IC[b]	46.5/51	37/51	10.5/51	27/51	20.5/51	24.5/51	39/51	24/51

[a] *100 × (EC-IC/IC).*
[b] *Number of littermate pairs in which value of EC rat exceeds that of IC rat.*

periods but where all rats in an experiment came from the same breeding and where the chemical analyses were done concurrently.

Results of the 1969 experiments demonstrated that, whereas brain weight effects and some AChE effects appeared in the 30-day groups, only the 80-day duration also yielded clear ChE effects (see Table 9). Whereas in 1964 we supposed that all of the cerebral effects we measured might be reflecting only different aspects of the same syndrome of changes, it now appears that various measures follow their own time courses and may eventually be shown to represent independent types of change. We had already shown in 1968 (Rosenzweig) that different measures may respond to different aspects of the environmental conditions, the effects in cortical weight being greater between EC and SC than between IC and SC, whereas the cortical ChE effect is larger between IC and SC.

It would appear worthwhile to make glial and neuronal counts in experiments of 30-day duration to determine whether the change in glial number is slow to develop as is the change in ChE activity. We have suggested (Rosenzweig et al., 1967, p. 50) that the EC-IC difference in ChE may reflect the change in glia, and making the counts would provide a test of this hypothesis. A more thorough study of this question would be to follow the development and migration of glia as functions of age and of duration and type of environmental experience, using radioactive tracer techniques.

We noted earlier that in the 25- to 105-day experiments the enzymatic effects occurred between IC and SC and not between EC and SC, but we stated that the relative roles of enrichment and impoverishment were different in young adult rats. Let us now consider in Table 10 the results for three ECT-SC-IC experiments run from 105 to 185 days of age. Body weight could exert relatively little influence in these experiments since ECT exceeded SC by only 6.6 percent, IC exceeded SC by 5.1 percent (and ECT exceeded IC by 1.4 percent). In brain weights, ECT was significantly greater than SC in most sections, whereas IC fell below SC but not significantly for most measures.

In AChE activity per unit weight, ECT was significantly below SC in remaining dorsal cortex, ventral cortex, and total cortex, whereas IC differed in the opposite direction but not significantly. In total activity of AChE, ECT was significantly greater than SC in both occipital and total cortex, but IC did not depart significantly from the SC values. For ChE activity, there were only two experiments, and more work is needed for conclusive results. Since the ChE results are thus preliminary, they are not shown in the table, but the following tentative statements can be made. In ChE/weight, enrichment did not produce any significant effects, but impoverishment caused losses that reached statistical significance in total cortex and total brain. In total ChE activity, there were both significant

TABLE 9. **EC-IC Percentage Differences[a] in 30-day and 80-day Experiments with S_1 Rats, 1969[b]**

	Weight			AChE/wt.			ChE/wt.			ChE/AChE		
	Occip. Cortex	Total Cortex	Rest of Brain	Occip. Cortex	Total Cortex	Rest of Brain	Occip. Cortex	Total Cortex	Rest of Brain	Occip. Cortex	Total Cortex	Rest of Brain
25-55	10.8***	3.0**	−2.0*	−4.8***	−2.6**	3.8***	1.2	1.9*	1.2	6.1**	4.5**	−2.6**
70-100	7.3***	5.2***	2.4*	−2.9**	−1.4	1.0	−2.8*	−1.9*	−0.3	0.0	−0.5	−1.3
25-100	7.4***	3.6***	−1.8	−6.6***	−3.4**	2.0**	3.0**	3.5**	2.2*	10.4***	7.2***	0.2

*$p<.05$, **$p<.01$, ***$p<.001$.

[a] 100 × (EC minus IC)/IC.

[b] Two replications, N=24 per group.

TABLE 10. Brain Weight and Brain Chemistry of Rats in ECT, SC or IC from 105 to 185 Days of Age*

	Cortex							
	Occip-ital	Somes-thetic	Rem. Dorsal	Ventral	Total	Rest of Brain	Total Brain	Cortex Rest
Weight								
EC vs. SC								
% diff.[a]	6.0	−0.4	3.3	5.5	4.3	1.0	2.4	3.2
p	<.001	N.S.	<.05	<.001	<.001	N.S.	<.001	<.001
IC vs. SC								
% diff.[b]	−3.6	−2.7	−2.1	−0.5	−1.6	−0.9	−1.2	−0.6
p	<.05	N.S.	N.S.	N.S.	<.05	N.S.	N.S.	N.S.
AChE/weight								
EC vs. SC								
% diff.[a]	−1.4	0.5	−1.7	−2.3	−1.7	−0.2	−1.1	−1.6
p	N.S.	N.S.	<.05	<.05	<.05	N.S.	N.S.	N.S.
IC vs. SC								
% diff.[b]	0.3	1.6	1.2	1.0	1.4	−0.5	0.1	1.8
p	N.S.	N.S.	N.S.	N.S.	N.S.	N.S.	N.S.	N.S.
Total AChE								
EC vs. SC								
% diff.[a]	4.5	0.1	1.4	3.0	2.4	0.8	1.2	1.8
p	<.05	N.S.	N.S.	N.S.	<.05	N.S.	N.S.	N.S.
IC vs. SC								
% diff.[b]	−3.4	−1.0	−1.0	0.4	−0.3	−1.6	−1.2	1.2
p	N.S.	N.S.	N.S.	N.S.	N.S.	N.S.	N.S.	N.S.

**34S$_1$ rats per group.*
[a]100 × (EC-SC/SC.
[b]100 × (IC-SC)/SC.

241 Table 10. Right-hand column heading should be Cortex/Rest rather than Cortex Rest.

increases with enrichment and significant losses with impoverishment. We conclude that among young adult rats enrichment of experience above the standard colony baseline has significant effects not only in brain weights but also in activities of AChE and ChE. In older adults run by Riege from 290 days on, enzymatic EC-IC differences were small, so perhaps they are slow to develop in old rats.

Effects on Histology

Histological depth studies are useful in bringing out more detailed regional changes in the cortex as a consequence of environmental manipulation. Such measures have been made on 30-day experiments starting at two different ages, 25 days and 60 days. Significant EC-IC differences in cortical depth were found in the somesthetic and motor cortices as well as in the occipital cortex (see Fig. 6A and 6B). Finding changes in the more anterior cortical areas was new, for with 80-day EC-IC experiments only the occipital cortex showed such differences, as was seen in Fig. 4. The 25- to 55- and 60- to 90-day experiments yielded similar EC-IC depth effects, with a few exceptions: in the occipital cortical samples, the Lateral Segments D and E showed significantly larger differences in the 60- to 90-day experiments. Also Section 3 showed significant effects on 60- to 90-days but not in the 25- to 55-day experiments.

The three experimental conditions, EC-SC-IC, have been compared above for wet weight and enzyme differences in 80-day experiments, and here effects of these conditions will be considered histologically for 30-day experiments. Both enrichment and impoverished are seen in Figure 7 to contribute to the overall EC-IC differences. Let us see how they contribute in the various sections and segments of the cortex. In Segment B both enrichment and impoverishment from SC cause significant effects at both age ranges. In Segment C and D the effects of isolation are more clearly evident, except perhaps in the occipital region of the 60 to 90-day group. The lateral areas are ones which myelinate late and, perhaps the detrimental effects of isolation on myelin formation are being manifested. These data indicate that one area of the brain can be more susceptible to increase in depth in response to a stimulating environment, while an adjacent area is more prone to decrease in depth due to an impoverished environment. As was shown in Table 3 for cortical weight effects in 25- to 105-day experiments, both enrichment and impoverishment contribute significantly to the overall effect.

The extent to which EC-IC effects in weight and depth covary deserves attention. A strict comparison between depth and weight effects cannot be made since the histological sections and the tissue samples taken for weight analyses do not include exactly the same areas. The size of samples removed for measures of weight is related to the surface area of the brain, so that its bulk varies with surface as well as with cortical thickness. To obtain a single value for depth relatively comparable to that for weight, we calculated an average for each rat for each region. That is, for each animal an average depth for the occipital region was obtained from the nine values for Segments B, C, and D of Sections 8, 9, and 10. For the somesthetic region, an average was obtained from the 12 values for Segments B, C, and D of Sections 3, 5, 6, and 7.

In spite of the problems of comparing weight and depth, Table 11 indicates a close correspondence between depth and weight difference for the 60- to 90-day group, and both are highly significant. In the 25- to 55-day group, the EC-IC differences for occipital cortex are twice as large as those for somesthetic, although the magnitudes of both weight effects are double those for the corresponding depth effects.

PREWEANING EXPOSURE TO EC. Our experiments described to this point have all involved treatments begun at weaning or later, but Malkasian (1969) has now found histological differences caused by exposure to EC before weaning. We had previously avoided preweaning treatments because our central interest has been in brain mechanisms of learning and, until recently, little evidence of learning had been shown in the suckling rat. Also, the rat is born in a very immature state, with eyes closed until the end of the second week and poor temperature regulation until the same age. Most behavioral experiments with suckling rats have studied effects of stress on later development, whereas our interest has been primarily in mechanisms of learning and memory. Recently learning, at least in regard to feeding behavior, has been shown in the first few postnatal days in the rat (Thoman and Arnold, 1968; Thoman et al., 1968) and in other mammals (Stanley et al., 1963; Rosenblatt, 1970). We had also noted (Rosenzweig et al., 1968, p. 295) that Forgays and Read (1962) found superior learning at 60 days in rats that had been kept for their first 21 days postpartum in an enriched environment with a lactating female. In view of the widely held belief that the young brain must be especially plastic, it seemed worthwhile to test for effects of preweaning experience. Malkasian, working with Dr. Diamond, therefore undertook to determine effects on cortical thickness and on perikaryon cross-section.

The experiment was designed so that six days after birth, mothers and pups of the Long-Evans strain were divided into the following groups:

1. One mother with three pups in each standard colony cage (unifamily environment).

2. Three mothers with three pups each placed in each large EC cage with toys (multifamily environment with EC).

3. Three mothers with three pups each in a large cage but with *no* toys (multifamily environment).

25–55 Days N = 31 Pairs

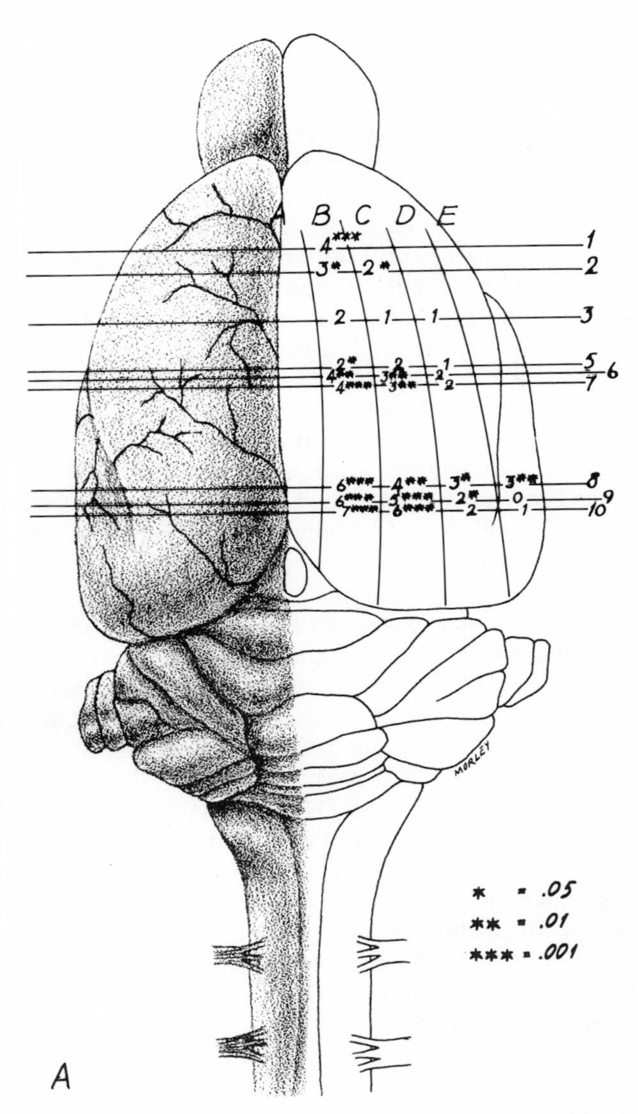

FIGS. 6A and 6B. Dorsal views of rat brain showing positions of transverse sections taken for cortical depth measurements. The numbers on the brain represent percentage differences in cortical depths between EC and IC. The measures were taken in both left and right hemispheres and were averaged to give the values shown. (Section 4 was discontinued after original experiments.) (From Diamond et al. In preparation.)

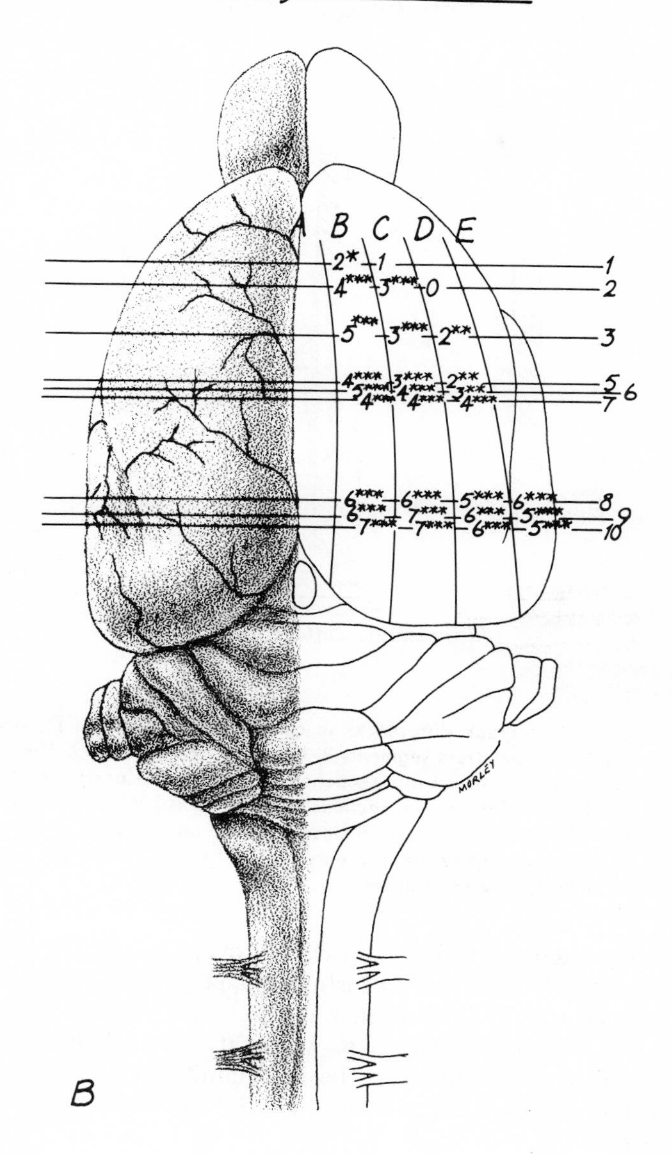
60 - 90 Days N = 50 Pairs
A B C D E
2* 1
4*** 3*** 0
5*** 3*** 2**
4*** 3*** 2**
5*** 4*** 3**
4*** 4*** 4***
6*** 6*** 5*** 6***
6*** 7*** 6*** 5***
7*** 7*** 6*** 5***
1
2
3
5
6
7
8
9
10
MORLEY
B

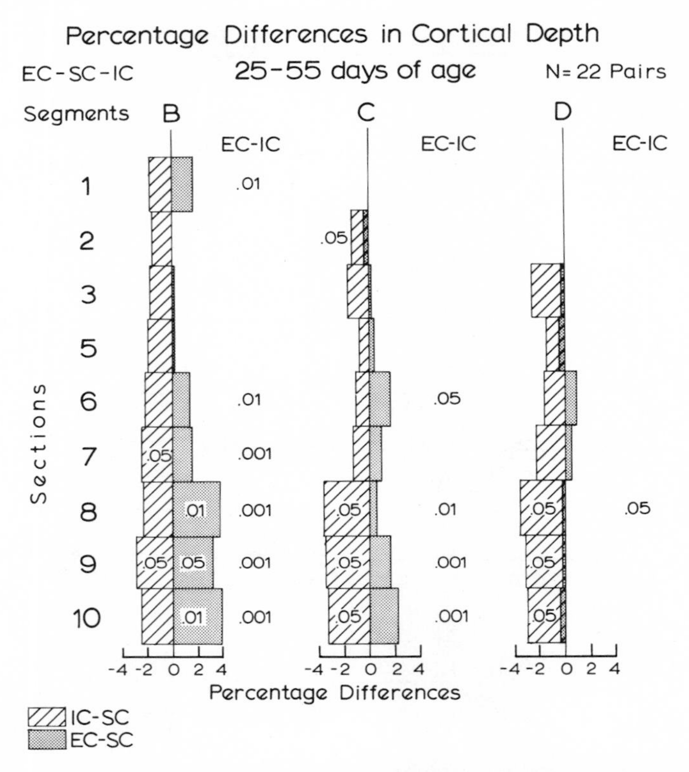

FIGS. 7A and 7B. **Percentage differences in cortical depth between EC and SC and between SC and IC. Values from segments B, C, and D are presented. The numbers at the far left designate transverse brain sections from anterior to posterior. (See Figures 4 and 6 for specific locations of the sections and segments.) Where differences were statistically significant, the level of significance is shown on the IC-EC or EC-SC blocks, or in the EC-IC columns. Figure 7A presents results for 25- to 55-day experiments; Figure 7B, for 60- to 90-day experiments.**

Pups were sacrificed at 14, 19, or 28 days of age.

Cortical depth measures on histological preparations of the somatosensory and occipital cortex indicated no differences between the unifamily and multifamily environment groups when rats were sacrificed at 28 days of age (see Table 12). However, the multifamily EC group had significantly greater cortical depth measure than did unifamily at all three preweaning ages.

In addition to cortical depth measures, cross-section areas of perikarya and nuclei from neurons in the somatosensory cortex and nuclear areas from the occipital cortex were significantly different between multifamily-EC and unifamily rats. Specifically, in the somatosensory cortex in Layers

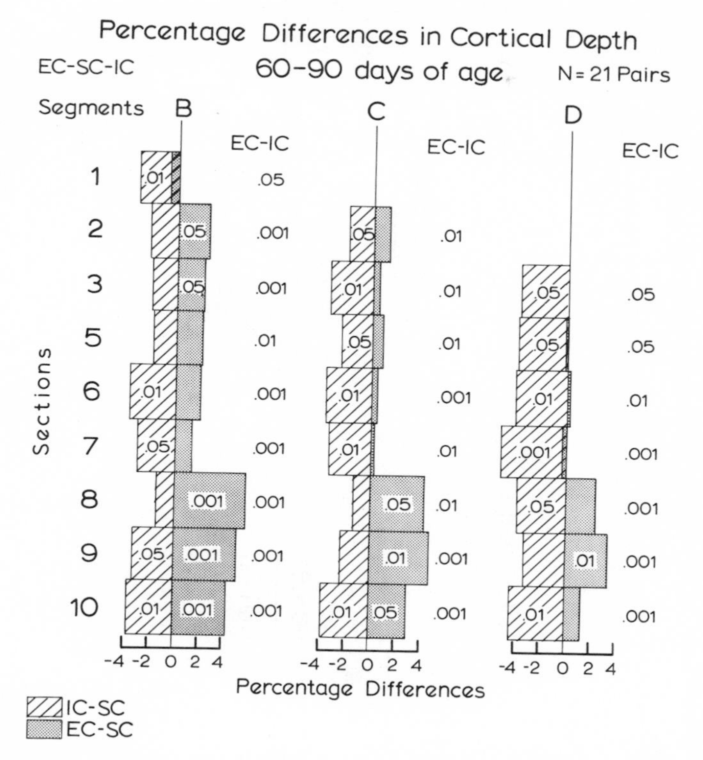

II and III the neuronal nuclear area was different by 19 percent ($p < .05$) and the perikaryon area was different by 16 percent ($p < .01$). In the occipital cortex of the same brains, the neuronal nuclear area was different by 25 percent ($p < .01$); perikaryon area was not measured in occipital cortex.

These findings complement the results obtained with EC and IC rats run postweaning. The preweaning brain shows changes in cortical depth that are comparable in magnitude to those of the postweaning brain in the dorsomedial aspect, but the preweaning EC-IC depth effects are considerably larger than postweaning effects in the lateral region of the occipital cortex.

New Related Effects

As well as replicating the previously-reported experiments and studying effects of varying the starting ages and durations of EC-SC-IC treatments, we have explored in several new directions, and results of these

TABLE 11. EC-IC Percentage Differences for Cortical Depths and Wet Weights

	25-55 days				60-90 days			
	Depth		Weight		Depth		Weight	
	%	No., EC>IC	%	No., EC>IC	%	No., EC>IC	%	No., EC>IC
Occipital	4.6***	23/30	10.4***	116/135	6.2***	45/70	7.8***	64.5/87
Somesthetic	2.2**	20/31	4.7***	99.5/135	3.6***	38/50	4.9***	66.5/87

**$p < .01$, ***$p < .001$.

TABLE 12. Percentage Differences in Cortical Thickness Measures between Rats Assigned to Differential Environments at 6 Days of Age†

		Somesthetic			Occipital Cortex		
	N	Medial	Intermed.	Lateral	Medial	Intermed.	Lateral
EC-multifamily envir. minus unifamily envir.							
14 days	7,7	10.6***	10.0***	7.0*	1.5	6.8	16.0**
19	6,9	5.5***	9.7***	7.6	6.8*	8.0*	14.1**
28	19,19	8.0**	6.9**	9.0***	9.1**	10.3***	11.9***
MFE minus UFE							
28 days	12,12	0.3	2.8	0.9	3.6	0.9	1.7

*$p < .05$, **$p < .01$, ***$p < .001$.

†*From Malkasian. Unpublished observations, p. 74.*

new probes will be reported in this section. The topics to be reported here include EC-IC effects on several further chemical measures (acetylcholine, choline acetyltransferase, and nucleic acids), effects on dry weight of brain compared with wet weight effects, effects of giving EC experience for only 2 hr per day, effects of stimulant and depressant drugs on the 2-hr EC effects, and species generality of the EC-IC brain effects.

EC-IC Effects on Varied Chemical Measures

Effects on Acetylcholine and Choline Acetyltransferase. Inasmuch as AChE is only one of the three main biochemical components of the cholinergic system, it was anticipated that differences in either the concentration of acetylcholine (ACh) or in the activity of choline acetyltransferase (ChAt) might result from differential experience.

In 1964 we did one 80-day ECT-IC experiment in which the concentration of ACh in whole brain (excluding cerebellum and medulla) was determined. The results are summarized in Table 13. No significant differences in either the concentration or total amount of ACh was found between the groups. It should be remembered that the concentration of ACh in the brain is much more responsive to the immediate physiological condition of the rat than are the enzymes concerned with its metabolism. The procedure for the removal of the brain and the extraction of ACh is more difficult and subject to more variation than the removal and dissection of the brain for enzyme assays. In addition, at the time that this work was done, only bioassay methods were available for the determination of ACh. Thus we believe that differences of at least 5 percent in ACh concentrations would have to result from EC-IC treatments in order to have been reliably determined. The analyses were made on whole brain, and none of our other measures has given an EC-IC effect as large as 5 percent on whole brain. Therefore it is not surprising that we have not found an EC-IC effect in ACh concentration in whole brain.

It seemed to us more reasonable to determine environmental effects

TABLE 13. Effects of 80-day ECT-IC on Total Brain ACh*

	Condition				
	ECT	SD	IC	SD	% Difference ECT minus IC
Total brain wt (g)	1.201±	.064	1.194±	.073	0.6
Total ACh (MμM)	32.3±	2.3	32.2±	2.8	0.3
ACh/g (MμM)	27.0±	2.6	27.0±	1.9	−0.2

**Eleven pairs of male S_1 rats were run in this experiment, but the table is based on complete data from 8 ECT and 9 IC rats.*

on another enzymatic component of the cholinergic system—ChAt. The procedures for dissection and weighing of brain used for AChE determination are directly applicable to ChAt, and presumably the activity of an enzyme does not change rapidly at the time of sacrifice.

During 1967 and 1968, ChAt activity was determined after 30 days of EC or IC. In the first experiment, the rats were placed in the appropriate conditions at 60 days of age; in the subsequent two experiments, the rats were assigned at 30 days of age. In each of these experiments, small decreases in ChAt specific activity of the EC rats were found in each of the cortical areas, but these decreases were essentially reflections of the increased weight of the cortical tissue produced by EC (Table 14). A significant but slight increase in total cortical ChAt activity of the EC rats was found. It would appear from these data that the EC-IC differences in ChAt follow a similar pattern to those found in AChE but not to those in ChE.

Garattini et al (1969) have reported a study by Consolo, who compared the ChAt activity of grouped versus isolated mice. They determined ChAt values in a number of selected brain areas and in whole brain. Although differences of up to 10 percent were found, they concluded (with an *N* of only 5) that no significant changes were observed. Their grouped—like our EC animals—had lower specific activity of ChAt. They point out that the method measures only enzymatic activity and does not measure the turnover rates of ACh in vivo of the differentially housed mice, and the same comment would apply to our differentially housed rats. Since feasible methods for study of ACh turnover have recently become available (Schuberth et al., 1969), such a study would appear worthwhile.

Effects of ECT-IC or dark-rearing on DNA and RNA. Biochemical speculation and experimentation on learning and memory have largely focused on the role of RNA, beginning with the pioneering investigations of Hydén in the 1950s. As indicated in our introduction, most of the experiments concerned the effects of a few training trials per day over a few days on RNA quantity or composition. More recently, it has been fashionable to investigate the short-term effects of training on the incorporation of precursors into RNA.

In our experiments on nucleic acids, done in 1964 and 1965, two different experimental designs were used to obtain rats with differing degrees of cerebral stimulation. The first of these was our usual 80-day ECT-IC experimental design beginning at weaning. The second experimental series used a three-group design with littermate triplets. One group (light) was raised in the usual EC condition under a 12-hr light, 12-hr dark cycle from weaning until 105 or 155 days of age. The second group (dark) was raised in EC but in complete darkness for the same period and was exposed to light for only a few seconds at the time of

252 Table 14. Right-hand column heading should be Cortex/Rest rather than Rest/Cortex.

TABLE 14. Percentage Differences in Weight, ChAt/Weight, and Total ChAt of Rats maintained from 30 to 60 Days in EC and IC

	Cortex							
	Occip-ital	Somes-thetic	Rem. Dorsal	Ventral	Total	Rest of Brain	Total Brain	Cortex Rest
Expt. 1[a]								
Weight	8.1*	4.4*	4.7	3.4	4.4*	−0.3	1.8	4.7**
ChAt/wt.	−5.8*	−5.3*	−3.1	−2.6	−3.4	2.1	−0.4	−5.1*
Total ChAt	0.9	−1.3	1.2	0.2	1.4	1.7	1.4	0.2
Expt. 2[b]								
Weight	7.6*	3.5	3.8	5.6*	4.8***	1.2	2.8*	3.6**
ChAt wt.	−6.3*	−5.2*	−5.3**	−2.9	−4.1*	1.7	−0.6	−5.8***
Total ChAt	0.7	−1.8	−1.8	2.5	0.5	2.9	2.1	−2.4
Combined Expts.								
Weight	7.8***	4.0**	4.3*	4.4*	4.6***	0.4	2.2	4.2***
ChAt/wt.	−6.1***	−5.2**	−4.2***	−2.8**	−3.7*	1.9	−0.6	−5.4***
Total ChAt	0.8	−1.5	−0.3	1.3	0.9	2.3	1.8	−1.0

*$p<.05$, **$p<.01$, ***$p<.001$.
[a] *30 days EC beginning at 30 days of age (N = 10 per group).*
[b] *30 days EC beginning at 30 days of age (N = 9 per group).*

sacrifice. The third group (dark-light) was kept in darkness for most of the experiment but was exposed to light for an interval before sacrifice—approximately 35 min in the first experiment, and from 3 to 8 hr in the second, and 18 days in the third experiment of this series.

The results of analyses for RNA and DNA in the occipital area of the cortex are presented in Table 15. No significant effect of treatment on the concentration of RNA was found in any of the experiments in either the occipital area or the somesthetic area (which is not shown in the table). Changes in the total amount of RNA were primarily a reflection of weight differences. On the other hand, in every comparison the amount of DNA per milligram was less in the brains of the more stimulated rats—that is, the ECT or light-raised groups. The ECT or light-raised rats had less DNA per milligram in 44 out of 52 comparisons in the occipital area and in 33 out of 41 comparisons in the somesthetic area.

Another measure that yielded consistent differences between the groups was the ratio of RNA/DNA, which was greater in the occipital cortex of ECT or light-raised rats than in the isolated or dark-raised rats. For the ECT-IC comparison this difference amounted to 5.9 percent ($p < .001$, ECT > IC in 19 of 23 pairs). For the light cycle versus dark comparison, the RNA/DNA difference was 11.0 percent ($p < .001$, light > dark in 24 of 26 pairs). The somesthetic cortex showed similar effects.

The results will be related in the *Discussion* to the increase with EC in perikaryon and nuclear size and to the other indices of cellular function.

The groups that had been removed from the dark and exposed to light for 8 hr or less were similar in brain values to the corresponding dark groups and differed significantly from the light-raised groups (see Section C in Table 15). These results thus corroborate those of the dark-raised animals. On the other hand, the group removed to the light for 18 days before sacrifice had brain values closely similar to those of the light group and differing significantly from the dark group (see Section D of Table 15).

EC-IC Differences in Dry Weights of Brain Tissue

We have often been asked whether the changes we observed in weights of fresh brain tissue samples might not reflect changes in fluid content of the brain rather than in formed tissue. In the 1964 *Science* paper we reported that total protein varied directly with weight of the fresh tissue. We further reported that enzymatic measures per unit of protein were, if anything, slightly more stable than measures per unit of weight, and so group differences in enzymatic activities were somewhat more significant when stated in this way.

TABLE 15. Environmental Effects on Weight, RNA, and DNA of Occipital Cortex

	Weight		DNA/mg		RNA/mg		RNA/DNA	
	% diff.	No.	% diff.	No.	% diff.	No.	% diff.	No.
A. ECT vs. IC								
	ECT–IC	ECT>IC	ECT–IC	ECT>IC	ECT–IC	ECT>IC	ECT–IC	ECT>IC
No. 1	10.0**	10/11	–3.2	3/11	0.0	6/11	3.0	9/11
No. 2	6.9*	8/12	–8.7**	1/12	–0.8	4/12	8.5**	10/12
Combined	8.4***	18/23	–6.1***	4/23	–0.7	10/23	5.9**	19/23
B. Light (L) vs. Dark (D)								
	L–D	L>D	L–D	L>D	L–D	L>D	L–D	L>D
No. 1	3.1	5/7	–9.6*	1/7	–1.0	4/27	10.8	6/7
No. 2	6.4*	7/9	–11.5***	0/8	–1.4	2/5	8.8*	5/5
No. 3	5.3	11/14	–8.3***	3/14	2.7	8/14	11.9***	13/14
Combined	5.0*	23/30	–9.5***	4/29	0.9	14/26	11.0***	24/26
C. Light (L) vs. Dark (briefly exposed to light) (DL)								
	L–DL	L>DL	L–DL	L>DL	L–DL	L>DL	L–DL	L>DL
No. 1	9.7*	6/7	–10.3*	0/7	0.0	2/7	11.0	6/7
No. 2	4.8	6/9	–10.1***	0/8	–4.2	1/5	5.4	3/5
Combined	7.1**	12/16	–10.2***	0/15	–1.8	3/12	8.7	9/12
D. Light (L) vs. Dark-Light (18 days of light) (DL)								
	L–DL	L>DL	L–DL	L>DL	L–DL	L>DL	L–DL	L>DL
No. 3	–0.4	5/14	–1.3	6/14	–1.1	6/14	0.2	6/14

*$p < .05$, **$p < .01$, ***$p < .001$.

TABLE 16. Percentage Differences between EC and IC Littermate S_1 Rats in Wet Weights and Dry Weights of Brain Tissues

	Original Experiment (N = 11 per group)		Replication Experiment (N = 23 per group)
Brain sample	Wet	Dry	Dry[a]
Cortex			
Occipital	12.7***	12.5**	9.2
Somesthetic	5.0*	4.8*	1.9
Remaining dorsal	8.4**	8.5***	9.4
Ventral	1.6	1.5	1.2
Total	5.7***	5.7***	5.8
Rest of brain			
Cerebellum, pons, medulla	−0.1	−0.4	−1.7
Remainder	1.2	1.0	0.4
Total subcortex	0.6	0.3	−0.7
Total brain	2.8*	2.6	1.9
Ratio: cortex to rest of brain	5.0***	5.3***	6.5

$*p<.05$; $**p<.01$, $***p<.001$.
[a]*Significance of differences not given for replication experiment because of pooling of samples, as explained in text.*

Questions about the validity of wet weight measures continued to arise, however, so in 1969 we compared EC-IC differences in both wet and dry weights. This was done for a 25- to 55-day experiment with 11 littermate pairs of S_1 rats. The usual brain samples were dissected, weighed, lyophilized, and reweighed. Both wet and dry weights were given in our report (Bennett et al., 1969). Here we reproduce in Table 16 only the EC-IC percentage differences in Columns I and II. Note that these are virtually identical. We concluded that our previous reports given in terms of wet weights of brain tissue would not have changed if dry weights had been taken instead.

A further recent experiment replicated the conditions of the previous one except that the tissues were frozen quickly upon dissection without taking time for obtaining wet weights. The samples were also pooled before lyophilization, so that the number of independent dry weights was reduced, thus decreasing the value of statistical analysis. Each mean in Column III is based upon 23 littermate pairs of S_1 rats, so these values are undoubtedly quite stable. Here again clear EC-IC percentage differences were found. The magnitudes of the effects, region by region, are comparable to those of the previous experiment and to the wet weight of other 25- to 55-day EC-IC experiments. We believe that there is no need to inquire further whether taking dry weights would either eliminate the effects or reveal something new; they are strictly proportional to the wet weights.

Effects of Two-hr Daily Exposure to EC

The fact that formal training seemed to produce little or no effect on our usual brain measures led William Love of our group to question whether a few hours a day of EC would have any effect. To our general surprise, two-hour a day in EC over a 30-day period proved to be sufficient to induce full-fledged effects in brain weights and in AChE/weight (Rosenzweig et al., 1968). In Experiments III and IV of that study the 30-min daily period in the Hebb-Williams apparatus, which had always previously been a part of the EC treatment, was dropped from the schedule without impairing the EC-IC brain effects. We have confirmed the effects of the two-hour schedule many times and have capitalized on it in drug experiments, which will be described next.

Drugs Modulate Effects of Two-hr EC

Finding the two-hour effects opened the way to work on reports of drug effects that had long intrigued us. Kennard and collaborators had observed in the 1940s that excitant drugs speeded recovery from effects of brain lesions in monkeys and that depressant drugs retarded recovery and led to a lowered eventual level of performance (Ward and Kennard, 1942; Watson and Kennard, 1945). We wondered if such drugs might also interact with effects of differential environmental experience.

A moderate dose of methamphetamine (2 mg/kg) given before the daily EC period was found to enhance significantly the EC-IC differences in brain weights. Except for the EC session, the EC rats lived the rest of the day in individual colony cages. The IC animals lived in the same room and in the same kind of cages as their EC littermates, and they received daily injections of saline solution. Animals given phenobarbitol (30 mg/kg) before the daily EC session showed little or no difference from the ICs; they were significantly below EC-saline animals in cortical weight measures. These drug effects occurred only or chiefly in the EC condition; giving the drugs to rats that then remained in their individual home cages produced only minor effects on brain weights. The drug effects have been described elsewhere (Rosenzweig and Bennett, 1968 and in preparation).

The potentiating effect of methamphetamine has, in turn, proved useful in our attempts to determine whether giving enriched experience to individual rats could produce cerebral effects. This will be taken up later.

Species Generality of EC-IC Effects

The 1964 *Science* paper was based entirely on work with rats. Although most of our experiments have continued to employ rats as subjects, we have since found basically similar effects with other rodents. La Torre (1968) used two inbred strains of mice and obtained 25- to 105-day EC-IC effects in brain weights and AChE activity that resembled those found in rats. Not enough ChE data were obtained for the mouse to compare effects on this enzyme with those found in the rat.

We then turned to the gerbil (*Meriones unguiculatus*) both to test an animal that had not been bred in the laboratory for many years and also to extend this work within *Rodentia* from the family *Muridae*, which includes rats and mice, to the family *Cricetidae*, to which gerbils belong. In two 25- to 55-day experiments, gerbils yielded EC-SC-IC effects in brain weights and AChE/weight that were quite similar to the familiar rat patterns (Rosenzweig and Bennett, 1969).

Possible Alternative Explanations or Interpretations of Cerebral Effects of Differential Experience

From the time that we first observed the cerebral effects produced by exposing rats to differential environments, we have attempted to determine the casual factors involved and to find how best to interpret these effects. Thus, in our initial publication on the ECT-IC differences in brain AChE (Krech et al., 1960), we reported results of control experiments indicating that neither handling nor locomotion accounted for the effects. In the 1964 paper, discussing effects on brain weights as well as on AChE and ChE, we reported evidence that the following additional factors were not responsible: (1) the combination of handling and locomotion, and (2) isolation stress. We also found by using animals kept under standard colony conditions that the cortical weight effects were not chiefly due to impoverishment from this baseline but that enrichment above it gave the greater part (1964, p. 615 and Figs. 5 and 6). Furthermore, the effects could be obtained in young adult rats as well as in weanlings (1964, p. 615 and Fig. 6). Even though the effects were largest in occipital cortex, they did not require vision, since they could be obtained in blind or light-deprived rats (1964, p. 617). We concluded by urging caution in interpreting the possible functional significance of these effects: "We wish to make clear that finding these changes in the brain consequent upon experience does not prove that they have anything to do with storage of memory. The demonstration of such changes merely helps to establish the fact that the brain is responsive to environmental pressure—a fact demanded by physiological theories of learning and memory" (1964, p. 618).

In seeking possible casual factors as well as alternative interpretations, we have been aided by suggestions of members of our research group and of colleagues, as well as by members of audiences to whom we have reported our findings. We have considered all suggestions seriously, even if some of them may have been advanced facetiously. An example of the latter sort was the comment, published in 1965, that we might well measure the thickness of the rats' skins to see whether environmental enrichment caused release of an ectodermal-stimulating factor that would promote growth of skin as well as of the CNS, which is derived embryologically from ectoderm. The person who made this comment ignored our reports that the growth effect in the subcortex is sometimes negative and never large, and that within the cortex there is considerably less effect in the somethetic area than in the occipital area. Since there appear to be specific regional brain effects rather than a general cerebral effect, it did not seem worthwhile to look for an even more general ectodermal effect. The factors that have been investigated further will be described under two main headings: (1) *Physiological Mechanisms*, including possible roles of stress, hormonal mediation, and maturation, and *Environmental Factors* that may contribute to development of EC-SC-IC cerebral differences.

Possible Physiological Mechanisms

CAN STRESS ACCOUNT FOR CEREBRAL EFFECTS IN EITHER EC OR IC? A frequent question addressed to us is whether differential stress might account for the EC-IC brain differences. IC might produce "isolation stress." EC might cause the stress of "information overload." One of the most often-cited papers on isolation stress is that of Hatch et al. (1963) in Ottawa. In our 1964 article we noted (p. 615) that our isolated rats of several strains did not show the slower weight growth, increased aggressiveness or scaly tail reported by Hatch et al. (Note that in our Fig. 3 the growth of weight is, in fact, the most rapid for IC rats.) A subsequent publication of the Ottawa group (Wiberg et al., 1966), presenting replication studies of organ weights of isolated and communally housed rats, fails to confirm several of the significant differences reported in their earlier paper. While not wishing to cast doubt on the importance of housing conditions and genetic factors in determining the physiological status of animals, we do have reservations about the generality and reproducibility of the syndrome of isolation stress as described by Hatch et al. We have nevertheless attempted to determine whether either EC or IC

258 Line 20. Before the word "Environmental" insert (2).

can be shown to be stressful and whether stress of various forms can alter the brain values that we usually measure. Adrenal weight was used as a convenient, although admittedly crude, index of stress; we shall see that it yielded clear effects whenever overt stressors were employed.

In the 1964 paper we reported briefly that adrenal weight did not differ between EC and IC groups. Later we analyzed the question more thoroughly on the basis of six experiments (64 littermate pairs) run that year, some of them performed after the paper had been written. The further analysis, based on more experiments, revealed that IC adrenals did weigh somewhat more than those of EC littermates, but only in proportion to the greater body weight of the IC rats. Adrenal weight as a percentage of body weight did not differ significantly between EC and IC.

In 1959–1960 we had tested whether daily sessions of unavoidable foot shock could affect adrenal weight and brain weights. Rats of the S_1 and K strains received shock or no shock for two-week periods (each of these experiments including eight littermate pairs); rats of the RCH and RDH strains received the same treatments for four weeks (11 and 12 littermate pairs, respectively).*

The shocked rats developed significantly greater ratios of adrenal weight to body weight than did their controls (14 percent, $p < .001$), but there were no differences in the brain measures. Taking these results with those of the groups in the paragraph above, we see that EC-IC affects brain values but not the adrenal/body weight ratio, whereas overt stress does not affect our brain measures but does significantly increase adrenal weight.

Riege and Morimoto (1970) ran three experiments (two with the S_1 strain and one with Fischer rats) with 30-day EC and IC groups, both with and without the stress of being tumbled daily in a revolving drum. The Fischer strain experiment also included SC and SC-tumbled groups. The results provide a further test of whether stress alters brain values in either the EC or the IC direction. Riege and Morimoto confirmed the previous findings: tumbling stress increased adrenal weight/body weight but did not affect our usual measures of brain weights or brain enzyme activity and, conversely, the differential EC, SC, or IC environments did not lead to significant differences in adrenal weight/body weight, but they did alter significantly the brain measures. In the same experiments, concentrations of serotonin, dopamine, and norepinephrine were measured. These amines did show significant effects with both stress and environmental treatment, but the patterns of change produced by stress and by environment differed from each other.

* The K, RCH, and RDH lines were developed in our laboratories and maintained there for several years. Each of these lines had been demonstrated to develop cerebral EC-IC effects.

More severe stress was then employed by Riege and Morimoto in two additional unpublished experiments. Here the stress included both 3- to 5-min daily tumbling and 3- to 5-min daily foot shock. Once again, stress increased the adrenal weights but did not affect brain weights, whereas the differential environments altered brain weight but not adrenal weights.

We conclude that neither the EC nor the IC situation is sufficiently stressful to induce changes in the adrenal weight/body weight ratio of the strains of rats that we have tested. Even if EC or IC were stressful, this could not explain the effects in cerebral weights or AChE, since overtly stressful foot shock or tumbling did not affect these brain measures.

HORMONAL MEDIATION OF ENVIRONMENTAL EFFECTS? Although the experiments of the preceding section have ruled out stress as the cause of the cerebral EC-SC-IC effects, it might still be possible that hormonal functions mediated these effects of experience. That is, the differential experience might lead to alterations in hormonal activity which could, in turn, affect the nervous system. We have therefore tested the hypothesis that the pituitary gland is essential to occurrence of the EC-IC effects. Choosing this gland allowed us not only to eliminate its secretions but also to diminish and cut off from environmental influence the secretions of these glands controlled by the pituitary—the thyroid, the adrenal cortex, and the testes.

These experiments were run, two with male Fischer rats and one with male Long-Evans rats. In each case, some animals were hypophysectomized shortly after weaning. Then, 5 to 10 days later, the animals were placed in one of four conditions (to which they had been preassigned) and kept there for 30 days. The conditions were these: EC-hypophysectomized, EC-control, IC-hypophysectomized, and IC-control. Results of operates were used only for those animals with verified complete hypophysectomy.

The results of these experiments demonstrated that, although hypophysectomy stunts bodily growth and checks brain growth somewhat, significant EC-IC brain differences nevertheless occur in brain weights and brain chemistry of the operated animals. The brain weight effects are summarized in Figure 8. If the operation had prevented the occurrence of EC-IC effects, then it would have been necessary to determine which gland or glands are essential, to verify this by the procedures of replacement therapy, and so on. With our present results, however, we do not plan to pursue the endocrine direction further unless other investigators produce clear evidence of hormonal intervention.

MATURATIONAL EFFECTS? It has sometimes been suggested that the EC-IC brain differences develop because the stimulation of the en-

260 Line 22. Replace "These" with "Three."

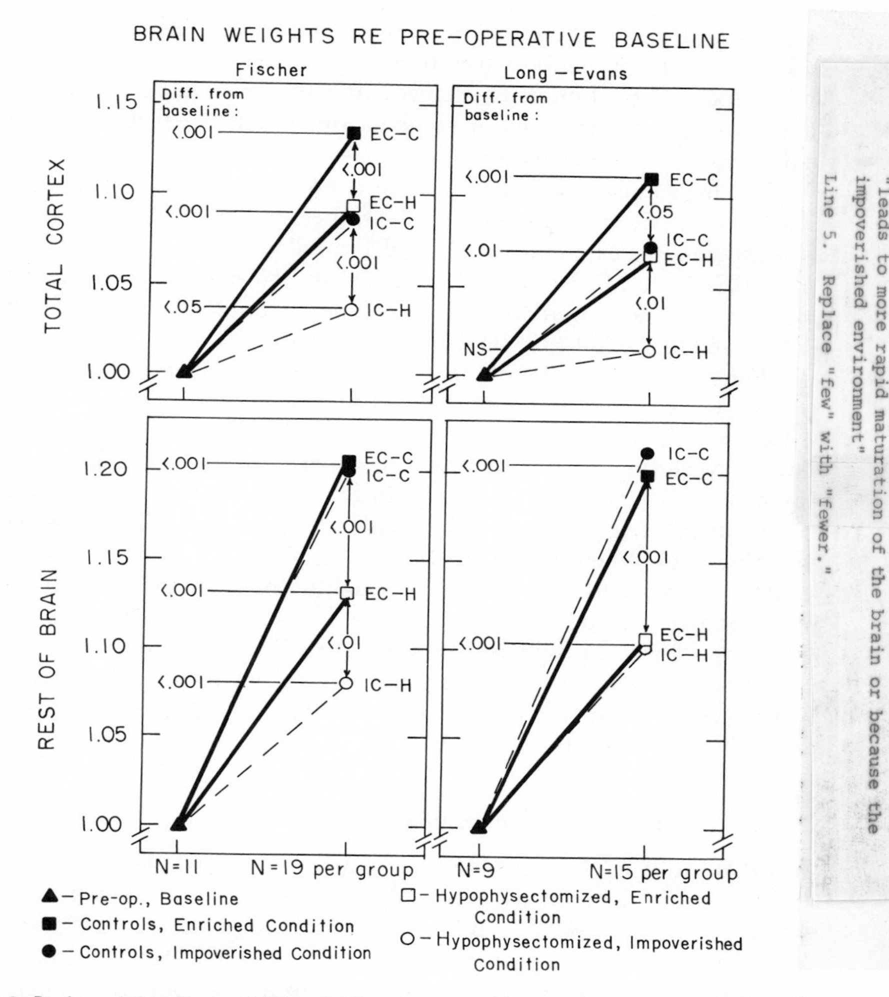

FIG. 8. **Brain weight effects of EC and IC among normal and hypophysectomized male rats of the Fischer and Long-Evans strains. Measures are taken from the baseline of like-strain rats sacrificed at the outset of the experimental period. Hypophysectomy is seen to impair growth of weight of both total cortex and the rest of the brain; nevertheless EC-IC effects occur among the hypophysectomized as among the control animals. (From Rosenzweig et al. In preparation.)**

riched environment retards brain maturation. It is true that certain EC-IC effects are similar to changes in the growing brain. For example, both the EC brain and the adult brain show, in comparison with the IC or the immature rat brain, greater cortical weight, a greater glial/neural ratio, and few neurons per unit of cortical volume. Nevertheless, two types of

261 In line 1 the following material was omitted after the word "environment," completely changing the meaning of the sentence: "leads to more rapid maturation of the brain or because the impoverished environment"
Line 5. Replace "few" with "fewer."

findings lead us to reject the interpretation that the effects reported in this chapter are primarily maturational: (1) the lack of a critical period and the possibility of inducing changes after the brain has attained mature values, and (2) the fact that on some measures EC-IC effects are opposite in direction to effects of maturation.

We have already noted that there is no critical period for producing EC-IC brain effects. In brain weights, rather similar effects are produced with experiments extending over all of the following age ranges: 25 to 55, 25 to 105, 60 to 90, 105 to 185, and 290 to 380 days. In cortical depth, rather similar effects were found in experiments running from 6 to 28 days: 25 to 55, 60 to 90, 25 to 105, and 105 to 185 days.

Three examples will serve to show that EC-IC effects can go opposite to changes with maturation. First, whereas cortical thickness is greater in EC than in IC, the thickness decreases with age, at least from 55 days onward. Nevertheless, in experiments begun at 60 days of age, cortical depth was significantly greater in EC than in IC. Secondly, the ratio of cortical to subcortical weight also decreases with age up to about 200 days, as we have shown elsewhere (Fig. 2 in Bennett et al., 1970). But EC rats show reliably greater cortical/subcortical weights than do IC rats. Thirdly, AChE activity per unit of weight increases with age, at least to about 100 days of age, but EC is significantly lower than IC on this measure.

Because of these types of evidence we are not inclined to agree with Sperry (1968, p. 324) when he suggests that our studies may concern primarily growth effects.

Possible Environmental Causes of EC-IC Brain Effects

EXTRACAGE ENVIRONMENT. Early in our work we supposed that stimuli outside the animals' cages might affect brain measures. It was for this reason that we put the IC rats in cages with solid side walls so that they could not see each other, although they could still get the social stimuli of sound and smell. The IC cages were also put in a quiet and dimly lighted room, whereas the SC and EC cages were in a brightly lighted room where caretakers came and went frequently. The belief that extracage stimulation might be important was based in part on a report of Forgays and Forgays (1952). They stated that rats confined to small mesh cages within a large "free-environment" cage containing other rats and "playthings" were better in subsequent problem-solving than rats in a similar environment but with no "playthings" outside the mesh cages. On the other hand, it had been reported that rats that have a circle and a triangle placed in their cage can later distinguish them better than controls, but only if the situation permitted direct contact with the stimuli. If the circle and triangle were surrounded by wire mesh so that they could

263 Line 16. Replace "Kretch" with "Krech."
Line 30. Replace "Richard" with "Ricard."

be seen but not touched, then their presence in the cage did not aid later discrimination performance (Bennett and Ellis, 1968).

Experiments in 1965–1966 demonstrated that our usual extracage environments did not affect either EC or IC animals. Rats kept in individual cages in the EC room had brain values virtually identical with those of littermates in the strict IC cages and room; an EC group kept in the quiet and dimly illuminated IC room developed brain values equal to those of littermates under usual EC. In the drug experiments described above, the IC rats are kept in colony cages on the same racks on which two-hour EC groups spend 22 hr a day, yet clear EC-IC cerebral differences develop.

Only when we went to an extremely impoverished environment did we find an extracage effect. For the extreme condition, we suspended each cage in a fiberglass padded box, and the boxes were placed in an audiometric test chamber. This condition did result in cerebral values that were even further removed than were IC from the SC baseline (Kretch et al., 1966).

Specific visual stimuli outside the cage have been reported recently to affect rats. Lavallee (1969, 1970) stated that rats given three two-hour daily slide shows from 21 to 80 days of age were subsequently superior in problem-solving to control rats that saw only light on a blank screen. Gilbert Ricard, in our laboratory, attempted in 1969 to replicate Lavallee's behavioral findings and to look for possible associated cerebral changes. In two experiments with Fischer rats, the animals given visual stimulation by slide programs were significantly inferior to controls on the Hebb-Williams maze ($N = 20$ pairs). Other animals kept concurrently in the experimental or control conditions were sacrificed, and no differences in brain weights were found between the groups that could see slides and their controls ($N = 20$ pairs). No reason is apparent for the reversal of behavioral effect between Lavallee's and Richard's results.

Singh et al. (1967) have reported that rats that could look out from their cages to a surface painted with vertical stripes developed both cerebral and behavioral differences from rats that see a blank surface. Rats were placed in the experimental conditions at 30 days of age and were kept there for 60 days. A stripe group, subsequently tested on frequency of bar-pressing for light reinforcement, showed significantly higher rates than controls. In the groups subjected to chemical analysis, the stripe group was said to show twice as great (A)ChE activity in posterior cortex as did the controls ($p < .01$). This is a far larger difference in AChE activity than we have ever seen; furthermore, the absolute values reported and their anterior-posterior pattern are quite different from those given by other investigators.

William Maki, who had collaborated with Singh et al. in North

Dakota, has attempted in 1969–1970 to replicate their results in our laboratories. An experiment with 12 male S_1 rats per group showed no cerebral differences between groups, but stripe-exposed rats did bar-press somewhat more for light reward. A subsequent replication was done with female Holtzman rats, the sex and strain used in the original experiment of Singh et al. Twelve rats were used in each of the four groups (stripes-brain analysis, blank surface-brain analysis, stripes-behavior test-brain analysis, blank surface-behavior test-brain analysis). No significant differences were found between stripe-exposed and blank-surface groups sacrified at the same time, either in brain measures or in behavior.

We conclude that it is difficult to alter the brain values we measure by extracage stimuli. The only significant effect of this sort that we have found to date is that of the extremely impoverished environment.

BRIEF DAILY EXPERIENCE IN ANOTHER APPARATUS. Recently there have been suggestions that the brain is so plastic that very little stimulation is required to change it measurably. In this vein, an article of 1968 referred to our work as follows, "There are indications in the data that if the daily monotony of the impoverished existence is relieved for only 15 minutes of handling during transfer to and from cages, this is in itself sufficient to cancel the growth differences" between EC and IC (Sperry, 1968, pp. 323–324). We do not believe that our data support such a statement.

The earliest experiment in our laboratories bearing on this point was done in 1964–1965 by Robert Slagle. It included an ICT (IC-training) group, that is, a group that was daily removed from IC, transported in individual compartments of a cart to a testing room, and given a few trials a day in various apparatus over a 50-day period. This ICT group showed almost no differences from a regular IC group, and the differences that occurred were not in the EC direction. ECT did differ significantly from ICT in this experiment, and along the same pattern that it differed from IC. Thus the daily handling and brief daily experience of test apparatuses given to ICT rats did not cancel the cerebral differences between them and ECT littermates.

In several other experiments, we have given one group daily runway trials as controls for a maze-running group. The runway groups, in spite of daily experience in an apparatus other than the cage, scarcely differed in brain measures from home-cage controls, and even the maze groups have shown disappointingly small cerebral differences from animals that experience only the home cage.

The misunderstanding expressed in the 1968 quotation may have arisen in regard to unpublished results showing that after 30 days of EC or IC experience, the EC-IC brain differences diminish during two subsequent weeks of pretraining and testing in the Lashley III maze. Further

265 Line 20. Replace "contradicted" with "contradict."
Bottom line. Replace "three" with "four."

work has shown that it is not primarily the maze testing that causes the EC and IC brain differences to decrease. Rather, it appears to be the food deprivation and the recaging of EC and IC rats in individual colony cages where we usually keep rats during testing; they experience this new cage environment for about 23 hr a day. In a control experiment we took rats from EC and IC, recaged them in individual colony cages, and put them on a food-deprivation schedule similar to that which we employ for pre-training and testing. At the end of two weeks in the new conditions the brain values of the groups taken from EC and IC were significantly less different than those of rats sacrificed directly after EC and IC experience. Food deprivation causes decreases in brain weights, making difficult a direct comparison between EC-IC effects in deprived and nondeprived groups. In an experiment in 1969, one set of animals remained in EC or IC except during the daily testing period, and some of their brain differences remained significant at the end of testing (cortical/subcortical weight ratio difference, 4.3 percent, $p < .01$). Groups tested concurrently, but housed in individual colony cages, showed distinctly lesser EC-IC differences (cortical/subcortical ratio difference, -0.4 percent N.S.). The results with the groups that remained in differential housing during the maze-test period contradicted the statement that 15-min handling of IC rats during transfer to and from cages could in itself cancel the brain differences. Further substantiation for our position will be found in the next section, in which it can be seen that putting single rats into an EC cage for two-hours per day over a 30-day period produces little or no cerebral effect, unless other treatments are given concurrently.

ROLES OF SOCIAL AND INANIMATE STIMULI IN DEVELOPMENT OF EC-IC EFFECTS. In a recent paper (Rosenzweig, 1970) we explored the relative importance of the social and the inanimate stimuli of the EC situation in producing the cerebral effects, and we can now add further findings. The previous study included Fischer male rats put for two-hours per day into bare large cages or large cages furnished with stimulus objects; the animals were put into these conditions either singly or in groups of 12. Neither social stimulation alone (12 rats in an empty cage) nor the stimulation of varied objects (a single rat in a furnished cage) produced significant brain differences from rats that remained all day in the individual home cage. Only the combination of social and inanimate stimulation (a group put into a furnished cage) produced cerebral effects. At the time, however, we noted that perhaps the social stimulation served principally to keep each rat in active commerce with the stimulus objects. We suggested that cerebral effects might be produced in single rats if they could be kept active rather than being allowed to spend most of their time quiescent, as is typical of isolated rats.

We have now tested this hypothesis in three experiments with S_1 rats activating the individual animals either by giving them their daily two-hour EC period in the dark part of the daily cycle, or by injection of methamphetamine, or by the combination of dark and methamphetamine. Summary results of these experiences are given in Table 17. The S_1 rats may respond more to the inanimate enriched environment than did the Fischer rats, since even individual S_1 rats given saline injections and put in the large furnished cages in the light (Column 2) developed cerebral differences from littermates kept in their home cages. Still larger cerebral differences from home cage rats are seen in rats given the enriched experience in the dark (Column 3), although only the cortical/subcortical ratio gave significant results in all three replications. The largest effects are found in the methamphetamine-treated individuals put into the complex environment in the dark (Column 4). In all three replications this treatment produced significant effects in weight of occipital cortex and of the cortical/subcortical ratio. Only a minor part of the effect can be attributed to the drug alone, since the rats given methamphetamine but kept in their individual home cages (Column 1) showed a significant effect in only one of the three experiments, and only small significant effects overall. The results of these experiments—demonstrating that social stimulation is not necessary to produce the brain changes and that interaction with a complex environment suffices—suggest that either complex and novel stimulation or learning (or both) are responsible for the cerebral effects of our experiments.

TESTS OF THE LEARNING HYPOTHESIS. We have also attempted to test directly the hypothesis that at least some of the EC-IC cerebral dif-

TABLE 17. Single Rats in Enriched or Home Cage Conditions: Percentage Differences of Brain Weights Measured from Baseline of Home Cage-Saline Group

Condition:	Home cage	2 hr/day in large furnished cage during		
		light hr	dark hr	
Injection:	Methamphetamine	Saline	Saline	Methamphetamine
N per group	44	46	46	46
Occipital cortex	0.6	4.0**	4.1***	10.6***
Total cortex	1.9*	2.2**	3.0***	5.5***
Rest of brain	0.8	0.6	0.6	2.0*
Cortex/rest	1.1**	1.6***	2.3***	3.4***

*$p<.05$, **$p<.01$, ***$p<.001$.

ferences are produced by prolonged learning in EC and therefore reflect the operation of mechanisms of learning or of memory storage. The findings that we have just reviewed are consistent with this hypothesis; that is, the observations that two-hours a day of exposure to a complex environment bring about cerebral changes and that these require active commerce with the stimuli. Even before making these recent findings, we had begun to test whether variation in formal training alone might produce measurable cerebral effects. Several experiments were done with operant conditioning devices. In these experiments, one animal of each pair mastered a succession of problems, mostly to visual stimuli in a Skinner box, while its yoked-control littermate received a pellet whenever the first rat earned one. No significant differences in cerebral weights were obtained, but some significant effects were found in chemical measures when data from three experiments were combined. For total cortex, experimentals exceeded controls in total AChE by 2.6 percent ($p < .05$) and in total ChE by 2.9 percent ($p < .05$). The effects obtained were relatively small, and their pattern did not resemble that obtained in the EC-IC case in that the increase in AChE was almost as great as that in ChE. Further experiments were run with various automatic mazes, at least 50 trials being given per day over 30 days. When maze rats were compared with runway-control and with food-control rats, only small and nonsignificant differences were found. Work in this direction is continuing, with use of more challenging learning tasks. If the more demanding learning tasks appear to produce cerebral effects, then we will be faced with the problem of devising fully adequate controls.

Discussion

The research of our group has, at the least, demonstrated the responsiveness of a number of brain measures to differential experience. A summary of such effects for occipital cortex of the rat, the region most studied in this regard, is given in Table 18.

Limitations of space preclude discussion of many of the findings surveyed above, so we will limit ourselves to one central question: What may be the importance of these anatomical and chemical brain effects for mechanisms of learning and memory from the point of view of neuroscientists—biochemists, anatomists, and psychologists? We believe that they may reveal not only the continuation into long-term learning of chemical changes seen in short-term and intermediate-term experiments but also the presence of structural changes that are presumably the result of earlier chemical synthesis.

TABLE 18. EC-IC Effects on a Variety of Brain Measures in Occipital Cortex*

	EC-IC period, days of age			
	25-105	25-55	60-90	6-28
Weight, fresh tissue	+	+	+	
Dry weight		+		
Depth	+	+	+	+
Number, neurons	=			
Number, glia	+			
Perikaryon size	+			
Nuclear size	+			+
AChE/weight	–	–	–	
Total AChE	+	+	+	
ChE/weight	+	+	=	
Total ChE	+	+	+	
Hexokinase/weight	=			
Total hexokinase	+			
ChAt/weight		–		
Total ChAt		=		
RNA/weight	=			
Total RNA	+			
DNA/weight	–			
Total DNA	=			
RNA/DNA	+			
Protein/weight	=			
Total protein	+			

**(+, EC significantly greater than IC; =, no significant difference; –, EC significantly less than IC).*

Anatomical Effects

The changes that we reported in 1964 in weight and thickness of cerebral cortex have been the most often replicated effects, and they can now be understood, in part at least, in terms of more specific cellular changes. For one thing, the cell bodies of cortical neurons have been found to be significantly larger in EC than in IC rats; a 12 percent difference in cross-sectional area indicates a difference of about 18 percent in volume. By itself this increase in perikaryon volume probably accounts for only a part of the effect in cortical bulk, since the cell bodies have been estimated to amount to only about 9 percent of the volume of adult rat cortex (Eayrs and Goodhead, 1959). The dendritic tree accounts for most of the bulk of cortical neurons, and with the greater depth of the EC cortex, the lengths of the dendrites presumably increase proportionately. In general, the cross section of a dendrite varies with its length, so an increase in length would amount to an increase in bulk. We have not, however, attempted to measure the mass of the dendrites. A further contribution to the greater bulk of EC cortex is the greater number of glial cells found there than in IC cortex.

Turning now to functional considerations, some anatomical indices support the conclusion that cortical neurons have greater functional loads in EC than in IC rats. The increased sizes of perikaryon and nucleus are themselves considered to indicate heightened function (Edström and Eichner, 1958; Gyllensten et al., 1966). In two preliminary experiments with Albert Globus, we have found a greater number of dendritic spines in EC than in IC. This also suggests the capacity to sustain a greater flux of neural impulses, and it may in turn be a response to increased functional demand.

Relating Chemical to Anatomical Effects

Let us now attempt to relate some of the chemical findings to the anatomical results. The finding of a reduction of DNA/weight in EC as compared to IC can be related to the greater volume of cells in EC. This follows from the premise that the amount of DNA is the same for all diploid cells of a given species, so that a lower concentration of DNA must mean fewer and therefore larger cells in a given volume of tissue. It must be remembered that the total cortical volume is greater in EC than in IC brains. The finding of a significantly increased ratio of RNA to DNA in EC is consistent with the anatomical results indicating greater functional activity in the EC cortex. Pevzner (1966) and Glassman (1969) have summarized several reports demonstrating increased ratios of RNA to DNA after neural stimulation. As we had already pointed out in the 1964 report, the percentage of protein does not vary between EC and IC brains, so that total protein increases with the weight of the EC cortex. Dry weight of the cortex has also been found to increase in the same proportion as wet weight, so we would expect that other components of the tissue such as polysaccharides and lipids would also increase in EC in proportion to tissue weight. These increases in a number of brain constituents apparently complement the short-term or intermediate-term results of Glassman, Hydén, and others. They have shown increased rates of labeling and presumably of synthesis in brains of animals undergoing training. It may be that our long-term results represent an integration of a continuing series of such pulses of synthesis with learning experience. We would anticipate that with properly designed experiments employing tracers, we could find differences in incorporation and/or turnover between EC and IC animals.

Our concepts about the meaning of effects in activity of AChE and ChE have changed during the course of our research. Early in our work we hypothesized that AChE might provide an index of synaptic number or activity. We were surprised when early analyses showed significantly lower AChE per unit of weight in EC than in IC cortex, but we then believed that this could be understood in terms of the increase in cortical weight and the small but significant increase in total AChE activity. The replication series of EC-IC experiments has shown, however, that this increase in total AChE activity is very small and of doubtful statistical significance. The finding of increased activity of ChE in EC brains was unexpected and has been abundantly confirmed in replication experiments. The differential pattern of changes in AChE and ChE activities suggested that differential cell counts might show an increase in glial cells in EC animals. This hypothesis was tested directly and confirmed by histological techniques. Although the finding of altered ChE/AChE activity led to the discovery of altered EC-IC glial number, we wish to be cautious about using ChE activity as an index to glial number. Capillary walls as well as glial cells are rich in ChE activity, and further chemical and histological observations are needed to characterize the cerebral changes. For example, it would be useful to employ other chemical indicators of types of brain cells. We have also suggested that a more thorough study of this question would be to follow the development and migration of glia as functions of age and of duration and type of environmental experience, using radioactive tracer techniques.

To date, at least, the enzymatic measures have led to more knowledge of changes in cortical structure than of changes in synaptic function as consequences of differential experience. The results obtained have provided definitive data on effects of the EC, SC, and IC conditions on AChE and ChE. The emphasis of our further chemical work will therefore be directed toward other types of analysis such as RNA, DNA, and incorporation of radioactive tracers.

Independence of Brain Measures

In trying to form an integrated picture of the anatomical and biochemical cerebral changes with differential experience, it must be borne in mind that the changes in various measures do not necessarily occur together and that some of the brain effects may be independent of others. The independence of effects is shown by the findings that some develop at different rates than others during the experimental period, that some are more influenced by enrichment and others by impoverishment of experience, and that some appear equally at all ages whereas others develop more readily in younger animals. Let us recall briefly some examples of these differences among measures; for many of our brain measures we do not yet have data that would allow comparisons of effects as experimental duration and ages of subjects are varied. Difference in rate of development of effects is seen in these examples: (1) EC-IC differences in cortical weights are well developed in 30-day experiments; (2) differences in AChE/weight are present after 30 days of EC or IC but become larger by 80 days; and (3) changes in ChE/weight are absent or small after 30 days but are clearly established by 80 days. Differential effects of enrichment versus impoverishment (both with regard to the standard colony baseline) are seen in the facts that cortical weight is affected more by enrichment whereas cortical ChE activity is affected more by impoverishment. Furthermore, depth of dorsomedial cortex is affected by enrichment whereas depth of lateral cortex is chiefly influenced by impoverishment. Finally, an example of the importance of age is that in experiments started at 290 days, EC-IC changes in weight of cortex appeared as in younger rats, but little or no effect on AChE or ChE was found even in such 90-day experiments.

A positive result of the interdependence of effects is that it should be possible to concentrate on effects of enrichment of experience rather than on those resulting from impoverishment. This can be done both by using behavioral enrichment procedures and by using anatomical and chemical brain measures that respond especially to enrichment.

Are the Cerebral Effects of Differential Experience Due to Learning?

In company with other investigators in this area, we have not been able to demonstrate conclusively that the brain changes induced by differential experience are due to learning rather than to other variables that are present in the experimental situation. Discussions of the problem of disentangling the effects of variables tied to learning are found in recent articles by the following researchers, among others: Miller (1967), Glassman (1969), Hydén and Lange (1971), Rosenzweig (1971).

Our own tactics have been two-pronged: (1) to look for brain changes with formal training versus control conditions, and (2) to test alternative hypotheses for causes of the brain effects. Concerning the first direction, the cerebral changes we have so far been able to induce with formal training have been small and have not closely resembled those caused by informal enriched experience. Perhaps this lack of result means only that the amount of formal training has been slight compared with the learning that occurs in the EC situation. For this reason we are working to give enhanced formal training.

Concerning alternative hypotheses, we have already ruled out several of the most obvious or most prominent. The brain differences induced by EC versus IC are not due to differences in handling, locomotion, stress, rate of maturation, alteration in secretion of several endocrines, or extracage environment. There is, however, no end to the alternative hypotheses that can be devised for any effect observed.

One further alternative hypothesis that may be considered briefly here is that of novelty. The novelty of stimulation or novelty of response to the enriched situation may call forth brain changes that are not related to the learning that may, in fact, be occurring at the same time. Miller (1967, p. 650) has pointed out that such responses to the situation should regress with disuse, whereas learning is not greatly affected by disuse. We had also pointed out: "Once learning was completed and the resultant processes of consolidation has subsided, then the changes needed to support this extra activity would regress, and only the correlates of long-term memory would remain to differentiate the trained from the untrained brain" (1968, p. 62). As a matter of fact, some EC-IC effects are present only in relatively short experiments, whereas others can be found even at the end of long durations. Thus differences in depth of anterior cortex are found in 30-day experiments but not in 80-day experiments. On the other hand, the persistence of differences in brain weights at the end of two 160-day EC-IC experiments (see Table 7) suggests that novelty is not required for these brain differences, since the rats have long ago become familiar with the set of stimulus objects. (Although it was not mentioned above in our discussions of results, these 160-day experiments also showed the usual pattern of significant EC-IC effects in AChE/weight, ChE weight, and ChE/AChE ratio.) To what extent the brain differences would persist if EC, SC, and IC animals were taken out of the differential environments and put into a common environment, we do not know. Testing this is not so simple as might appear, since it would involve changes of environment—and thus novelty—for at least two of the three groups. We have undertaken further research on this question.

Studies with Primitive or Simplified Neural Systems

Because of the complexities of the mammalian nervous system, many investigators in recent years have attempted to study mechanisms of learning in neurally isolated preparations or in invertebrates with comparatively small numbers of neural units. Interesting results are being obtained, although defining learning in such preparations and applying results obtained to more complex animals then provide new sets of problems. We believe that results of studies on the mammalian brain show that much can be accomplished with an intact complex organism. Even electrophysiological studies, which have been the special focus of work on simplified systems, can be made on learning in the mammalian brain (John, 1967; Olds, 1969). Recently Leon Dorosz in our laboratories has recorded differences in transcallosal transmission between EC and IC rats. While not wishing to contest the value of research on "learning" in simplified neural systems, we believe that results justify optimism about research on learning and memory in the mammalian brain.

Qualitative or Quantitative Cerebral Changes?

A key problem of memory mechanisms is whether the increased synthesis seen in the brain with learning represents only a quantitative change in brain function or a qualitative change as well. That is, are new and unique molecules synthesized as a result of "new memory," or does the increased synthesis of RNA and protein and probably other molecular classes as well represent only "more of the same"? We believe that there is little evidence to date for the formation of new and unique molecules as a result of learning. The "memory transfer" phenomenon, if adequately confirmed among independent laboratories, would of course constitute excellent evidence for formation of novel molecules, but at the present time, in our view at least, evidence for "memory transfer" is still open to question.

Bonner (1966) has raised the possibility of synthesis of unique RNA molecules during learning and memory storage. Machlus and Gaito (1968, 1969) provided some evidence that this might in fact occur. Work in our laboratory (von Hungen, 1970), however, indicates that present competitive hybridization techniques may not be adequate to test critically Bonner's hypothesis. Further developmental work on hybridization techniques as well as further behavioral work seems warranted to follow up this exciting possibility.

In histological studies, too, quantitative changes with experience are being found in the number and size of cellular structures. It may well be that further modifications of such structures as synapses, axon terminals, and dendritic spines will be identifiable with electronmicroscopic techniques which we have begun to use. In fact, encouraging results have been published indicating that quantitative differences in structures studied with the electron microscope in rat visual cortex can be measured as functions of altered visual experience (Cragg, 1969; Fifková, 1970).

In sum, we are tending to interpret our findings as showing a brain that, as a consequence of increased behavioral demands, is adapting both anatomically and chemically to increase functional capacity. The neural cells are larger, they form more interconnections, and their function is supported by greater numbers of glia. The cells are also more active in chemical synthetic processes. Some of the observed changes are transitory and presumably reflect only short-term demands, whereas others persist and may support long-term memory. To begin to spell out the ABCs of these complex processes, we need our own ABCs: Anatomy, Behavior, and Chemistry. Any comprehensive interpretation of brain processes in learning requires at least all of these approaches.

An Invitation

The results described in this chapter show relatively large cerebral effects of experience—some changes of 10 percent in cortical weight and chemical ratios, and 20 percent changes in perikaryon volume. We believe that such results demonstrate both the feasibility of work in this field and also the need for many more minds and hands (and animals!) to be devoted to such research. The mammalian brain has now been found to change in a number of ways as a result of differential experience, but undoubtedly many new findings of this sort remain to be made, and the most critical and significant changes for learning and memory must still be singled out. Furthermore, the behavioral procedures needed to produce the cerebral effects are now seen to be much less time-consuming or arduous than those described in 1964, and they are effective with a wider range of subjects (both as to age and species). We hope therefore that further investigators will choose to orient their endeavors in this direction. To aid this development, we wish here to renew the invitation extended in our 1964 article to offer any help possible to other investigators. This includes making available detailed descriptions of procedures and welcoming visitors to our laboratories to see first-hand any apparatus or procedures of interest and to discuss research in this area.

Acknowledgments

This investigation was supported by Research Grants GB-5537 and GB-8011 from the National Science Foundation and by Grant 09-140398 from the Office of Education. It also received support from the United States Atomic Energy Commission. We wish to acknowledge warmly the help of many collaborators in this work over the past several years, and especially the following: Chemists Marie Hebert and Hiromi Morimoto; anatomists Bernice Lindner, Lennis Lyon, Alma Raymond and Carol Ingham; behavioral technicians Clarence Turtle, Don Gassie, and Todd Grant; secretary Jessie Langford and statistician Ann Muto.

References

Adair, L. B., Wilson, J. E., Zemp, J. W., and Glassman, E. (1968). Proc. Nat. Acad. Sci. U.S.A., 61:606–613.

Altman, J., Wallace, R. B., Anderson, W. J., and Das, G. D. (1968). Develop. Phychobiol., 1:112–117.

274 Bottom line. Replace "Phychobiol." with "Psychobiol."

Beach, G., Emmens, M., Kimble, D., and Lickey, M. (1969). Proc. Nat. Acad. Sci. U.S.A., 62:692–696.

Bennett, E. L., Crossland, J., Krech, D., and Rosenzweig, M. R. (1960). Nature (London), 187:787–788. ——— Diamond, M. C., Krech, D., and Rosenzweig, M. R. (1964). Science, 146:610–619. ——— and Rosenzweig, M. R. (1970). In: Handbook of Neurochemistry, Vol. 6, Lajtha, A., ed. New York: Plenum Press. ——— Rosenzweig, M. R., and Diamond, M. C. (1969). Science, 163:825–826. ——— Rosenzweig, M. R., and Diamond, M. C. (1970). In: Molecular Approaches to Learning and Memory, Byrne, W. L., ed. New York: Academic Press.

Bennett, T. L., and Ellis, H. C. (1968). J. Exper. Psychol., 77:495–500.

Bondy, S. C. (1966). J. Neurochem., 13:955.

Bonner, J. (1966). In: Macromolecules and Behavior, 1st ed., Gaito, J., ed. New York: Appleton-Century-Crofts.

Bowman, R. E., and Strobel, D. A. (1969). J. Comp. Physiol. Psychol., 67:448–456.

Cragg, B. G. (1967). Nature (London), 215:251–253.

Crossland, J., (1961). In: Methods in Medical Research, Vol. 9, Quastel, J. H., ed. Chicago: Year Book Medical Publishers.

Dellweg, H., Gerner, R., and Wacker, A. (1968). J. Neurochem., 15:1109–1119.

Diamond, M. C. (1967). J. Comp. Neurol., 131:357–364. ——— Law, F., Rhodes, H., Lindner, B., Rosenzweig, M. R., Krech, D., and Bennett, E. L. (1966). J. Comp. Neurol., 128:117–125. ——— Rosenzweig, M. R., and Krech, D. (1965). J. Exp. Zool., 160:29–36.

Eayrs, J. T., and Goodhead, B. (1959). J. Anat., 93:385–402.

Edström, J. E., and Eichner, D. (1958). Z. Zellforsch, u. Mikroskop. Anat., 48:187–200.

Ferchmin, P. A., Eterovic, V. A., and Caputto, R. (1970). Brain Res., 20:49–57.

Fifková, E. (1970). J. Neurobiol., 1:285–295.

Fonnum, F. (1966). J. Biochem., 100:479–484.

Forgays, D. G., and Forgays, J. W. (1952). J. Comp. Physiol. Psychol., 45:322–328. ——— and Read, J. M. (1962). J. Comp. Physiol. Psychol., 55:816–818.

Garattini, S., Giacalone, E., and Valzelli, L. (1969). In: Aggressive Behavior, Proc. Symposium on the Biology of Aggressive Behavior, Garattini, S., and Sigg, E. B., eds. Amsterdam: Excerpta Medica.

Geller, E., Yuwiler, A., and Zolman, J. (1965). J. Neurochem., 12:949–955.

Giacalone, E., Tansella, M., Valzelli, L., and Garattini, S. (1968). Biochem. Pharmacol., 17:1315–1327.

Glassman, E. (1969). Ann. Rev. Biochem., 38:605–646.

Gold, A. M., Altschuler, H., Kleban, M. H., Lawton, M. P., and Miller, M. (1969). Psychon. Sci., 17:37–38.

Gyllensten, L., Malmfors, T., and Norrlin, M. L. (1966). J. Comp. Neurol., 126:463–470.

Hatch, A., Wiberg, G. S., Balazs, T., and Grice, H. C. (1963). Science, 142:507.

Heppel, L. A., Hurwitz, J., and Horecker, B. L. (1957). J. Am. Chem. Soc., 79:630–633.

Hydén, H., and Lange, P. W. (1971). In: Handbook of Neurochemistry: Alteration of Chemical Equilibrium in the Nervous System, Vol. 6, Lajtha, A., ed. New York: Plenum Press.

Jacob, M., and Mandel, P. (1966). In: Protides of the Biological Fluids, Vol. 13, Peeters, H., ed. Amsterdam: Elsevier. ——— Stevenin, J., Jund, R., Judes, C., and Mandel, P. (1966). J. Neurochem., 13:619–628.

John, E. R. (1967). In: The Neurosciences, Quarton, G. C., Melnechuk, T., and Schmitt, F. O. eds. New York: Rockefeller University Press.

Kahan, B. E., Kriginan, M. R., Wilson, J. E., and Glassman, E. (1970). Proc. Nat. Acad. Sci. U.S.A., 65:300–304.

Kimble, D. P., ed. (1965). The Anatomy of Memory. Palo Alto, Calif.: Science and Behavior Books.

Krech, D., Rosenzweig, M. R., and Bennett, E. L. (1960). J. Comp. Physiol. Psychol., 53:509–519. ——— Rosenzweig, M. R., and Bennett, E. L. (1966). Physiol. Behav., 1:99–104.

La Torre, J. C. (1968). Exp. Neurol., 22:493–503.

Lavallee, R. J. (1969). Psychon. Sci., 17:21–22. ——— (1970). Develop. Psychol., 2:257–263.

Leslie, I. (1955). In: The Nucleic Acids, Vol. 2. Davidson, J. N., and Chargoff, eds. New York, Academic Press.

Logan, J. E., Mannell, W. A., and Rossiter, R. J. (1952). Biochem. L., 51:470–479.

Machlus, B., and Gaito, J. (1968). Psychon. Sci., 10:253–254. ——— and Gaito, J. (1969). Nature (London), 222:573–574.

Mahler, H. R., Moore, W. J., and Thompson, R. J. (1966). J. Biol. Chem., 240:2122–2128.

Malacarne, V. G. (1819). Memorie storiche intorno alla vita ed alle opere di Michele Vincenzo Giacinto Malacarne. Padova, Tipografia del Seminario, p. 88. (See also notice in Journal de Physique, 43:73, 1793.)

Malkasian, D. R. (1969). Morphological Effects of Environmental Manipulation and Litter Size on the Neonate Rat Brain. Unpublished doctoral thesis, University of California, Berkeley.

May, L., and Grenell, R. G. (1959). Proc. Soc. Exptl. Biol. Med., 102:235–239.

Mayer, J., Marshall, N. B., Vitale, J. J., Christensen, J. H., Mashayekhi, M. C., and Stare, F. J. (1954). Am. J. Physiol., 177:544–548. ——— Roy, P., and Mitra, K. P. (1956). Am. J. Clin. Nutr., 4:169–175.

McCaman, R. E., and Hunt, J. M. (1965). J. Neurochem., 12:253–259.

Miller, N. E. (1967). In: The Neurosciences, Quarton, G. C., Melnechuk, T., Schmitt, F. O., eds. New York: Rockefeller Univ. Press.

Munro, H. N., and Fleck, A. (1966). In: Methods of Biochemical Analysis, vol. 14, Glick, D., ed. New York: Interscience Publishers.

Olds, J. (1969). Am. Psychologist, 24:114.

Pevzner, L. A. (1966). In: Macromolecules and Behavior, 1st ed., Gaito, J., ed. New York: Appleton-Century-Crofts.

Rappoport, D. A., Fritz, R. R., and Myers, J. L. (1969). In: Handbook of Neurochemistry: Chemical Architecture of the Nervous System, vol. 1, Lajtha, A., ed. New York: Plenum Press.

Riege, W. H. (1971). Develop. Psychobiol., in press. ——— and Morimoto, H. (1970). J. Comp. Physiol., 71:396–404.

Rose, S. P. R. (1969). FEBS Letters, 5:305–312.

Rosenblatt, J. S. (1970). In: Biopsychology of Development, Tobach, E., ed. New York: Academic Press.

276 In Leslie reference, add "E." after "Chargoff."
In Logan reference, replace "Biochem. L." with "Biochem. J."
In Rosenblatt reference, replace "1970" with "1971."

Rosenzweig, M. R. (1971). In: Biopsychology of Development, Tobach, E., ed. New York: Academic Press. ——— (1968). Atti Accad. Naz. Lincei (Italy), Quaderno, 109:43–63. ——— (1966). Am. Psychol., 21:321–332. ——— and Bennett, E. L. (1968). Proc. 76th Ann. Conv. APA, 3:269–270. ——— and Bennett, E. L. (1969). Develop. Psychobiol., 2:87–95 ——— Bennett, E. L., Diamond, M. C., Wu, S. Y., Slagle, R. W., and Saffran, E. (1969). Brain Res., 14:427–445. ——— Krech, D., and Bennett, E. L. (1958). In: Neurological Basis of Behavior, CIBA Foundation Symposium. London: J. & A. Churchill. ——— Krech, D., and Bennett, E. L. (1958). In: Biological and Biochemical Bases of Behavior, Harlow, H. F., and Woolsey, C. N., eds. Madison: Univ. of Wisconsin Press. ——— Krech, D., Bennett, E. L., and Diamond, M. C. (1962). J. Comp. Physiol. Psychol., 55:429–437. ——— Krech, D., Bennett, E. L., and Diamond, M. C. (1968). In: Early Experience and Behavior, Newton, G., and Levine, S., eds. Springfield, Ill.: Thomas. ——— and Leiman, A. L. (1968). Ann. Rev. Psychol., 19:55–98. ——— Love, W., and Bennett, E. L. (1968). Physiol. Behav., 3:819–825.

Santen, R. J., and Agranoff, B. W. (1963). Biochim. Biophys. Acta, 72:251–262.

Schuberth, J., Sparf, B., and Sundwall, A. (1969). J. Neurochem., 16:695–700.

Shapiro, H. S. (1968). In: Handbook of Neurochemistry, Sober, H. A., ed. Cleveland: Chemical Rubber Co.

Singh, D., Johnston, R. J., and Klosterman, H. J. (1967). Nature (London), 216:1337–1338.

Sperry, R. W. (1968). 27th Symp. Soc. Dev. Biol., Developmental Biology Supplement 2, 306–327.

Stanley, W. C., Cornwell, A. C., Poggiani, C., and Trattner, A. (1963). J. Comp. Physiol. Psychol., 56:211–214.

Thoman, E., and Arnold, W. J. (1968). J. Comp. Physiol. Psychol., 65:441–446. ——— Wetzel, A., and Levine, S. (1968). Anim. Behav., 16:54–57.

U.S. Department of Health, Education and Welfare (1968). Public Health Service Publication No. 1024.

Ward, A. A., Jr., and Kennard, M. A. (1942). Yale J. Biol. Med., 15:189–288.

Walsh, R. N., Budtz-Olsen, O. E., Penny, J. E., and Cummins, R. A. (1969). J. Comp. Neurol., 137:261–266. ——— Budtz-Olsen, O. E., and Torok., A. (1970). Develop. Psychobiol., in press.

Watson, C. W., and Kennard, M. A. (1945). J. Neurophysiol., 8:221–231.

Welch, A. S., and Welch, B. L. (1970). In: Physiology of Fighting and Defeat, Eleftheriou, B. E., and Scott, J. P., eds. Chicago: Univ. of Chicago Press. ——— Welch, B. L., and Welch, A. S. (1969). In: Proc. Int. Symp. Aggressive Behavior, Garattini, S., and Sigg, E. B., eds. Amsterdam: Excerpta Medical Foundation.

Wiberg, G. S., Airth, J. M., and Grice, H. C. (1966). Fd. Cosmet. Toxicol., 4:47–55.

Yajima, A. (1966). Tohuka J. Exptl. Med., 89:235–244.

Zamenhof, S., Bursztyn, H., Rich, K., and Zamenhof, P. J. (1964). J. Neurochem., 11:505–509.

277 Under Rosenzweig, the following reference was omitted:

Rosenzweig, M. R., Bennett, E. L., and Diamond, M. C. (1967). In: Psychopathology of Mental Development, Zubin, J., and Jervis, G., eds. New York: Grune and Stratton.

Chapter 13

Macromolecular Change and the Synapse

J. A. DEUTSCH

University of California (San Diego)
La Jolla, California

The Problem of Finding a Change

An organism capable of learning must be able to change some internal state whenever it lays down a memory. To learn one habit or to lay down one memory the organism must change at least one state. To learn another habit or to lay down another memory the organism must be able to change at least one different state in addition to the preceding one. For it is difficult to see how an identical change could form the substrate of two different memories. Given this necessity of a different state for each memory, we must suppose that there must be at least as many different changes of state inside the organism as there are memories.

A particular piece of learning therefore selects one state out of a large number of possible states the organism is capable of assuming. If we suppose that a particular organism can learn, say, 200 different habits, then that organism must have, before any learning has taken place, at least 200 states which it is capable of assuming. Further, for each habit that the organism will learn there will be many habits that it will not learn because the learning of one piece of one habit excludes others. For instance, I can learn that the sign for "stop" is red. I was capable of learning that such a sign was any of a number of different colors. Thus states appropriate to these other possibilities could have been selected, and must still exist as possibilities. We can therefore assume that if an organism is capable of learning 200 habits, a very much larger number of possible states must exist within it. The very reason why an organism learns is that it allows whatever environment it finds itself in to select the appropriate repertoire of habits. If the number of habits it was capable of learning equalled the number of possibilities within its nervous system, then the organism has those habits genetically preselected and no learning can take place. The number of possibilities must be at least twice the number of habits.

It seems, therefore, that when a habit is learned—to name a figure which seems conservative—something like one out of a thousand possibilities is selected. The proportion might very well be much smaller. If we are looking for some change as a result of a particular piece of learning, then this thought is somewhat discouraging. It means that even if we are right about the kind of change to expect, our chances of finding it are very low. For instance, we may believe that memory consists of a change at a synapse. Even if all the synapses in the nervous system were modifiable by learning, we would expect only a very small proportion to be modified through the learning of one habit. In fact, we can say that if a relatively massive difference is found as a result of learning, such a change is most unlikely to represent the substrate of learning. It is, instead, most probably some other physiological correlate of the situation. A similar argument applies if we look for biochemical differences. The changes that form the substrate of memory can only be so small as to be below the resolution of our observational methods. If, by any chance, we were to observe some change due to learning we would be likely to dismiss it because it would be so small and infrequent as to qualify as some random event or a part of the "noise" inherent in our observations.

If it is unlikely that we can detect the substrate of memory in a large nervous system, then perhaps we should look at some small nervous system or a small portion of a large nervous system. The problem here is that the system or subsystem we select should be capable of learning. It seems likely that memory storage requires at least some degree of specialization on the part of nervous tissue. Any kind of learning scheme that is consistent with the behavioral facts involves the postulation of characteristics of the components which have not yet been observed by neurophysiologists. We may therefore doubt the wisdom of turning to unspecialized nervous systems to study learning. Even if we can force modifications on pieces of spinal cord or neurons of Aplysia, such systems do not display any conspicuous talent in the area of learning and our manipulations are therefore unlikely to be related to memory storage. We can burn out resistors in a computer by passing large currents across them. This does produce long-term alterations but it would be perilous to assume that this is the way that the computer stores information in its memory banks. We must somehow make sure that the tissue we study has the properties we are interested in. If, for instance, we are interested in the information transmitting properties of tissue, a minute study of liver cells will not be very enlightening. Such a study is only likely to generate wrong ideas.

The above arguments indicate strongly that an approach to the substrate of memory through use of the methods of direct observation is unlikely to succeed. There are, of course, other arguments which can be made against the direct approach. It seems, therefore, that any approach to the problem of the substrate of memory must use methods which are less direct. Being less direct, such methods must give answers which are less certain. Any answers given by indirect methods have more of the status of hypothesis than fact. On the other hand, it is through the use of such hypotheses that the field of observation for a direct approach may be narrowed to manageable proportions and the stage set for a direct frontal assault. One of the methods available for an indirect investigation of the problem is furnished by pharmacology. By means of intervention with drugs of known activity, and by the use of other drugs with similar or related action, it is possible to infer things about various aspects of neural functioning.

The Synapse and Memory

Almost as soon as the hypothesis of the synapse was enunciated, it was suggested that memory resided in synaptic modification (Tanzi, 1893). Indeed, so plausible was this idea that it has not been seriously challenged since. However, no good evidence for the idea has been forthcoming. Given that there are pharmacological agents which interfere with synaptic transmission, it should be possible to obtain at least some kind of confirmatory evidence for the idea that synaptic modification forms the substrate of memory. Our strongest evidence concerning amnesia comes from human clinical material. In retrograde amnesia the memories of events before an accident are lost. However, memories of events prior to those lost are retained. The span of time for which there is no memory may cover minutes, hours, days, or even weeks. The gap in memory usually shrinks in a gradual manner, the events farthest in time before the accident normally returning first. This suggests that whatever change is triggered at the time of the registration of a memory is not instantly complete but that the change is slow and continuous. If the change is one which is due to synaptic alterations in conductance, it should be possible to track such a change by pharmacological methods. For instance, if transmission across a synapse modified by learning is initially poor, such transmission could be facilitated by the same dose of the same drug which would block the synapse if conduction across the synapse were strong. Not only is such intervention possible, but it is actually in medical use in the treatment of myasthenia gravis. In this disease the patient becomes gradually incapable of movement. The efferent nerves and the muscles themselves are intact. However, an insufficient amount of transmitter (acetylcholine) to produce muscle contraction is ejected at the neuromuscular junction. An anticholinesterase is used to boost the amount of acetylcholine and this enables the patient to move. The anticholinesterase inactivates the enzyme cholinesterase. Cholinesterase destroys acetylcholine. In the normal case, the destruction of acetylcholine is crucial, because if the acetylcholine is not destroyed there is a block at the neuromuscular junction. The amount of anticholinesterase which permits the accumulation of sufficient acetylcholine to produce muscular contraction in the case of a myasthenic produces paralysis in a normal.

A very similar situation exists in the case of a cholinergic synapse, as in the case of a neuromuscular junction. The arrival of an electrical disturbance at the presynaptic junction causes the emission of a large number of packets of the transmitter acetylcholine. This transmitter spreads rapidly across the synaptic cleft to the postsynaptic membrane. There it produces depolarization. The acetylcholine is rapidly destroyed by the enzyme acetylcholinesterase. Pharmacological agents are available which modify the process of transmission in various ways.

The anticholinesterases (diisopropyl flurophosphate or DFP, physostigmine) inactivate the enzyme acetylcholinesterase either partially or wholly, depending on the dose. With partial inactivation of acetylcholinesterase the destruction of acetylcholine is slowed down. This slowing down of the destruction of acetylcholine leads to a temporary increase of acetylcholine in the synaptic cleft. If the amount of acetylcholine at a synapse is inadequate to produce sufficient depolarization of the postsynaptic membrane, the pile-up of acetylcholine due to the inactivation of acetylcholinesterase will increase the amount of depolarization, leading to a facilitation of synaptic activity. Such facilitation may occur either because the amount of transmitter at a given synapse is subnormal or because the sensitivity of a given postsynaptic junction to transmitter is subnormal. If, on the other hand, the amount of acetylcholine is adequate to produce sufficient depolarization at the postsynaptic membrane, then the pile-up of acetylcholine will produce synaptic block. Due to causes at present obscure, the synapse ceases to transmit.

Another class of agents, the anticholinergics (scopolamine, atropine) appears to occupy receptor sites on the postsynaptic membrane in the same way as acetylcholine but does not produce depolarization. Because of this, anticholinergics reduce the effectiveness of transmitter. In a synapse where depolarization of the postsynaptic membrane is only just sufficient to excite a spike potential, an anticholinergic will block transmission. On the other hand, transmission across another synapse where postsynaptic depolarization is higher could occur as usual given the same dose of anticholinergic.

Another class of agents, the cholinomimetics (carbachol, oxotremorine) mimics the action of acetylcholine. Unlike acetylcholine, however, they are not destroyed by acetylcholinesterase. As a result, these agents have a persistent action. In low doses they produce a small depolarization and thus a facilitation at cholinergic synapses. At higher doses they have a blocking action in a manner similar to acetylcholine when it is not destroyed by acetylcholinesterase.

Now let us assume that when learning occurs a synapse is modified in some way so that it begins to conduct. Previous to such a modification, when an electrical disturbance reached the presynaptic ending no depolarization of the postsynaptic membrane occurred. After learning, such depolarization occurs with the consequence that the synapse conducts the message which reaches it. As we saw above, it seems that whatever change is initiated at the time of learning, such a change increases spontaneously with time. At least this is an interpretation consistent with the facts of retrograde amnesia. If the "learning" change is an increase of conductance across a synapse, we would then expect such conductance to increase with time, and then after coming to a maximum perhaps to decline with time, as forgetting sets in. A change in conductance across a synapse might be due to either of two factors. Either the amount of transmitter emitted by the presynaptic ending increases, or the sensitivity of the postsynaptic ending to transmitter increases. In either case, as conductance increases differential effects on transmission should be seen as we apply a uniform dose of the pharmacological agents described above. At low levels of conductance an anticholinesterase should increase the amount of depolarization at the postsynaptic membrane and facilitate memory immediately after learning and when forgetting normally sets in. On the other hand, anticholinesterase should block memory at a longer time after learning, when conductance across the synapse is high. An anticholinergic, on the other hand, should block memory soon after learning when depolarization of the postsynaptic membrane is low, but leave memory intact when conductance is high, that is, at a longer time after learning.

The test of the hypothesis that the substrate of memory consists of a gradual change in synaptic conductance is experimentally simple. We teach animals a habit at a certain point in time. The animals are then divided into different groups. Each group waits a different time before retest. Each group is injected with a drug at the same time before retest. Any difference in performance during retest (our measure of retention) is then due not to differential effects of the drug on performance but to the differential effects of time of initial learning. This is because the time between drug injection and retest is invariant whereas the time between initial learning and retest is varied.

Experimental Investigations

The first two experiments (Deutsch et al., 1966; Deutsch and Leibowitz, 1966) show that effects of facilitation or block of a memory can be obtained with the same dose of anticholinesterase simply as a function of time of injection since the time of original learning.

The Ss in these experiments were male albino rats of the Sprague-Dawley strain. They were 250 to 300 at the start of the experiment.

Experiment I

In the first experiment, rats were trained on a simple task. Then an intracerebral injection of anticholinesterase was made at different times after initial training, the time being varied from one group of subjects to another. After injection, all rats, irrespective of the group to which they were assigned, were retested 24 hr after injection. Thus, what was varied was the time between training and injection. The time between injection and retest was kept constant. Any difference in remembering between groups was therefore due to the time between initial training and injection.

Rats were dropped on an electrified grid in a Y-maze. The lit arm of the Y was not electrified and its position was changed from trial to trial. The rats therefore learned to run into the lit arm. When they had chosen the lit arm 10 trials in succession, the criterion of learning was met and training was concluded.

Then, at various times after training, the rats—under nembutal anesthesia—were placed in a stereotaxic instrument and four intracerebral injections were made. These were aimed at the hippocampus. (The placement of the bilateral injection was anterior 3, lateral 3, vertical +2, and anterior 3, lateral 4.75, vertical −2, according to the atlas of DeGroot.) Peanut oil, 0.01 ml containing 0.1 percent of diisopropyl fluorophosphate (DFP), was injected in each locus. After 24 hr from time of injection, the rats were retrained to the same criterion of 10 successive trials correct. The number of trials prior to the first of the 10 successive correct trials in this retraining session gave a measure of retention.

The first group was injected 30 min after training. Its retention was significantly worse than that of an equivalent group injected simply with peanut oil. In contrast, a group injected with DFP three days after training showed retention as good as that of the peanut oil control group. Up to this point it seems that memory is less susceptible to DFP the older it is. However, Group 3, injected five days after training, showed only slight recollection at retest and Group 5, trained 14 days before injection, showed complete amnesia. The score of Group 5 on retest was the same as the

TABLE 1. The Effects of DFP on Memories of Different Age

Group	N	Substance injected	Training-injection interval	Initial score	Retest score	P of difference*
1	13	DFP	30 min	27.1	14.3	<.001 (groups 1 and 2)
2	9	DFP	3 days	22.1	5.0	
3	10	DFP	5 days	25.1	20.4	
4	6	DFP	14 days	23.3	32.5	
5	13	DFP	trained 24 hr after injection		35.4	<.001 (groups 4, 5 and 6)
6	10	Peanut oil	30 min	19.9	5.0	
7	7	Nothing	15 days training to retest	20.4	3.0	
†						
a†	5	DFP	1 day	27	7.2	
b†	9	DFP	2 days	30.7	5.9	
c†	6	Peanut oil	2 days	35	6.8	

*t test. †*Light in safe alley was dimmer.*

score of a group which had not been trained before but had simply been injected with DFP 24 hr before initial training. The amnesia of group 5 was not due to normal forgetting as other controls showed almost perfect retention over a 15 day span. (A subsidiary experiment [Deutsch and Stone, unpublished data] has established that injections of DFP on habits that are one and two days old have no effect, showing that the initial stage of vulnerability lasts less than 1 day; see Table 1). The same results have been obtained by Hamburg (1967) with intraperitoneal injections of the anticholinesterase physostigmine, using the same escape habit.

To make sure that we were not observing some periodicity in fear or emotionality interacting with the drug, another experiment was conducted in which rats were taught to run to a reward of sugar water the position of which correlated with the lit arm of a Y-maze (Wiener and Deutsch, 1968). The results show a very similar pattern of amnesia as a function of time of learning before injection. It is therefore most likely that we are in fact studying memory.

Taken together, the results of the experiments establish a *prima facie* case for the notion that an agent known to interfere with synaptic conduction (depending on the level of transmitter) does have differential effects on memories of different age. Except for a short period at the outset, the older a habit the greater the vulnerability of this habit to the anticholinesterase DFP. This suggests that after an initial period of decrease, the level of conductance at a synapse storing a habit increases spontaneously after learning.

Experiment II

The first experiment here dealt with the effects of the anticholinesterase DFP on habits which were normally well retained. The effects of the drug were to decrease the retention of a habit, depending on its age. Thus one of the predicted effects of an anticholinesterase was verified. However, the other predicted effect, facilitation, was not shown. The reason for this is that the habit which was acquired was so well retained without treatment over 14 days that one could not, on methodological grounds, show any improvement of retention subsequent to injection of the drug. It may be the case that 1-, 2-, and 3-day-old habits were facilitated instead of merely being unaffected, but the design of the experiment would not allow us to detect this because there is an effective ceiling on performance. Consequently, an attempt was made to obtain facilitation where it was methodologically possible to detect it—namely, where retention of the habit by a control group was imperfect. It was found that 29 days after learning the habit described above was almost forgotten by a group of animals injected with peanut oil 24 hr earlier. Consequently, the following experiment was devised. Rats were divided into four groups. The first two were trained 14 days before injection; the second two, 28 days before injection. One 28-day group and one 14-day group were injected with the same dose of DFP; the other 28-day group and the other 14-day group were injected with the same volume of pure peanut oil. The procedure and dosage were exactly the same as in the first experiment.

On retest the 14-day DFP group and 28-day peanut oil group showed poor retention. In contrast, the 28-day DFP group and 14-day peanut oil group showed good retention. The results (Table 2) show a large and clear facilitation of the otherwise almost forgotten habit by an anticholinesterase injection, while they confirm the obliteration of an otherwise well-remembered habit when it is 14 days old, which was already shown in Experiment I. The findings also lend strong support to the notion that forgetting is due to a reversal of the change in synaptic conductance which underlies learning. The same result has been obtained by Wiener and Deutsch (1968), with use of an appetitive habit. Here, forgetting occurs at 21 days and the maximum of block with anticholinesterase occurs after 7 days. A large facilitation of the 21-day-old forgotten habit was obtained after anticholinesterase injection.

Experiment III

So far it has been shown that the anticholinesterase drug DFP has different effects on memories of different age. Though its actions on memory are consistent with, and plausibly interpreted by, its anticholinesterase

TABLE 2. Effects of DFP on Memory of a Well-Remembered and Almost-Forgotten Memory

Group	N	Substance injected	Training-injection interval (days)	Initial learning scores		Retest scores		P of difference*
				Means	Medians	Means	Medians	
1	10	DFP	14	44.5	46	44.2	47	
2	7	Peanut oil	14	45	49	6.0	1	<.01; <.001
3	8	DFP	28	47.4	50	14.5	12	<.001
4	9	Peanut oil	28	51	56	46	46	<.001

The training scores are much higher than in Table 1 because a much dimmer light was used to indicate the correct alley.
**Mann-Whitney U test.*

action, some other property besides its indirect action on acetylcholine could in some unknown manner produce the same results. A check on the hypothesis that the effects observed are due to an effect on acetylcholine can be provided by the use of an anticholinergic drug. An anticholinergic drug (such as atropine or scopolamine) produces an effective lowering of the level of acetylcholine at the synapse. It does this apparently by occupying some of the receptor sites on the postsynaptic membrane without producing depolarization. It thus prevents acetylcholine from reaching such receptor sites and so reduces the effectiveness of this transmitter. We would therefore expect an anticholinergic to block conduction at a synapse where the level of acetylcholine is already low, while simply diminishing conduction at synapses where the level of acetylcholine is high. If the interpretation of the effects of DFP is correct, we would then expect the reverse effect with the administration of an anticholinergic drug. That is, we would expect the greatest amnesia with anticholinergics where the effect of anticholinesterase was the least, and we would predict the least effect where the effect of anticholinesterase on memory was the largest. It will be recalled that the least effect of DFP was on habits one to three days of age.

In the third experiment (Deutsch and Rocklin, 1967) the anticholinergic agent chosen was scopolamine, and it was injected dissolved (0.58%) in peanut oil, using precisely the same amount of oil and location as in the previous experiments. The same experimental procedure was also used. A group injected 30 min after training showed little if any effect of scopolamine. However, a group injected one day after training showed an intermediate amount (Table 3). Groups injected 7 and 14 days after training showed little if any effect. As far as the experimental methodology allows us to decide, then, the effect of an anticholinergic is the mirror image of the effect of the anticholinesterase. This strongly supports the initial hypothesis that variations of level of acetylcholine during transmission occur at certain synapses as a function of time after the modification of such synapses through learning. Further, the experiment mirrors the two phases of drug sensitivity found with DFP. There is an increase of sensitivity between 30 min and 24 hr, followed by a decrease of sensitivity to an anticholinergic. This further confirms the notion that there are two phases present in memory storage. The result with scopolamine is reliable and has been repeated by Wiener and Deutsch (1968) with use of an appetitive habit. However, the maximum effect of scopolamine in causing block occurred at three days, though the general shape of the curve was very similar. Finally, it is of interest to note that amnesia can result in man from anticholinergic therapy for Parkinsonism (Cutting, 1964).

TABLE 3. The Effects of Scopolamine on Memories of Different Age

Group	N	Time injected after training	Initial learning (medians)	Retest scores (medians)	P of difference*
1	8	30 min	21.5	4.75	<.007
2	8	1 day	20.5	16.5	
3	12	3 days	29.5	10.3	<.004
4	8	7 days	20.5	4.25	
5	8	14 days	16.5	7.5	

*t test.

Experiment IV

It can be seen from the preceding experiments that amnesia can be produced by manipulating the level of transmitter with suitable pharmacological agents. As such amnesia is attributed to synaptic block due to alterations in the effective level of acetylcholine after it has been emitted, the amnesia produced should be temporary in nature. According to the hypothesis, amnesia here is not due to interference with the capacity of the presynaptic ending to emit transmitter. The amnesia is assumed to be due simply to an increase or effective decrease of transmitter brought about by the presence of the injected drug. We would therefore expect the amnesia to be temporary and to disappear as the injected drug wears off. Two experiments were carried out to see if the amnesia obtained was indeed temporary. As the curve of amnesia effect is apparently made up of two components, a sample was obtained from both of these components. Animals were injected with DFP 30 min after learning. One group was retested 24 hr after injection, another after 48 hr, and a third five days after injection. Return of memory was complete after five days (Table 4). In a second study, animals were injected 14 days after training. One group was retested one day after injection and another five days after injection. There was considerable return of memory in the five-day group. Taken together the results support the prediction that anticholinesterase induced amnesia is only temporary in nature. It is of clinical interest to note that most human traumatic amnesias are also temporary. It is possible that they represent temporary states of cholinergic insufficiency.

The original hypothesis we introduced has been well supported by the effects of cholinergic drugs on memory. There seems to be a cycle of altering sensitivity to cholinergic agents. However, if the hypothesis is

TABLE 4. The Recovery of Memory After DFP Injection

Group	*N*	Time injected after training	Time retested after injection	Initial learning (means)	Retest Scores (means)	*P* of difference*
1	13	30 min	1 day	27.1	14.3	
2	5	30 min	2 days	20.4	10.0	<.001 (groups 1–3)
3	10	30 min	5 days	21.7	2.5	
4	10	14 days	1 day	44.5	44.2	<.001 (groups 4–5)
5	9	14 days	5 days	44.2	17.8	

* *Mann-Whitney U test.*

correct, we would expect an alteration in the strength of memory with time. As there is an increase of susceptibility to anticholinesterase block with time, we would expect memory to improve with time if this is due to an improvement with synaptic transmission. Yet in our undrugged control groups we saw no such improvement with time. So it seems that there is no obvious correlation of drug effect with normal memory at the beginning of the life of a memory trace. We have seen that there is a good correlation at the end, where forgotten habits show strong facilitation. The reason for an absence of correlation at the beginning may very well be methodological. The rats are initially trained to a very high criterion (10 out of 10 trials correct) and it would therefore be impossible to see any improvement. We therefore designed a situation in which such an improvement could be seen (Huppert and Deutsch, 1969). Rats were trained for 15 trials only, then allowed to wait from 30 min to 17 days and then retested. During retest we counted the number of trials it took the animals to come to criterion. If training to criterion (10 consecutive correct trials) was given in one session, the rats took 35.9 trials. If the second session was given one day later, the rats took 37.7 trials; three days later, it required 38.6 trials. Here there was no strengthening of the initial memory with time. On the other hand, if the second session took place seven days later the rats took only 27.15 trials and 10 days later only 24.9 trials. This was a significant improvement over the other scores. If, on the other hand, the second session took place 17 days later, the total number of trials to criterion was 48.9 trials. This score shows that the original 15 trials had almost been forgotten after 17 days. Similar results have now been obtained (Huppert, personal communication) in an appetitive task in which sugar pellets were used as reward. Taken together, these results support the idea that the cycle of sensitivity seen when we administer drugs reflects an aspect of normal functioning and change in the memory trace.

Experiment V

The experiments already outlined support the idea that at the time of learning some unknown event stimulates a particular group of synapses to alter their state. Synapses which do not transmit when an impulse arrives are modified in such a way that with the passage of time they transmit more and more effectively when an impulse arrives. We may ask ourselves if such a modification of a synapse represents an all-or-none process or whether it is graded. What is meant by this is the following. The amount of transmission when an impulse arrives at a certain time after modification by learning may be the same independent of the amount of learning. In the above case a synapse can be modified only once during learning or any further learning after the first has no effect on the further development of the capacity of one synapse to emit transmitter. On such a model increases in "habit strength" would be due to a progressive involvement of fresh synapses and so a spread involving more parallel connections in the nervous system. The second possible hypothesis is that each successive learning trial modifies the same synapses in a cumulative way and thus produces an increase both in the rate at which transmissive capacity increases and in the upper limit of such transmissive capacity. In other words, it is possible that with more training a synapse will come to transmit more efficiently per impulse.

If, with increased training, a synapse transmits more efficiently, then a habit should become increasingly more vulnerable to anticholinesterase with increased training. Furthermore, the memory of the same habit should be facilitated when its level of training is very low. In other words, we should be able to perform the same manipulations of memory by varying level of training as we were already able to perform when we varied time since training.

If, on the other hand, increases in training simply involve a larger number of synapses but no increase in the level of transmission at any one synapse, then increases in training should not lead to an increased vulnerability of a habit to anticholinesterase. Rather, the opposite should be the case. As the number of synapses recruited is increased, some of the additional synapses will by chance variation be less sensitive to a given level of anticholinesterase. Thus a larger number of synapses should be left functional after anticholinesterase injection when we test an overtrained habit.

In the first experiment done to permit us to choose between these possibilities we used two groups of animals. The first group was trained to criterion and took a mean of 44.5 trials to criterion. The group therefore received a total of 54.5 trials in initial learning. The group of 10 rats was then injected 14 days later with the same dose of DFP and in precisely the same way as has already been described. On retest, the rats in this group took a mean of 44.2 trials to criterion. It can be seen that here was a considerable loss of memory. A second group was given only 30 trials on the same habit during initial training and was therefore given approximately half the practice of the first group. Very little learning was evident by the thirteenth trial. Fourteen days later the rats in this group were injected with DFP in the same way as the first group. On retest 24 hr later they took a mean of 17 trials to relearn, significantly different beyond the 0.005 level (two-tailed test) from the retest score of the first group. In fact, the total score to criterion—adding the 30 trials of initial training and then the trials to criterion on retest (47)—is hardly higher than the initial score to criterion of the first group (44.5). It can safely be concluded that the group with low initial training suffered less of a memory impairment than the group trained to a good criterion. This result favors the hypothesis that synaptic modification during learning is a graded process and that increased learning is stored by an increase in capacity to transmit at a particular set of synapses. A higher level of transmission during activity should be more vulnerable to anticholinesterase.

A more thorough study in which we employed different degrees of training and baseline peanut oil control groups at each level of training was carried out to study this phenomenon further (Deutsch and Lutzky, 1967). The task used was the same as in the previous studies except that the light used to illuminate the lit alley was made dimmer to make the task more difficult and thus to make graded levels of training easier to achieve. The first two groups were trained for 30 trials only. One group was then injected with the dose of DFP five days after initial learning and the other, with peanut oil. At retest 24 hr after injection the peanut oil group scored 15.6 percent lower in the first 10 trials than in the last 10 trials of the initial 30, whereas the DFP-injected group scored 29.6 percent better during the first 10 trials than in the last 10 of the 30 trials during initial training (Table 5).

In spite of the many instances in which the same DFP injection caused either amnesia or no change, it was logically possible that the result in this instance was due to faster learning under DFP than peanut oil during retest or to an enhanced attractiveness of the light due in some way to the injection. To investigate this possibility two large groups of animals were injected, one with DFP and the other with peanut oil, and tested for rate of learning 24 hr after injection. The rates of learning of the two groups were almost identical, ruling out the possibility that some response bias or enhanced learning rate could be responsible for the result.

The conclusion that there was no enhancement for the attractiveness of the light is supported also by the results of the next two groups which

TABLE 5. Degree of Facilitation or Amnesia Depending on Degree of Initial Training of Habit*

Group	*N*	No. of trials initial training	Substance injected	Mean correct in last 10 before injection	Mean correct in last 10 after injection	First 10 to last 10
1	12	30	DFP	6.75	8.75	+2.00
2	12	30	PO	8.00	6.75	−1.25
3	12	70	DFP	8.43	8.52	+0.09
4	12	70	PO	8.25	7.92	−0.33
5	12	110	DFP	9.75	7.50	−2.25
6	12	110	PO	8.92	8.00	−0.92

**All rats were injected five days after initial training. Task was made difficult by dimming light in safe alley.*

were trained to 70 trials. One was then injected with peanut oil and the other, with DFP. Only small and insignificant difference between the two groups emerged on retest. This again makes it unlikely that response bias or enhanced rate of learning accounted for the earlier result in this experiment.

In a third part of the experiment rats were trained 110 trials in initial learning. Again, five days after initial training one group was injected with peanut oil and the other with DFP. Here the DFP injected animals showed worse retention than animals simply injected with peanut oil, and less than animals trained only for 30 trials and injected with DFP. Such results fully support the notion that the amount of transmission at a synapse at a certain time after learning is a function of amount of initial learning.

Similar results have been obtained by Leibowitz et al. (unpublished data) with the use of habits of varying difficulty but with the number of trials kept the same. The easy habit was, of course, well trained and suffered a block. The poorly trained habit was hardly learned at all but showed large facilitation after anticholinesterase injection.

So far the hypothesis that the drugs we are using are acting at a synapse is supported by the fact that all the three drugs (DFP, physostigmine, or scopolamine) have the same or reverse actions on the phenomenon we are studying. Another way to show that the phenomena observed are synaptic in origin is to take advantage of Bacq and Brown (1936) that at a neuromuscular junction treated with physostigmine the higher the rate of neural stimulation the greater the depression of muscle contraction. This is due to the fact, mentioned above, that physostigmine in submaximal dosage inactivates only some of the acetylcholinesterase. The remaining acetylcholinesterase takes a longer time to destroy acetylcholine.

If stimuli are repeated quickly, then each stimulus arrives before the last emission of acetylcholine has been destroyed. Acetylcholine will then accumulate to produce a synaptic block. But if stimuli are spaced so that acetylcholine from the last stimulus is cleared up before that from the next stimulus arrives no block takes place. If we are right about the explanation of memory block with physostigmine in terms of synaptic block, we should observe an analogous phenomenon in anticholinesterase amnesia. We should expect amnesia when trials during retest under physostigmine are massed, but retention when they are spaced.

Four groups of rats were trained on a shock escape task to a criterion of 10 successive correct trials (Rocklin and Deutsch, unpublished data). Seven days later, two groups were injected with physostigmine and two with saline. One of the physostigmine and one of the saline groups were retested with 25 sec between trials. The other two groups were retested with 50 sec between trials. There were 16 rats per group. (One half of each group had been originally trained at 25-sec intervals and the other at 50-sec intervals.) The results are presented in Table 6A. The physostigmine massed group took 31.8 trials to criterion. All the other groups took less than nine trials. The physostigmine spaced group took 4.8 trials, thus showing retention as good as or slightly better than the two saline groups. There were no significant differences between the saline and spaced physostigmine group scores. When these groups are combined and compared with the physostigmine massed group, the difference is highly significant ($p < .001$, $f = 48.52$, d.f. = 1/60).

To make sure that the result was due to amnesia and not to some effect of performance we altered the retest condition. In the design described above it is possible that rats retrained with massed trials under physostigmine are somehow disabled. This could be due to some inability

TABLE 6 (A, above; B, below). Retest Trials to Criterion Under Massed and Spaced Physostigmine or Saline Injection*

	Saline injection	Physostigmine
Massed (25-sec interval)	8.1	31.8
Spaced (50-sec interval)	7.2	4.8
Massed (25-sec interval)	58.6	34.4
Spaced (50-sec interval)	49.7	51.7

**A = Mean trials to relearn original habit (N = 16 per cell); B = Mean trials to learn reversal of original habit (N = 16 per cell).*

to recover between trials when these are closely spaced and not due to an inability to remember. However, if we make the rats reverse during retraining the two explanations of the initial result can be separated. If the rats remember they take longer to learn the reversal habit than if they had not been taught the original habit. If the massed physostigmine group remembers and is somehow disabled its score should be even higher than that of the saline groups. On the other hand, if the massed physostigmine group is amnesic and not disabled, then it should take fewer trials to learn the reversal than the saline groups because the reversal to this group should be like an entirely new habit. The design of the experiment was entirely similar to the experiment in which retest was on the original habit. The scores are presented in Table 6B. It is clear that the massed physostigmine group reached criterion in a much smaller number of trials than the other three groups. The result therefore confirms the hypothesis that we are dealing with an effect on memory rather than on performance. The whole experiment, taken together, supports the notion that we are blocking memory by interfering with transmission across a cholinergic synapse by producing an accumulation of acetylcholine which dissipates with time.

As we have seen above, the anticholinesterase drugs physostigmine and DFP produce amnesia for habits that are seven days old but not for habits that are three days old. It is possible to apply this property to an analysis of extinction. In extinction a previously rewarded habit becomes unrewarded and the animal stops performing the habit. Such a cessation of responding can be regarded as a weakening by nonreward of the bond or connection underlying the habit, or it can be viewed as the acquisition of an opposing behavior tendency or counter-habit. Such a counter-habit can be thought of as increasing in strength after each nonrewarded response until it equals in strength and thus cancels out the previously rewarded habit. Such a theory was proposed by Miller and Stevenson (1936).

To test these hypotheses, we trained rats seven days before an injection of physostigmine and retest. In this way amnesia for the original habit should always result. However, we varied the time between training and extinction. If training and extinction are separate habits, then extinction three days before injection and retest should lead to an amnesia for the original habit learned seven days previously but leave the memory for the extinction intact. We should, therefore, expect rats in this situation to take longer to learn in the retest than animals that had been taught seven days previously, never extinguished, and then retested after an injection of physostigmine. Both groups would have forgotten the original habit, but the group extinguished three days earlier would remember the extinction session. The effects of this memory of extinction would have to be overcome during retraining. An increase in the number of retraining trials would be the prediction of the theory of extinction which regards the event as the acquisition of an opposing habit. What of the prediction of the theory that regards extinction as a weakening of the original habit? Here we would expect that extinction at any time after training within the seven days should actually facilitate relearning under physostigmine. It has been shown (Deutsch and Lutzky, 1967; Leibowitz et al., unpublished data) that weak habits and almost forgotten habits (Deutsch and Leibowitz, 1966; Wiener and Deutsch, 1968) are actually extremely well facilitated by the same dose of acetylcholinesterase which blocks the retrieval of a well-learned habit.

The *Ss* were trained, extinguished and retained in a Y-maze. Centered and 3 inches from the end wall of each arm was a small plastic cup mounted on a metal platform. An automatic pipeting device was connected with plastic tubing to each cup. *E* could deliver .05 cc of a 20 percent sucrose solution per reinforcement.

Three days after being received at the laboratory the rats were placed on a 23-hr water-deprivation schedule which was maintained for the duration of the experiment.

Five days after being placed on the deprivation schedule *S* was introduced into the maze with 22-hr thirst. After 10 min of adaptation to the unlit maze one arm was lit and .05 cc of sucrose solution was available in that arm. *S* was allowed up to 30 min to initiate drinking although typically rats drank within 1 or 2 min of light onset. The arm remained lit for 8 sec after licking commenced, then another arm was lit and sucrose solution was available there. This procedure continued until *S* had run toward the light 10 successive times without making an error; learning was then assumed to have occurred and the training session was concluded.

Ss were extinguished immediately after reaching criterion or after one, three, five, or six days in the home cage *Ss* were run as in the training session except that they received no reinforcement in the lit alley. However, instead of extinguishing the light in the lit alley and lighting another 8 sec after licking commenced, this was done 8 sec after *S* entered the final 9 inches of the lit arm. When the extinction criterion of not entering the lit alley within 2 min was reached, *Ss* were removed from the maze and returned to their home cage until Day 7.

On Day 7 *Ss* were intraperitoneally injected with .25 cc of a 0.05 percent physostigmine salicylate solution and 20 min later were reintroduced into the maze. Control *Ss* were injected with the same volume of normal saline but otherwise were identically treated. Except for the reversal groups, *Ss* were retrained following the same procedure as in original training. Retention loss was measured by the number of trials to reach criterion on retest.

The reversal groups were retested in the same manner as the other experimental groups except that instead of the sucrose reinforcement's being available in the lit arm, with the other two remaining dark, the reinforcement was in the unlit alley, the others remaining lit. Retention loss was measured by the number of trials to criterion. In this situation retention loss is greater if it results in fewer trials to criterion.

Preinjected controls were run to determine drug effects, if any, on learning. Two control groups were trained and seven days later injected and retrained or reversed to measure the amnesic effects of the drug on ordinary retest and reversal performance. The results are presented in Tables 7 and 8. As has been found in other experiments, rats injected with physostigmine seven days after training show considerable memory deficit. Retraining in this experiment took a mean of 21.6 trials to criterion and reversal took 40.8 trials to criterion. The analysis of variance of the experimental group relearning measure indicated a significant training-extinction or extinction-retraining interval effect ($p < .01$, $f = 3.67$, d.f. = 5/78). Duncan's multiple range test showed that the four-day extinction group was significantly worse on retest than the no-extinction drug group and the groups extinguished on Days 1 and 2.

Analysis of variance of the experimental group reversal data reveals that the difference among the groups was significant ($p < .05$, $f = 3.63$, d.f. = 2/21). The difference between the no-extinction drug group and the Day 4 group was significant ($p < .002$, $v = 6$, Mann-Whitney test).

It must be noted that control groups subjected to the extinction procedure showed no effects of this during retraining or reversal. All control groups demonstrated near-perfect retention of the discrimination habit.

The possibility that amnesia can be induced for the reinforcement discrimination habit and not for the nonreinforced extinction learning was tested by injecting physostigmine at a time when the drug would normally disrupt memory for the conditioning session and minimally disturb the memory for the nonreinforcement experience. Previous research has indicated that the anticholinesterases block memory seven or 14 days after original learning while the same drug produces little or no amnesic effect if injected one to three days after learning.

The fact that extinction one and two days before retraining or reversal produces a behavioral effect suggests that extinction is also subserved by cholinergic synaptic modification. In this experiment it was also possible to induce amnesia for both what was conditioned and the effect of nonreinforcement. Amnesia for the conditioned habit was induced while the memory subserving extinction responses remained operative. Although there was no nondrug behavioral evidence that extinction had occurred, manipulation of the interval between extinction and retraining or reversal produces a behavioral effect if the rat is injected with physostigmine.

TABLE 7. **Trials to Criterion for Retraining Groups***

Day of extinction training	Retraining-injection	Training (trials to criterion)		Extinction (trials to criterion)		Retraining (trials to criterion)		
		M	Mdn	M	Mdn	M	Mdn	S.D.
†	Physio.	33.5	30.0			21.6	20.5	13.3
1	Physio.	36.0	36.5	16.9	14.5	21.6	22.5	13.3
2	Physio.	36.4	38.5	21.1	21.0	21.9	24.5	14.4
4	Physio.	36.3	35.5	22.2	22.0	42.9	44.0	16.5
5	Physio.	35.2	35.5	28.6	29.5	34.9	42.0	16.9
6	Physio.	36.1	37.5	27.4	25.0	28.6	27.5	15.2
1	Saline	37.5	37.5	15.8	15.0	8.0	8.5	7.2
4	Saline	33.3	34.5	26.4	26.5	4.5	5.0	4.4

**In each group, $N = 14$.*
†No extinction.

TABLE 8. Trials to Criterion for Reversal Groups*

Day of extinction training	Injection for reversal training	Training (trials to criterion)		Extinction (trials to criterion)		Restraining (trials to criterion)		
		M	Mdn	M	Mdn	M	Mdn	S.D.
†	Physio.	34.6	32.5			40.8	40.5	12.4
2	Physio.	34.9	34.5	24.8	25.5	33.8	43.5	19.8
4	Physio.	31.2	34.0	28.0	26.0	21.9	21.5	7.4
†	Saline	33.9	35.5			69.5	72.0	4.5
4	Saline	34.6	33.0	31.0	29.5	70.2	72.0	3.3

**In each group, N = 8.*
†No extinction.

The main results confirmed those of a substantial pilot study and strongly support the theory that during extinction a separate habit is learned. The results of the reversal experiment indicate something about the nature of the habit learned in extinction. Reversal is actually hastened so that it seems as if some aversion to the previously rewarded habit is acquired.

Discussion

The experiments described above point to a synaptic locus of memory. Let us summarize some of the findings which lead us to that conclusion.

1. Synaptically acting drugs (DFP, physostigmine, scopolamine) can block memory.
2. The effect of such a drug varies with the age of the memory and on its strength (as manipulated by amount of initial learning).
3. Anticholinesterases can, under select conditions (e.g., low amount of initial learning, forgetting of habit), selectively facilitate memory in the same dose as they block it under different conditions (e.g., high amount of initial learning).
4. Physostigmine produces memory block only when retest trials are close together.

Looking at this picture it would seem to be an unlikely coincidence if the situation were produced by some nonsynaptic action of the drugs. Not only do these various drugs converge in their effects in the manner to be predicted from their synaptic action but also the type of action observed looks like a synaptic action. For instance, the effect of spacing on recall under physostigmine is just what would be expected from the drug's action at a synapse. Nor can the effects be explained by some general effects which might produce amnesia by producing some type of malfunction. Such an idea might account for the observed amnesia. However, the facilitation of memory obtained by the same dose of drug which, under other circumstances, produces amnesia seems inconsistent with the idea of a general malfunction. This is especially so because it is the "weak" habits which are facilitated. A general malaise or malfunction would be expected on general grounds to eliminate the "weak" habits rather than the strong habits. On the other hand, such facilitation is directly predictable from the synaptic action of anticholinesterases on poorly conducting synapses.

If the action of the drugs is indeed synaptic then we may infer that there is a cycle of conductance of the memory synapse. When such a synapse is first activated through learning, conductance is at first relatively poor. There is a relative absence of block by anticholinesterases and block by anticholinergics. (This is if we exclude what happens 30 min after learning.) After this there is an increase of conductance with time, peaking at 7 to 14 days after learning. The precise position of this peak in time probably depends on the amount of initial learning. After this peak there is a gradual decline of conductance, leading to behavioral forgetting. When conductance has declined sufficiently, transmission across the synapse is facilitated by anticholinesterases. Besides the time of occurrence of the peak, the height of the peak is determined by the amount of initial learning. This is indicated by the fact that, with the same dose of anticholinesterase, facilitation occurs for poorly trained habits and amnesia is manifested for well-trained habits. A simple explanation can be given for a rather complex and surprising set of results if we assume that variation in synaptic conductivity forms the substrate of memory.

Various theories can be advanced to explain the data. The first has already been dealt with—namely, that the drugs affect directly the synapses which are modified through learning. The second alternative is that the effects observed are not due to the synaptic effect of the drugs used and that no real conclusions can be drawn, except that the substrate of memory changes with time and amount of initial learning. The fact that different pharmacologic agents all have effects on memory predictable from their synaptic effects would, on this alternative, have to be dismissed as a fantastic coincidence.

A third explanation which is available is that the pharmacological effects are indeed synaptic but the synapses affected are not concerned with memory storage. Instead, these target synapses interact with the true substrate of memory. This true substrate of memory does have different stages connected with time since learning and amount of learning. A uniform synaptic change produced by a particular type of drug then interacts with the different stages of the true substrate of memory to produce either block or facilitation. As anticholinergics cannot facilitate synaptic transmission but only block it to varying degrees, we would have to assume that their effects on the target synapses must be to block them. Further, the effect of such block should then produce memory block depending on the state of the true substrate of memory. As the anticholinesterases produce reverse effects in time, their effect on the target synapses cannot also be a block, because a block cannot produce amnesia and no effect simultaneously. (We are, of course, assuming that the information concerning what agent caused a block is not relayed to the true substrate of memory but only that there is a block at the target synapse.) We would therefore have to assume that anticholinesterases produced facilitation of the target synapses, and that this facilitation produced block of a memory when block of the same synapses by anticholinergics produced no effect on the true substrate of memory. In other words, we would have to believe that there were synapses whose block produced a block of memory by an interaction with a certain state of the substrate of memory. Further, we would have to believe that the facilitation of such synapses would produce amnesia or facilitation of memory depending on the state of the substrate of memory. Such a theory would also have to explain why there was facilitation of the target synapses when trials were close together but no facilitation when trials were far apart. While it is at present impossible to exclude such an explanation absolutely, it is to be noted that its postulations are quite complex and *ad hoc*.

On the empirical level, a theory postulating an indirect effect of synaptic block on some other substrate of memory would, as we have seen, have to postulate that the effect of anticholinesterases is to produce a facilitation of synaptic activity. (If the dose of anticholinesterase used in our experiments produced a block, then the effect of anticholinergics and anticholinesterases on the true substrate of memory would have to be the same.) It is known that increasing the dose of anticholinesterase changes synaptic facilitation to no effect and then to block. We should then expect, as we raise the dose physostigmine or DFP, to observe a lessening effect on memory and then to see an effect on memory similar to that obtained with anticholinergics. While we have undertaken no systematic study of this type, we have frequently had occasion to establish dosage by trial and error as a preliminary to an experiment. No hint of the relation expected on the indirect hypothesis has been seen.

It is evident that pharmacological methods can never produce a direct demonstration or proof of the synaptic substrate of memory. However, such methods can be important in establishing the most plausible hypothesis to guide our further search. In this way the scope of inquiry can be rationally focused. Macromolecular approaches can then concentrate on the most probable areas.

Another question which arises is whether there is any relation between the amnesic effects found with some protein-synthesis inhibitors and cholinergic agents. The work of Burkhalter (1963) and Koenig (1965, 1967) suggests that some at least of these agents (such as puromycin and actinomycin D) have effects on acetylcholinesterase. Dahl and Leibowitz (unpublished data), working in conjunction with the author, found that DFP and puromycin produce similar results when their effectiveness on memories of different age is compared (Deutsch and Deutsch, 1966). It is therefore possible that the work on cholinergic agents and protein-synthesis inhibitors in the area of memory is in fact related.

References

Bacq, Z. M., and Brown, G. C. (1936). J. Physiol., 89:45.
Burkhalter, A. (1963). Nature (London), 199:598.
Cutting, W. C. (1964). Handbook of Pharmacology—The Actions and Uses of Drugs. New York: Appleton-Century-Crofts.
Deutsch, J. A., and Deutsch, D. (1966). Physiological Psychology. Homewood, Ill.: Dorsey Press. ——— Hamburg, M. D., and Dahl, H. (1966). Science, 151:221. ——— and Leibowitz, S. F. (1966). Science, 153:1017. ——— and Lutzky, H. (1967). Nature (London), 213:742. ——— and Rocklin, K. (1967). Nature, 216:89.
Hamburg, M. D. (1967). Science, 156:973.
Huppert, F. A., and Deutsch, J. A. (1969). Quart. Exp. Psychol., 21:267.
Koenig, E. J. (1965). J. Neurochem., 12:343. ——— (1967). J. Neurochem., 14:429.
Miller, N. E., and Stevenson, S. S. (1936). J. Comp. Psychol., 21:205.
Tanzi, E. (1893). Riv. sper. Freniat., 19:149.
Wiener, N. I., and Deutsch, J. A. (1968). J. Comp. Physiol. Psychol., 66:613.

Chapter 14

Autoradiographic Examination of Behaviorally Induced Changes in the Protein and Nucleic Acid Metabolism of the Brain

JOSEPH ALTMAN

Purdue University, Lafayette, Indiana

When a peripheral nerve or central nervous tract is cut, the cell bodies of the severed axons undergo a complex of pathological changes referred to as retrograde degeneration. Among other things, retrograde degeneration is characterized by the dispersion or dissolution of the chromatin material in the soma of nerve cells, called chromatolysis. Chromatolytic changes in neurons are most easily demonstrated histologically with Nissl stains. Soon after the introduction toward the end of the last century of various Nissl stains as an aid for the microscopic investigation of normal, pathological, and experimental brain tissue, it was also discovered that, under certain conditions, chromatolytic changes would occur in nerve cells that were not directly traumatized. Thus, following sectioning of a fiber tract, chromatolytic changes may be observed in those nerve cells which have synaptic relationships with, or receive their signal input from, the severed fiber tract. This is the phenomenon of transneuronal degeneration, which is most easily obtained in cells of the lateral geniculate nucleus some time after removal of the eye or cutting of the optic nerve.

Transneuronal degeneration was generally interpreted to represent cellular atrophy produced by disuse. Accordingly, the question arose immediately whether the Nissl staining technique would be suitable for revealing the "functional state" of neurons. Will, for instance, the functional "deafferentation" of the eye by blindfolding or its overstimulation with light produce differential effects in the stainability of retinal ganglion cells? The problem was studied around the turn of the century by several investigators (Mann, 1895; Bach, 1895; Birch-Hirschfeld, 1900; Carlson, 1902). Unfortunately, the results were either unreliable or highly controversial, and to this day the problem has remained unresolved (see Bech,

1957). Since the Nissl staining technique may not be sufficiently reliable or sensitive to reveal the more delicate "functional" changes of neurons, the entire problem has been reinvestigated recently with various modern techniques. The chromatin material reacting with Nissl dyes is now identified with the RNA- and ribosome-rich endoplasmic reticulum, the major site of cellular protein synthesis. Accordingly, attempts have been made to determine whether changes occur in the concentration or metabolic turnover of nucleic acids and proteins in nerve cells as a consequence of physiological and behavioral manipulation of experimental animals. Most notable among such efforts has been the continued work of Hydén (1962), who has used a variety of biophysical and biochemical techniques (including ultraviolet microspectroscopy, microincineration, and microchemical analyses of single cells) for this end.

Similar considerations have also prompted our use of fine-resolution autoradiography to study changes in the protein and nucleic acid metabolism of the brain as a consequence of physiological or behavioral manipulations. In these studies, to be described below, animals were injected with radioactively labeled precursors of proteins and nucleic acids and the distribution and rate of utilization of these tagged precursors in the central nervous system, both under normal and experimental conditions, were determined with the aid of autoradiography.

The Technique of Autoradiography

In the simplest application of autoradiography, a photographic plate is brought into contact in the dark with a material containing radioactive isotopes and the photographic emulsion is exposed to bombardment by the particles emitted by the material. The radioactive particles produce a latent image—that is, they reduce the silver grains lying in their path in the emulsion—in the same way as light rays do. After an appropriately selected exposure period, and following photographic development of the plate, the site and concentration of the radioactive particles is revealed in the form of blackened grains in the emulsion. In the technique's modern biological application (Bélanger and Leblond, 1946; Kopriwa and Leblond, 1962) glass slides with histologically prepared tissue sections are dipped into a vessel containing melted, specially designed nuclear emulsion. The emulsion is dried, which forms a thin film over the tissue section, and in this way intimate contact is established between the two. Since the tissue section and the nuclear emulsion adhere to one another, the two can be examined simultaneously under the microscope, and the location of the radioactive sources with respect to the regional components of the tissue is easily determined.

Perhaps the greatest impetus to the development of modern autoradiography was given by the introduction of tritium (^3H) as an isotopic tracer of chemicals. A simple and relatively inexpensive technique (Wilzbach, 1957) is available for the tagging of a variety of organic chemicals with tritium, and due to the optimal half life of tritium (12.3 years), radiochemicals of high specific activity (high concentration of labeled over unlabeled molecules) are easily obtained. Moreover, the weak beta particles or electrons emitted by tritium penetrate only about 1 to 2 μ into the photographic emulsion, revealing with that degree of accuracy the regional source of radiation; hence the term fine-resolution autoradiography.

Figure 1 shows an unstained autoradiogram of a brain section from a rat injected intraperitoneally with a tritium-labeled amino acid (2 millicuries of leucine-^3H dissolved in isotonic saline; specific activity 3.6 curies/millimole; amount of radiochemical injected 0.094 mg). The animal was killed two hours after the injection by cardiac perfusion with neutral formaline. (Due to this procedure the free or unbound amino acid is washed out and the amino acid incorporated into proteins is preserved as a consequence of the denaturation of protein by formaldehyde.) The brain was subsequently embedded in paraffin and processed for cutting. The sections were left unstained and, after removal of the paraffin, were coated with Ilford G-5 nuclear emulsion. The dried, coated sections were exposed for five months in sealed lightproof boxes at 5°C and were developed in the usual manner. The great regional differences in the binding of the administered amino acid are indicated by the variable grain concentration or density over different brain regions. Regions rich in nerve cells show much higher "protein metabolism" than is seen in fibrous tracts, but there are also great differences in the grain density over neurons in different nuclei and of different cells with a single nucleus.

Grain concentration over a given tissue area is directly proportional to the concentration of radioactive particles in that region. Therefore, a quantitative estimate of the regional utilization of the tagged precursor can be made by counting under a microscope the number of blackened grains in that region. However, in most cases this is a very laborious and time-consuming task. An easier and more reliable method of estimating grain density is by photometry or densitometry, where optical absorption by the opaque grains is used for the determination of the concentration of such grains in the path of a regulated light beam. With its aid we can easily and accurately establish the grain density of any selected brain region, including that of single neurons. Figure 3 shows the relationship found between visual grain counting and photometric density in a section coated with Kodak NTB-3 nuclear emulsion. Table 1 gives the grain density of a few selected structures from one of our early experiments (Altman, 1963a).

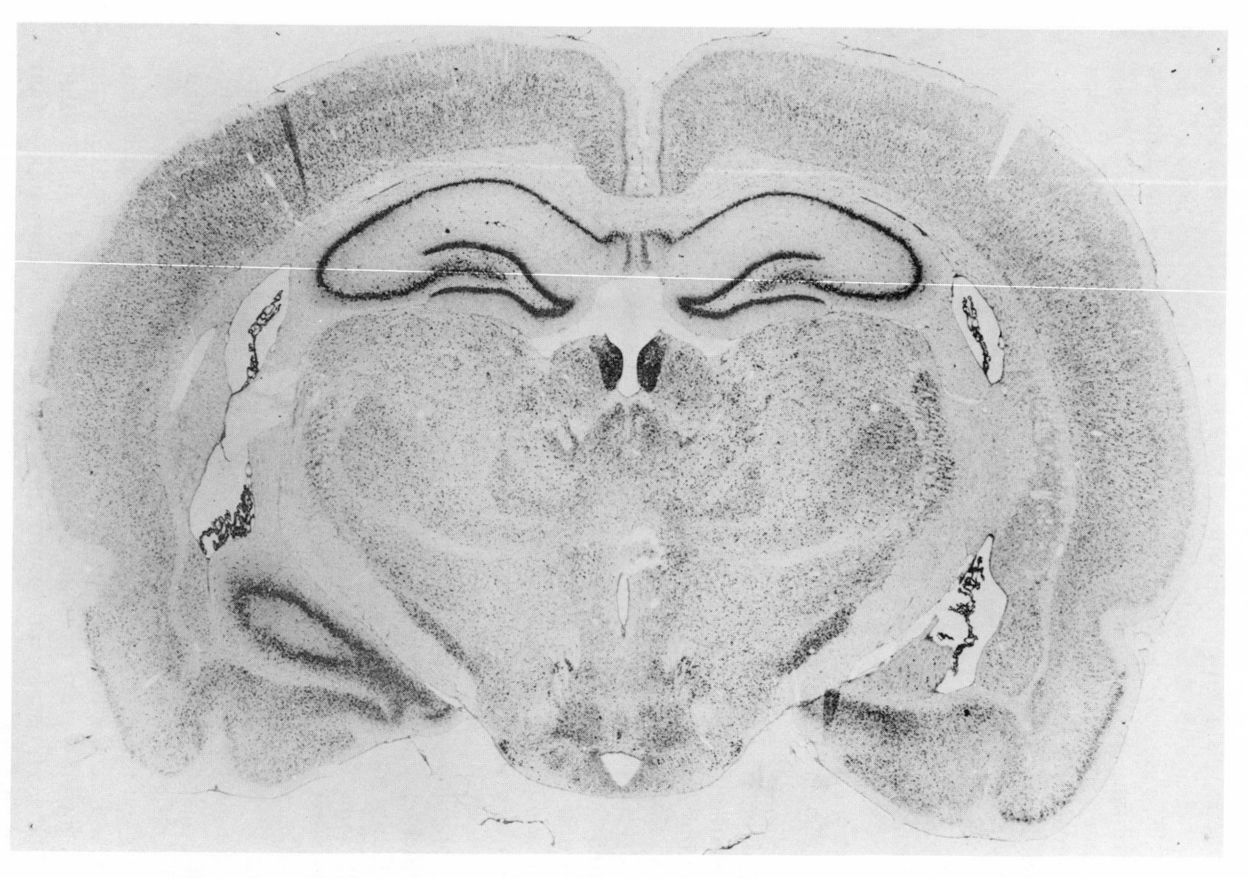

FIG. 1. Low-power photomicrograph of an unstained autoradiogram of a section of the rat brain. The animal was injected intraperitoneally with leucine-^{3}H.

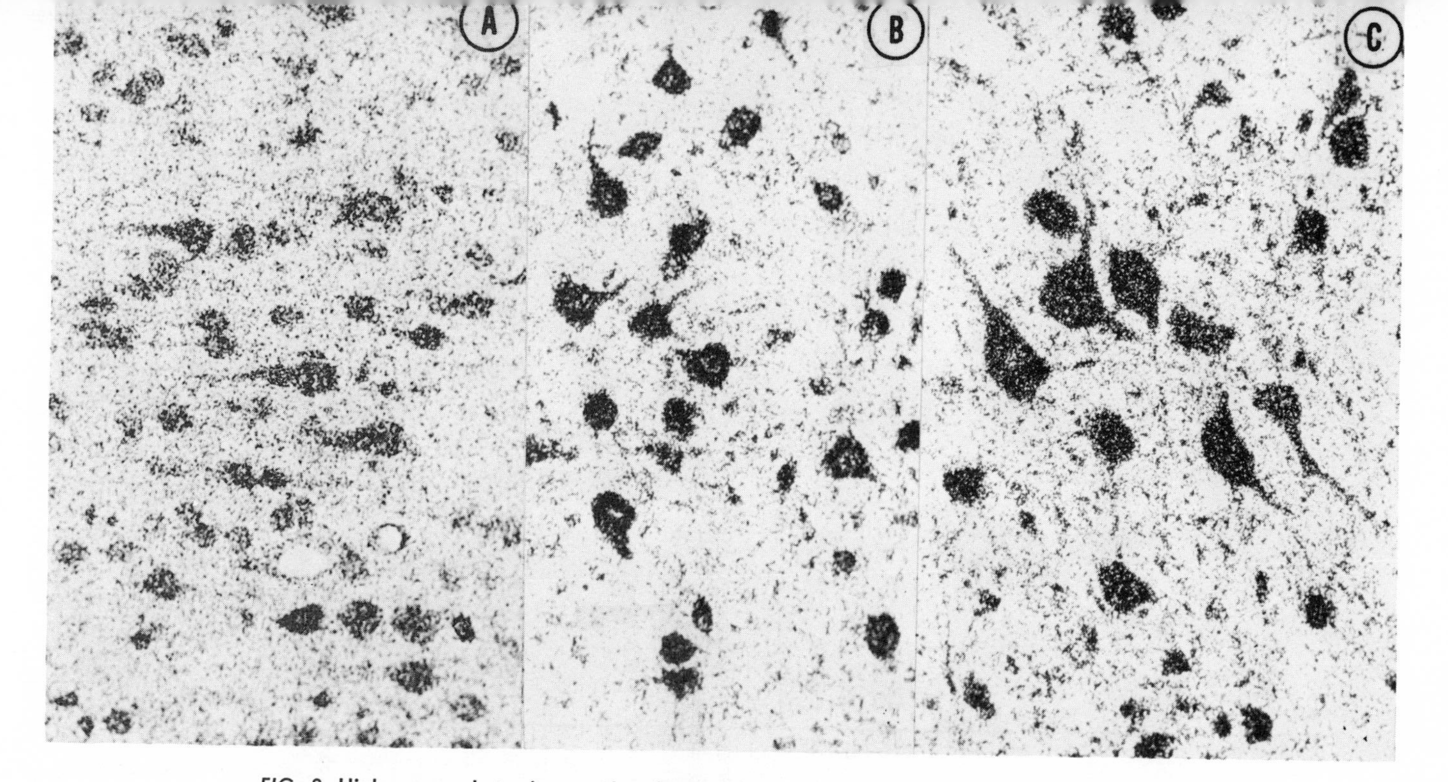

FIG. 2. High-power photomicrographs of autoradiograms of the rat brain. A. Pyramidal cells of the motor cortex. B. Cells of the dorsal cochlear nucleus. C. Ventral horn cells in the spinal cord. Note differences in grain density over different cells in different brain regions and also the differences in grain density over the neuropil.

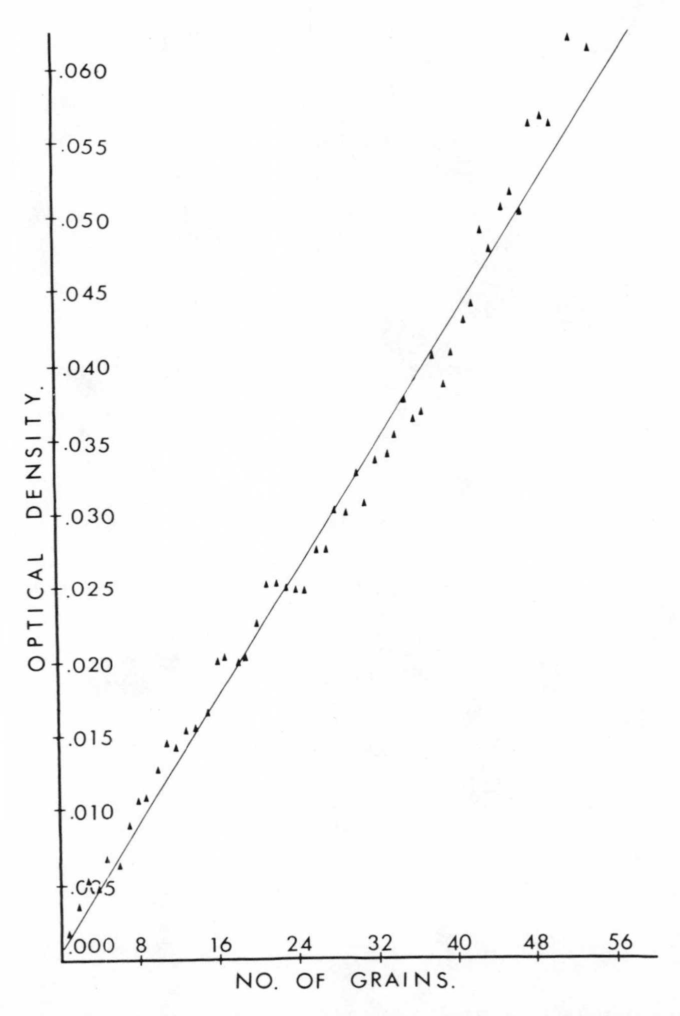

FIG. **3. Relationship between visual grain counting and optical density measurements for Kodak NTB-3 nuclear emulsion. All points represent the means of 10 measurements.**

Changes in Protein Metabolism

Proteins constitute about one-half of the dry weight of brain tissue, and nerve cells show an exceptionally high rate of protein metabolism. This high protein concentration, and the very rapid turnover of many of the nerve proteins, is thought to be due to the intense work output of

TABLE 1. The Optical Density of Autoradiograms in Several Brain Regions on the Right and Left Side, within and between Sections.*

Region	Type	Sections, Group A		Sections, Group B	
		Right	Left	Right	Left
Cord, white matter	fiber tract	.05	.06	.05	.07
Cortical radiation	fiber tract	.07	.07	.06	.06
Fimbrium of hippocampus	fiber tract	.07	.08	.08	.09
Corticospinal tract	fiber tract	.08	.08	.10	.09
Corpus callosum	fiber tract	.08	.08	.08	.09
Superior colliculus	neuropil	.09	.09	.08	.08
Periaqueductal gray	neuropil	.09	.09	.09	.10
Lateral geniculate nucleus	neuropil	.11	.11	.15	.11
Cortex, 5th layer	neuropil	.13	.13	.15	.16
Cerebellum	granular layer	.14	.09	.11	.13
Pontine nucleus	neuropil	.25	.25	.19	.20
Third nerve nucleus	neuropil	.26	.27	.24	.26
Cord ventral horn	neuropil	.27	.21	.20	.18
Striate cortex, 2nd layer	neuropil	.28	.28	.30	.29
Cerebellum	molecular layer	.32	.30	.28	.30
Ventral cochlear nucleus	neuropil	.30	.33	.19	.24
Superior colliculus	neurons	.38	.39	.43	.40
Cortex, 2nd layer	neurons	.44	.42	.49	.46
Hippocampus, granule layer	neurons	.47	.44	.47	.49
Lateral geniculate nucleus	neurons	.49	.47	.63	.44
Medial habenular nucleus	neurons	.56	.58	.52	.56
Hippocampus, pyramidal layer	neurons	.68	.67	.66	.66
Cortex, 5th layer	pyramidal neurons	.69	.57	.62	.77
Pontine nucleus	neurons	.94	.96	1.12	1.04
Cord ventral horn	motor neurons	.98	.90	1.04	1.02
Third nerve nucleus	neurons	1.00	.98	1.03	.99
Cerebellum	Purkinje cells	1.11	1.11	.98	1.04
Reticular formation	neurons	1.29	1.30	.94	.98
Ventral cochlear nucleus	neurons	1.47	1.39	1.13	1.24

**All density figures represent the mean of 10 measurements in circular areas 12 μ diameter.*

nerve cells. If this were the case, it could then be further assumed that alterations in the level of neural activity would be associated with changes in the rate of protein metabolism and, thus, the demonstration of such alterations could be used as indicators of the "functional state" of neurons. In our first experiment, carried out in 1959 (Altman, unpublished data), we occluded one eye in a group of 10-day-old chicks and the animals were then placed for varying periods (ranking from 2 to 18 hr) in an experimental box furnished on all sides with continuously flickering light bulbs. The animals were injected during the stimulation sessions with leucine-^{3}H intraperitoneally and were killed after different periods (ranging from 10 min to 24 hr). Since in birds there is a total decussation of the optic tracts, unilateral visual stimulation could be expected to produce differen-

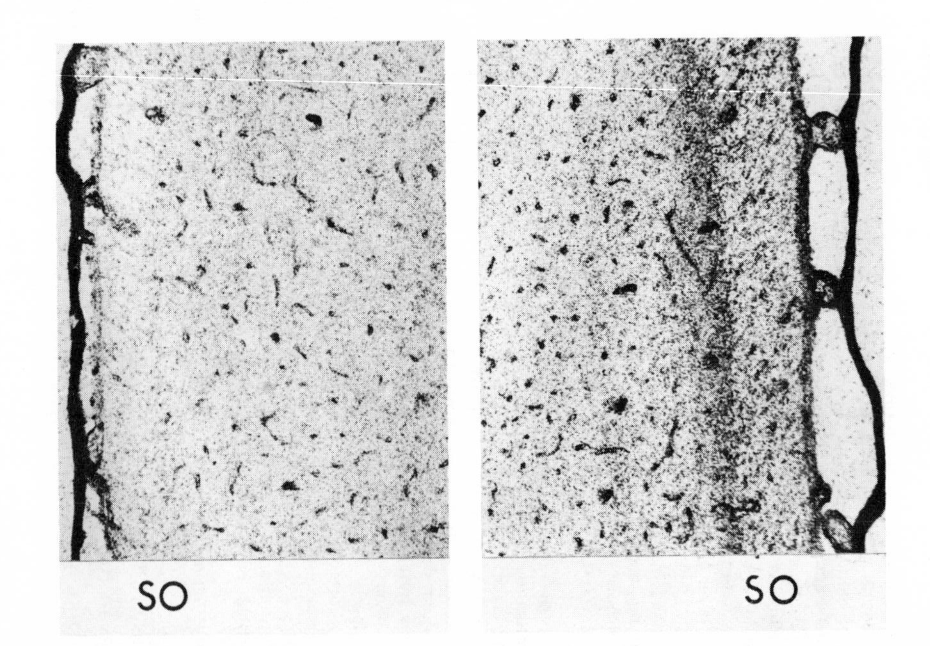

FIG. 4. Unstained autoradiograms of the superficial layers of the optic lobes of a pigeon. Left, normal optic lobe. Right, optic lobe contralateral to the removed eye. Note highly increased uptake of glycine-^{3}H in the stratum opticum (SO) undergoing degeneration.

tial rates of protein metabolism in the two optic lobes. However, in these experiments no differences could be detected in the autoradiographic grain density of the various layers of the "stimulated" and "unstimulated" optic lobes in any of the animals studied. In a subsequent study, we occluded one eye of adult pigeons for one month and injected the animals with glycine-^{3}H. Again, we failed to obtain differences in the stainability or autoradiographic grain density of cells in the two optic lobes (Altman and Altman, 1962).

In the latter study we also removed one eye in several animals to investigate the effects of morphological rather than functional deafferentation. With this more drastic procedure we obtained considerable differences in the optic pathways in the rate of incorporation of the injected glycine-^{3}H. There was a great increase in grain density in the severed optic nerve and tract, and in the contralateral stratum opticum of the optic lobe, the layer which contains the incoming optic tract fibers (Fig. 4). That is, there was increased "protein metabolism" on the "deafferented" side of the optic system. This seemingly paradoxical finding was clearly related to a great pathological increase in glia cells in the severed optic pathway. Indeed, no differences could be detected in grain density over the deeper layers of the optic lobes. In a similar later study (Altman and Das, 1964a) we removed one eye in a group of rats and after varying periods of postoperative survival the animals were injected with leucine-^{3}H. Again, a considerable increase in the utilization of the radiochemical was established microdensitometrically in the severed optic nerve and tract, but no differences were obtained in grain density over single cells of the lateral geniculate nucleus as long as four months after the deafferentation procedure (Fig. 5).

In another study (Altman et al., 1966), rats at the age of six weeks were divided into two groups. One group was reared for 10 weeks under conditions of relative visual deprivation while the other, in addition to exposure to normal visual stimulation, underwent intense training for an hour every day on a series of visual pattern discrimination tasks. The two groups of animals were then injected with leucine-^{3}H and killed one hour later. Contrary to prediction, there was no increase in the incorporation of leucine-^{3}H in neurons of the visual system in the group of animals undergoing intensive visual training; indeed, in at least some of the animals a slight decrease was observed.

The autoradiographic studies so far considered suggest that while radical changes may occur in the rate of amino acid utilization in the brain under pathological conditions, no appreciable changes are produced by such functional manipulations as blindfolding, differential visual stimulation, or intensive visual training. In addition to these studies, we may now consider two sets of experiments in which we did obtain autoradiographic

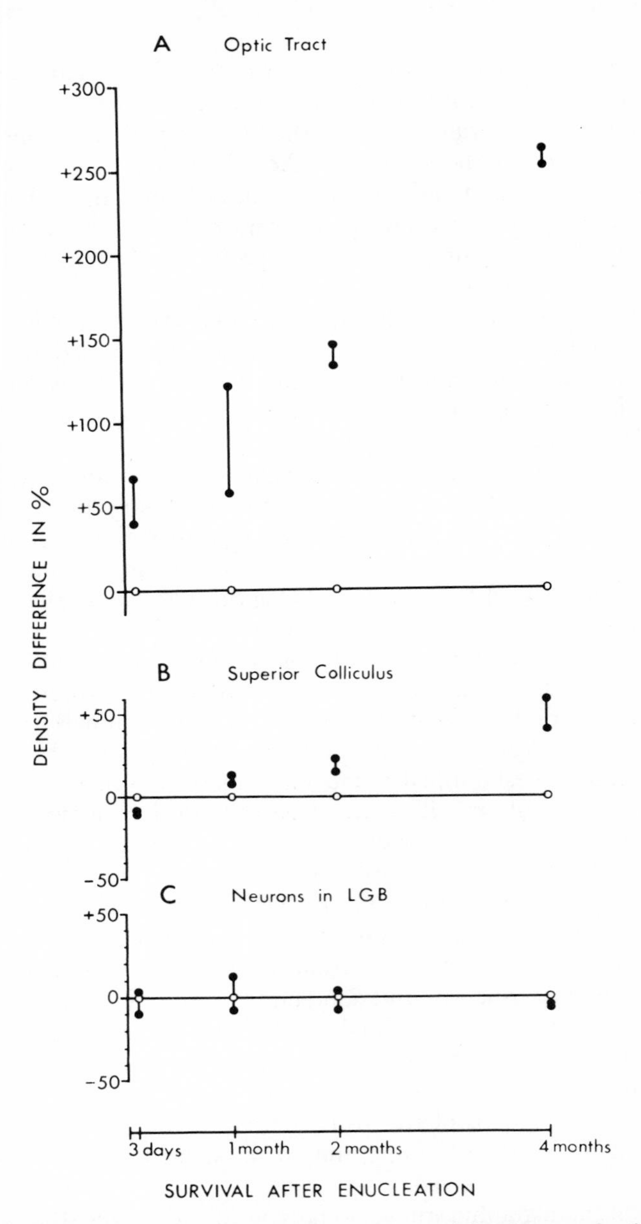

FIG. 5. Diagram of percent changes in grain density following enucleation in three visual structures. A. Optic tract. B. Stratum griseum superficiale of the superior colliculus. C. Single neurons in the dorsal nucleus of the lateral geniculate body. Open circles on baseline indicate the grain density on the "normal" side, considered as zero; closed circles show percent change on the side contralateral to the removed eye. Animals were injected with leucine-^{3}H.

evidence of changes in protein metabolism as a consequence of behavioral manipulations. In the first of these experiments (Altman, 1963b) we placed adult rats into a motor-driven, slowly rotating activity wheel and studied the rate of utilization of leucine-^{3}H in animals injected during exercise, animals injected after exercise, and control animals that were not exercised. Significant increases in the incorporation of the labeled amino acid by single neurons were obtained in the animals injected during exercise. There were considerable differences in this increase over different brain regions (range 19 percent to 66 percent), but all structures were affected to a greater or lesser degree (Table 2).

Since the placement of previously little-handled rats into an activity wheel may have represented a stressful situation to the animals, the obtained increases in protein metabolism in the animals injected during exercise could be attributed to nonspecific autonomic or "arousal" effects. Accordingly, we replicated this experiment (Altman and Das, 1966a) by using animals which were handled from infancy and were raised in an "enriched environment" (see below). In addition, in order to adapt these animals to the test situation, they were, from an early age, exercised daily in an activity wheel. After three months of such treatment the pretrained animals were injected with leucine-^{3}H during exercise and killed. Unlike the case in the previous study, we failed to obtain consistent differences in grain density between the two groups of animals. The latter result would suggest, then, that the increases in the incorporation of leucine-^{3}H

TABLE 2. Mean Grain Density of Single Neurons in Several Brain Structures in the Control Group and the Animals Exercised at 7 RPM and Injected During Exercise

	Control	Exer-cised	% increase	t	P
Motor cortex: neurons in 2nd layer	.185	.307	66	13.6	<.001
Cervical cord: ventral horn, neuropil	.063	.099	57	7.2	<.001
Motor cortex: pyramidal neurons	.282	.421	49	12.6	<.001
Posterior cortex: neurons in 2nd layer	.215	.315	47	10.0	<.001
Cervical cord: ventral horn cells	.476	.671	41	16.2	<.001
Posterior cortex: neurons in 5th layer	.274	.374	36	10.0	<.001
Lumbar cord: ventral horn cells	.512	.693	35	8.6	<.001
Mammillary body: neurons	.331	.442	34	7.4	<.001
Cerebellum: Purkinje cells	.458	.612	34	11.8	<.001
Medulla: large reticular neurons	.512	.674	32	10.1	<.001
Choroid plexus: 3rd ventricle	.333	.426	28	7.2	<.001
Hippocampus: pyramidal cells	.262	.330	26	7.6	<.001
Deiters' nucleus: neurons	.573	.718	25	13.2	<.001
Hippocampus: granule cells	.176	.218	24	6.0	<.001
Cochlear nucleus: neurons	.639	.760	19	6.7	<.001

in the exercised animals obtained in the first experiment may have been due to the "stress" produced by exercise rather than to the engagement of senior motor circuits by the task itself.

In another study (Das and Altman, 1966), rats at the age of about three weeks were weaned and separated into two groups, to be raised differentially in "restricted" and "enriched" environments. The restricted animals were raised singly in small, dark cages, provided *ad libitum* with food and water. The enriched animals were raised communally, males and females together, in a very large cage in which, in addition to normal opportunities of social interaction, they had to engage in increasingly more complex behavioral tasks (searching for food, climbing over ropes, jumping) in order to obtain food and water. Contrary to our prediction of an increase in the protein metabolism of single nerve cells in the brain of the enriched animals, we obtained autoradiographic evidence of highly significant decrease in the incorporation of leucine-^{3}H in various brain structures in the enriched group. Grain density over single neurons in the restricted animals exceeded that found in the enriched animals, ranging from 34 percent (in spinal ganglion cells) to 75 percent (in ventral horn motor cells) (Table 3). Similar differences in grain density were found over the neuropil in gray matter and over some fibrous tracts, such as the corpus callosum. We may mention here that the decrease in protein metabolism in the enriched animals was associated with morphological changes, i.e., an increase in the rate of glial multiplication (see below) and an increase in the size and volume of various brain structures. The latter finding has been reported previously by Rosenzweig, Krech, and their collaborators (Bennett et al., 1964).

In summary we may say that our studies to date do not provide support for the hypothesis that the presumed "activation" of brain structures by behavioral manipulations, as by optic stimulation or visual training, will affect the rate of incorporation of labeled amino acids into proteins in cell systems known to be associated with that function. The only be-

TABLE 3. Mean Grain Density of Single Neurons in Several Brain Structures in the Animals Raised in an Enriched and Restricted Environment

	Enriched	Re-stricted	% increase	t	P
Motor cortex: pyramidal neurons	.232	.328	41	63.13	<.001
Motor cortex: neuropil	.052	.087	66	16.97	<.001
Ammon's horn: pyramidal cells	.197	.313	59	16.01	<.001
Cerebellum: Purkinje cells	.266	.437	64	16.34	<.001
Inferior colliculus: neurons	.278	.437	57	15.53	<.001
Cervical cord: ventral horn cells	.272	.476	75	15.44	<.001
Spinal ganglion: neurons	.302	.404	34	10.63	<.001

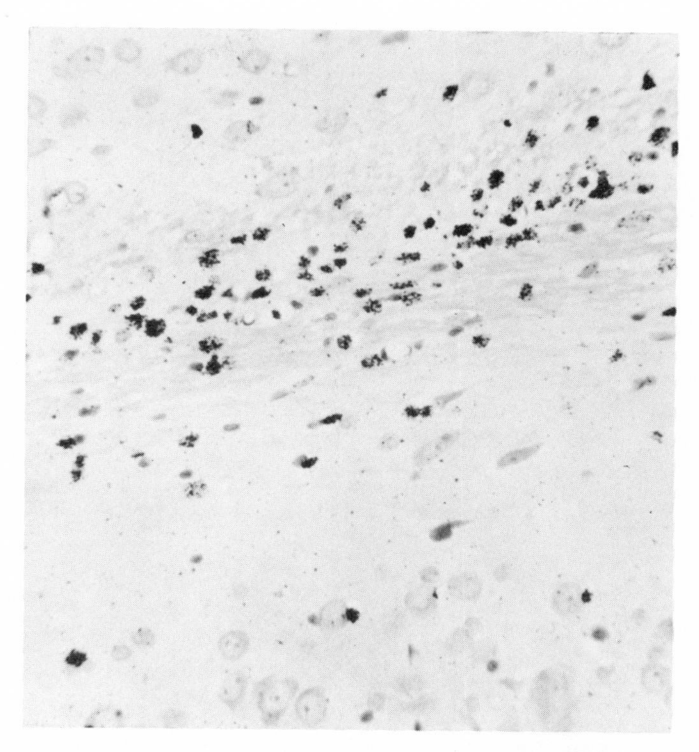

FIG. 6. Gallocyanin chromalum stained section showing labeled glia cells in large number in the cortical radiation of rat injected with thymidine-^{3}H following lesioning of the lateral geniculate hody.

havioral manipulation which could be shown to produce an increase in the protein metabolism of neurons was motor exercise in animals not adapted to the task. Insofar, as this effect was not obtained in the handled animals that were adapted to the task, the increased protein metabolism in the unadapted animals could be interpreted to represent a stress reaction. Such an explanation would also account for the increase in the incorporation of radioleucine in the unhandled, restricted animals in which the injection procedure may have produced a greater arousal effect than in the handled, enriched animals.

Changes in the Rate of DNA Metabolism or Cell Multiplication

It is generally maintained that there is little or no proliferative activity in the central nervous system of adult mammals. Neurons do not appear to multiply in the adult mammalian brain and the multiplication

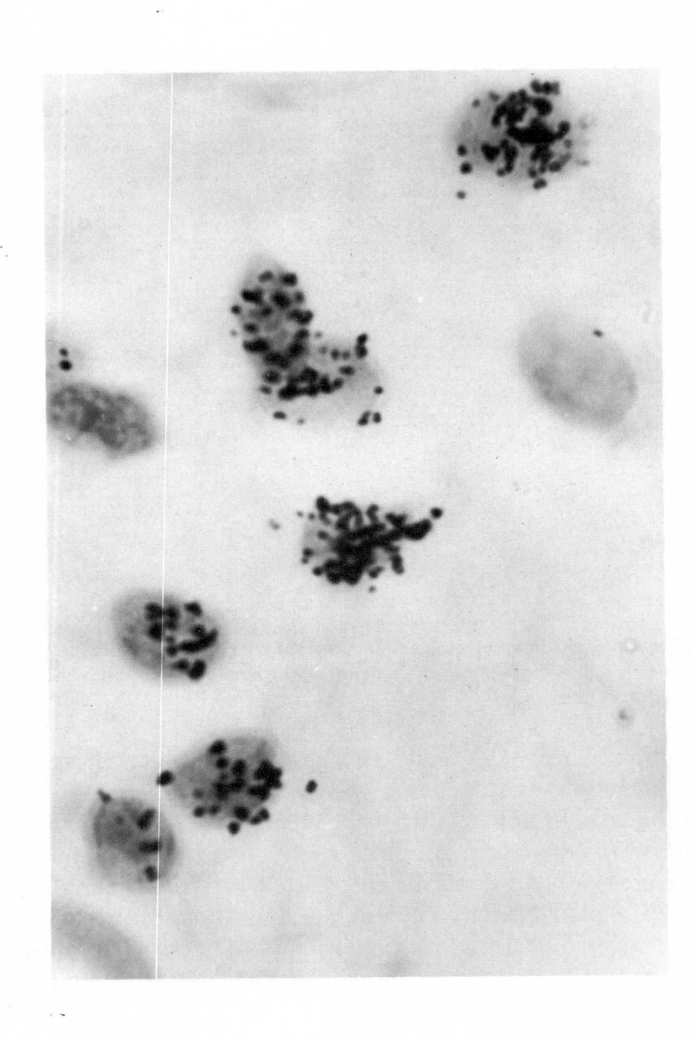

FIG. 7. High-power photomicrograph (oil immersion) of some labeled glia nuclei from the region shown in Figure 6.

of glia cells was for long considered to be essentially a pathological phenomenon. This presumed inactive property of the central nervous system was proved wrong by discoveries made with the use of thymidine-^{3}H autoradiography. Thymidine, a specific precursor of DNA, is known to be utilized exclusively by the nuclei of cells preparing for multiplication (Taylor et al., 1957; Hughes et al., 1958; Leblond, et al., 1959). In a pilot study (Altman, 1962a) we made lesions in the lateral geniculate nucleus of rats and then injected small doses of thymidine-^{3}H into the lesion area. Autoradiographic investigation of this material showed labeling of many glia nulcei in brain areas structurally or functionally associated with the lesion site (Figs. 6 and 7). This indicated induced multiplication of glia cells by the brain lesion. However, since glia nuclei were also found labeled in brain regions not associated with the site lesion, we subsequently injected thymidine-^{3}H into normal adult rats intraperitoneally and into adult cats introventricularly to investigate glial multiplication in the normal brain (Altman, 1963c). This study showed clearly that glia cells proliferate at a low rate at virtually all brain sites in adult animals. We were subsequently able to establish that the rate of glial multiplication is very high in neonates, declines rapidly, but remains at a low appreciable level through adulthood.

Glia cells are considered to be responsible for the myelination of nerve fibers and for the nourishing of neurons. What is the possible functional significance of continued glial multiplication in the adult brain? Could the rate of glial proliferation be used as an index of the "functional state" of given brain regions? In another experiment (Altman and Das, 1964b) we injected adult rats raised in restricted and enriched environments with thymidine-^{3}H, the animals were killed one week after injection, and their brains were prepared for autoradiography. In evaluation of this material, the numbers of labeled glia cells in different brain regions in the two groups of animals were compared. We obtained a highly significant increase in the number of labeled glia cells in the neocortex of the enriched group, with trends of nonsignificant increases in various subcortical structures. As we mentioned earlier, we also found a significant increase in the thickness of the neocortex in the enriched group, which may at least partly be attributed to the increase in the number of glia cells (Tables 4 and 5). A higher number of glia cells in the cortex of enriched animals was also reported by Diamond et al. (1964).

Our autoradiographic studies with tymidine-^{3}H indicated that not only the nuclei of glia cells but also those of occasional neurons may be labeled in the brains of adult rats and cats (Altman, 1962b; 1963c). Since neurons are seldom if ever seen to multiply, this finding appeared paradoxical. In these earlier studies we observed that several granule cells in the dentate gyrus of the hippocampus were always labeled in all sections and

TABLE 4. Mean Number of Labeled Glia Cells in Circular Areas 310μ in Diameter in the Brains of Rats Reared in Enriched and Impoverished Environments

Structure*	Enriched group	Impoverished group	% increase in enriched group	p t test
Neocortex	.44	.28	59	<.001
Amygdala and pyriform cortex	.38	.34	12	n.s.
"Dorsal thalamus"	.28	.24	17	n.s.
Hypothalamus	.25	.25	0	n.s.
Inferior colliculus	.34	.28	21	n.s.
Medulla	.20	.18	11	n.s.

**In all instances the designated structures were scanned in their entirety bilaterally.*

all the animals investigated (Fig. 8). Accordingly, we undertook to investigate specifically the significance of postnatal DNA metabolism in the hippocampus (Altman and Das, 1965a). Rats of different ages (from neonates to mature adults) were injected with thymidine-^{3}H and the animals were permitted to survive for varying periods after the injection (from four days to eight months). In addition, the brains of a large series of non-injected rats of different ages were prepared for histological evaluation.

TABLE 5. Mean Total Number of Labeled Glia Cells at Different Coronal Levels of the Neocortex

Coronal sections of neocortex at level of:	Stereotaxic coordinate (De Groot)	Enriched group	Impoverished group	% increase in enriched group	t-test p
(a) Inferior colliculus	–	57.5	23.8	142	<.10
(b) Oculomotor nucleus	A 1.4	68.8	39.7	73	<.001
(c) Parafascicular nucleus	A 3.0	79.7	38.8	105	<.05
(d) Posterior mammillary nucleus	A 3.4	76.5	43.8	75	<.005
(e) Mediodorsal nucleus of thalamus	A 4.6	83.5	47.2	69	<.005
(f) Paraventricular nucleus of hypothalamus	A 6.2	68.5	41.3	65	<.10
(g) Caudate nucleus	A 9.4	68.8	53.7	28	<.30
Neocortex as a whole	–	71.9	41.2	75	<.001

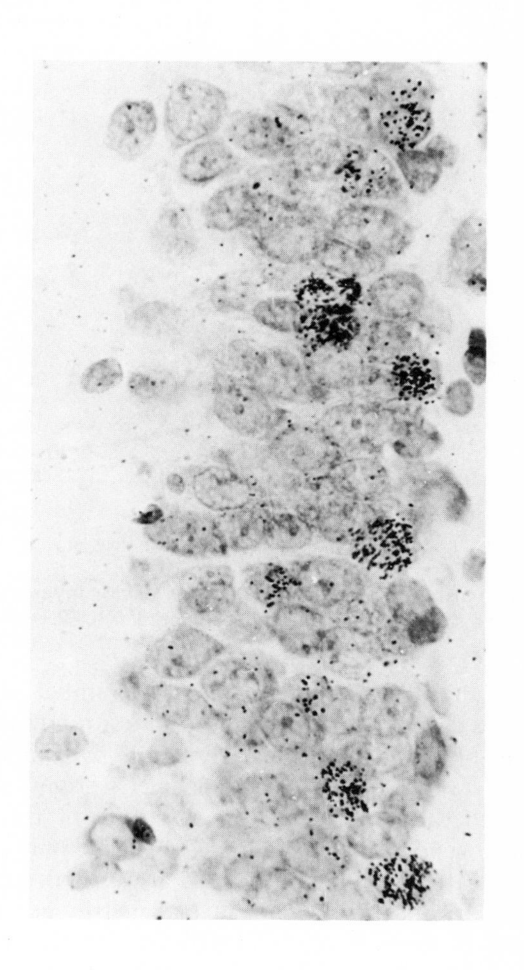

FIG. 8. Labeled granule cells in the dentate gyrus of the hippocampus in a young rat injected with thymidine-^{3}H and killed 2 months later.

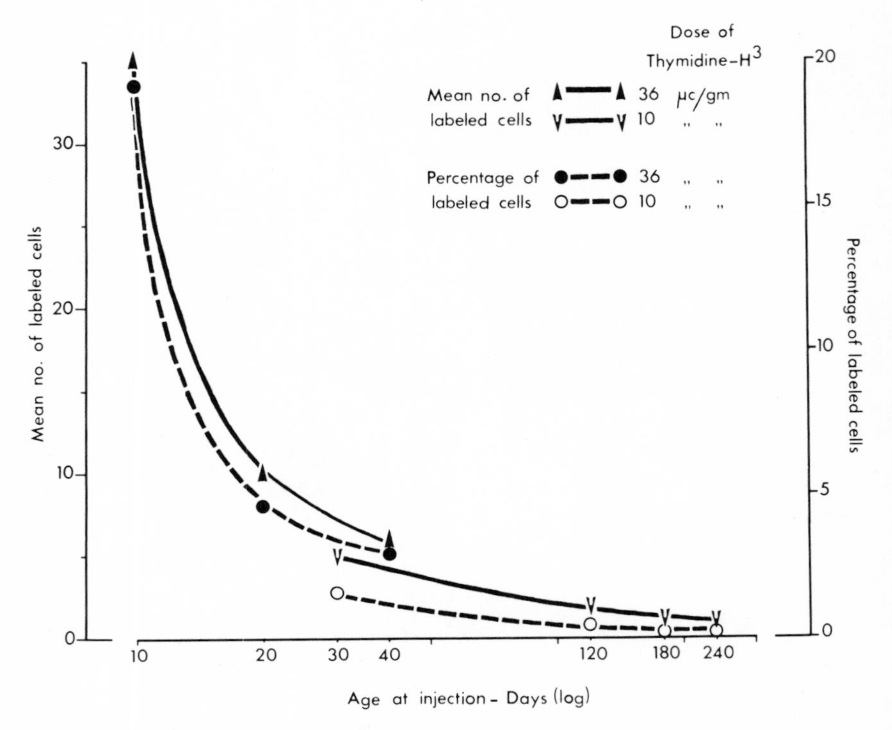

FIG. 9. **Mean number of labeled cells in the granular layer of the dentate gyrus as a function of age at the time of injection (material from 2 separate experiments).**

This study (Fig. 9) showed that the number of labeled cells in the dentate gyrus declined rapidly from neonates to infants, and reached a low but appreciable level in adults. Cell counts of granule cells, furthermore, showed a sixfold increase in homologous sections of the dentate gyrus from birth to three months of age. In spite of this evidence of hippocampal neurogenesis, mitotic granule cells were only exceptionally seen in this region. In this material our attention was drawn to the presence of an extensive ependymal and subependymal matrix around the lateral ventricles, in which the majority of cells were darkly staining, labeled with thymidine, and many also displaying mitotic activity. We observed many such dark, apparently undifferentiated cells also in and around the dentate gyrus of the hippocampus in all the younger animals. Quantitative work showed (Fig. 10) that the number of cells and the areal extent of the germinal matrix of the lateral ventricle declined rapidly from birth on, with a transient rise at about 15 days. During this period the number of undifferentiated cells in or near the granular layer of the dentate gyrus

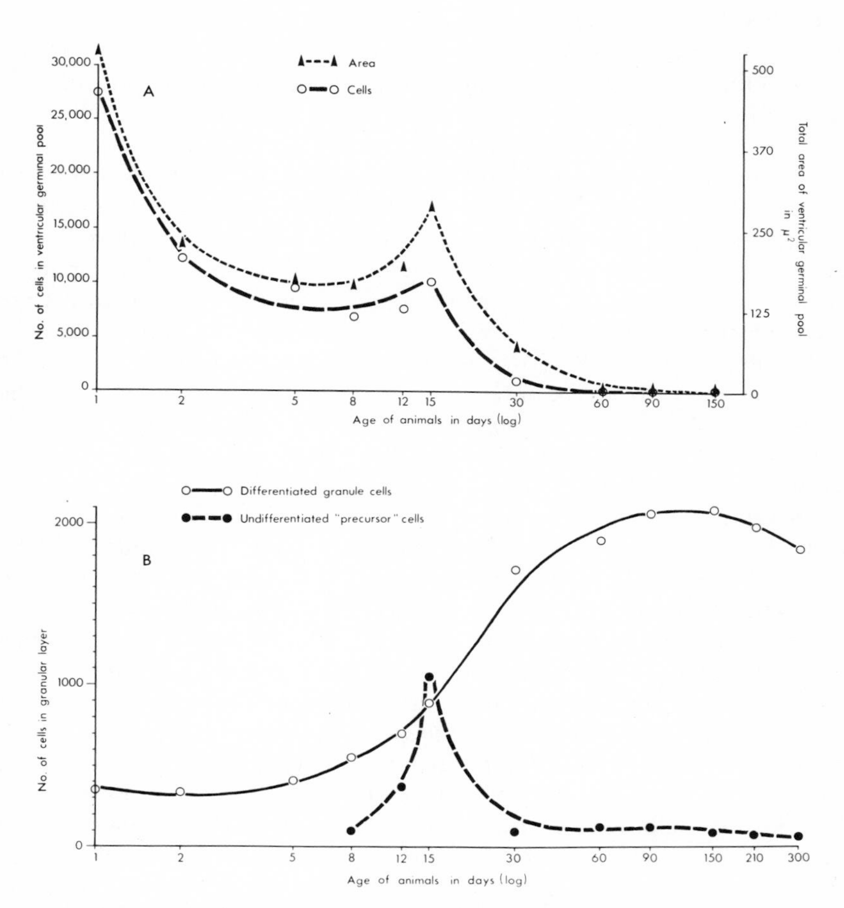

FIG. 10. **A. Graphs showing changes in cell population and in area occupied by the germinal matrix of the lateral ventricle in animals of different ages. B. Mean number of "undifferentiated" granule cells in the granular layer of the dentate gyrus of the hippocampus in animals of different ages. Note decline of "undifferentiated" cells, and concomitant increase in differentiated granular neurons after 15 days of age.**

showed a rapid rise, followed by a decline. The decline in the number of these undifferentiated cells, in turn, was accompanied by a corresponding increase in the number of differentiated granule cells. On the basis of this evidence we postulated that undifferentiated cells multiply near the lateral ventricles postnatally, migrate from there to the hippocampus, and become subsequently differentiated into granule cells.

The hypothesis of this experiment was confirmed in an extensive study completed in our laboratory. In these investigations (Altman and

Das, 1965b), we obtained unequivocal evidence that a large proportion of the granule cells of the dentate gyrus are formed postnatally in the rat. Hippocampal neurogenesis was established to be dependent primarily on the multiplication and subsequent migration of undifferentiated cells from the ependymal and subependymal layers of the forebrain ventricles, and to a lesser extent on local cell proliferation (Altman, 1966b; Altman and Das, 1966b). Moreover, this study also shows the occurrence of postnatal neurogenesis in a variety of brain structures in which short-axoned neurons, as the granule cells in the hippocampus, form distinct granular layers. Thus the granular layer of the olfactory bulb is formed to a large extent postnatally through migration of cells multiplying around the wall of the olfactory ventricles (Altman, 1966b; 1969b). Similarly, a large proportion of the cells of the internal granular and molecular layers of the cerebellar cortex is formed postnatally by local multiplication and the migration of newly formed cells from the subpial external granular layer (Altman, 1969a).

Several investigators reported the presence in adult animals and man of a mitotically active "subependymal layer" (Kershman, 1938) around the ependymal wall of the anterior lateral ventricle. Because techniques were not available for tagging these cells, these earlier investigators could only speculate about their fate. The technique of thymidine-^{3}H autoradiography was first addressed to this problem by Smart (1961). In a combined histological and autoradiographic investigation, Smart examined the morphology of this subependymal layer in infant and adult mice. He established that the proliferative subependymal layer extended in adult mice from the anterior wall of the lateral ventricle rostrally into the olfactory bulb. Smart's results indicated that these cells give rise, in infant mice, to glia and neurons, but he failed to obtain evidence of migration in adult mice. He postulated, in agreement with the hypothesis of previous investigators, that the mitotic activity of this layer in adults is an abortive phenomenon and that the newly forming cells degenerate. Our demonstration of cell migration in various brain regions in infant and adolescent rats, with evidence of their differentiation into neurons and glia, raised the possibility that cell production on the subependymal layer of the anterior horn of the lateral ventricle is not an abortive phenomenon but that the cells that are produced there are incorporated into the architecture of the adolescent and adult brain.

In a recent study (Altman, 1969b) we examined the autoradiographic procedure the brains of rats that were injected systematically with thymidine-^{3}H at 30 days of age and were killed 1, 6 and 24 hr, and 3, 6, 20, 60, and 120 days after injection. Our observations established that the cells produced in large numbers in the anterior wall of the inferior horn of the lateral ventricle migrated to the olfactory bulb (Fig. 11).

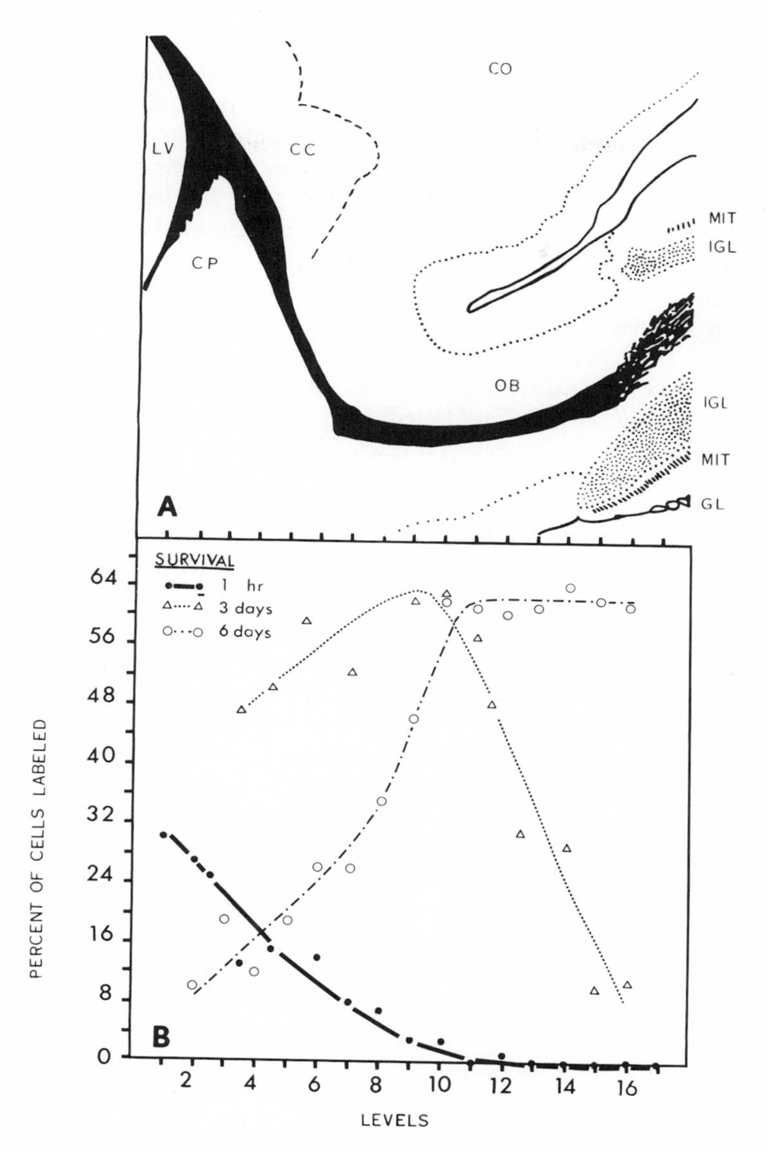

FIG. 11. A. Tracing of the anterior forebrain from a young adult rat at a magnification designed to match the scale of the abscissa, with nominal distances of 180 μ between levels. CC, corpus callosum; CO, cerebral cortex; CP, caudateputamen; GLO, glomerular layer; LV, inferior horn of lateral ventricle; MIT, mitral cell layer; OB, olfactory bulb. B. Percentage of labeled cells obtained in coronal autoradiograms in the subependymal layer and rostral migratory stream, at the coronal levels indicated, in rats that survived for 1 hour, 3 and 6 days after injection. The continued multiplication of cells near the lateral ventricle and their subsequent migration into the olfactory bulb is indicated.

Regional cell proliferation (as determined in animals killed shortly after injection) in the subependymal layer showed a steep caudorostral gradient during this time, with the high rate of cell proliferation near the ependymal layer of the lateral ventricle (with about 30 percent of the cells labeled) falling to zero in the olfactory bulb. The labeled cells that were present in high proportion near the lateral ventricle in the rats killed 1 to 24 hr after injection had further multiplied and moved to the middle portion of the "rostral migratory stream" by the third day, being located in the subependymal layer of the olfactory bulb by the sixth day after injection. By Day 20 the labeled cells disappeared from the subependymal layer of the olfactory bulb and were distributed throughout the internal granular layer (Fig. 12). The differentiated cells were tentatively identified as granular nerve cells and neuroglia cells. These results established that the major target structure of cell production in the subependymal layer of the lateral ventricle in young adult rats is the olfactory bulb.

In another study (Altman, 1969a) three experimental approaches were used to study the postnatal development of the cerebellar cortex in the rat: (1) a gross morphological investigation of the growth, in the sagittal plane, of the cerebellum as a whole, and of the various layers of the cerebellar cortex, from birth to maturity; (2) histological and cytological analyses of developmental changes in the cell composition of the cortical layers and in the structure of the cells themselves; and (3) autoradiographic analyses of the location and identity of labeled cells in the cerebellar cortex of adult rats that were injected with single or multiple doses of thymidine-^{3}H early in life. With the first technique we established that in the sagittal plane the cerebellum increases more than 20-fold between birth and 21 days. The second technique, the analysis of differential laminar contribution to the growth of the cerebellar cortex, gave the following results. The molecular layer, because of its thinness, could not be measured before five days. There was minimal increase in its area between five and nine days, but there occurred thereafter a rapid growth phase, which lasted until 23 days, leading to a nearly ten-fold increase. Little growth was indicated for the subcortical regions of the cerebellum (which in this plane is made up largely of the medullary layer) between 10 and 21 days, but there was some increase in the area of the subcortical regions between 21 and 90 days, presumably reflecting a growth due to myelination. Finally, the autoradiographic study established that the growth of the cerebellar cortex is directly or indirectly related to the migration of cells produced at a high rate in the subpial, external granular layer. During the first week the growth of the different layers—except for the proliferative external granular layer—is sluggish. During this period, the cells of the external granular layer do not differentiate, but they do provide stem cells to this growing proliferative matrix. These proliferating

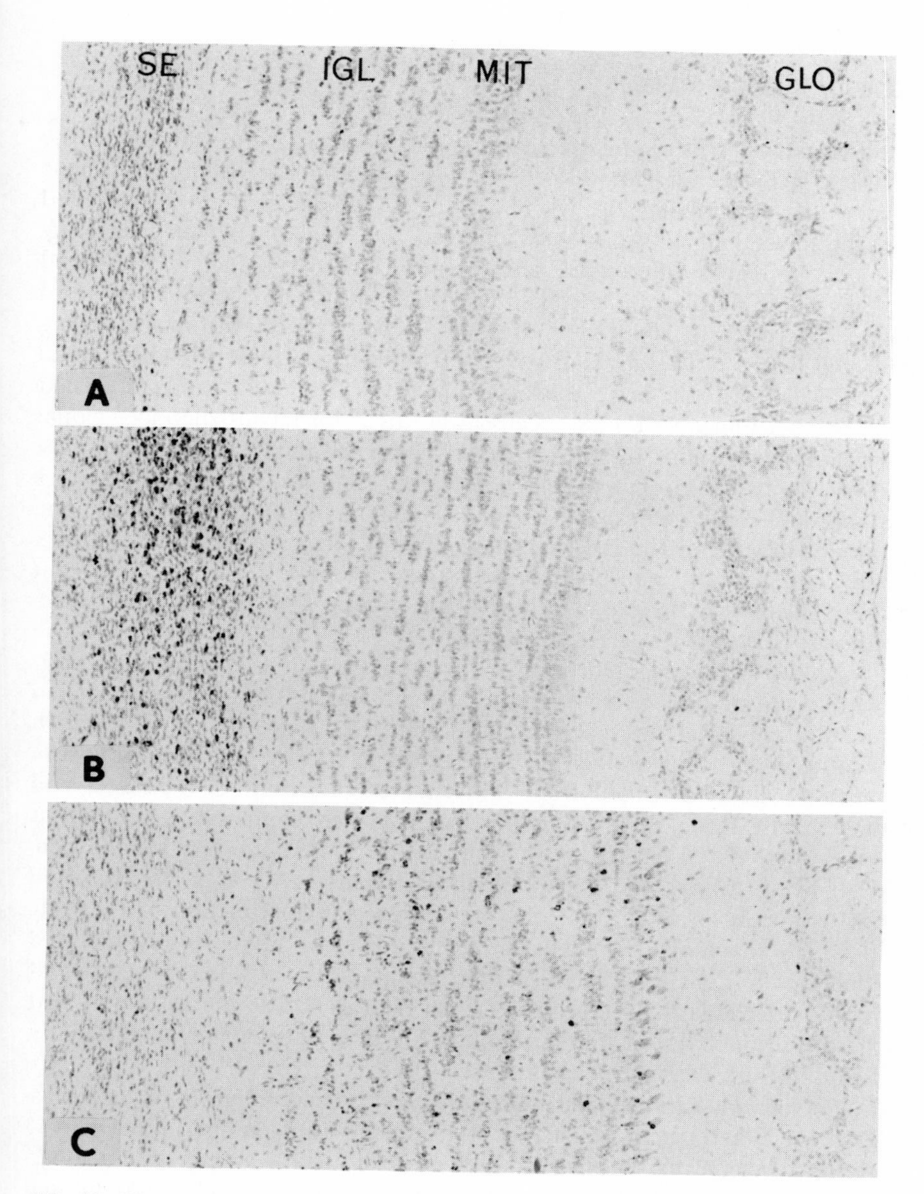

FIG. 12. Photomicrographs of autoradiograms of the laminated cortex of the olfactory bulb in rats that were injected at 30 days and lived for different periods. GLO, glomerular layer; IG, internal granular layer; MIT, mitral cell layer; SL, subependymal layer. A. From a rat that was killed 1 hour after injection. A few labeled cells are seen in the glomerular layer, but in general the bulb is free of labeled cells. B. From a rat that was killed 3 days after injection. Note the high concentration of labeled cells in the subependymal layer. Labeled cells are not seen in the internal granular layer. C. From a rat that was killed 20 days after injection. Labeled cells have disappeared from the subependymal layer and moved radially into the internal granular layer.

and migrating cells of the external granular layer are the precursors of the basket, stellate, and granule cells of the cortex; glia cells probably arise from cells multiplying locally. The first cells to differentiate are situated in the lower half of the molecular layer; they include basket cells (Fig. 13). Stellate cells differentiate later, with a peak at the end of the second week. The bulk of granule cells differentiate during the second and third weeks, with 25 to 80 percent of them, depending on the region, being formed between 11 and 21 days. These differences, together with several histological criteria (e.g., thickness of external granular and molecular layers, appearance of Purkinje cells) were used for reconstructing regional developmental maps of the cerebellar cortex. Granule cells differentiate in the depth of vermian fissures before they do over the exposed surfaces of the lobes; the ventral lobes (lingula and nodulus) mature before the anterior lobes; and the last maturing vermian lobes are the tuber, declive, and culmen. The hemispheres, with some exceptions, mature later than the vermis, with the paraflocculus being among the last maturing structures.

What is the possible functional significance of the delayed or postnatal origin of short-axoned neurons (microneurons)? It has been known for some time that the supragranular layer of the cortex (Layer II), which is composed of short-axoned neurons, grows in relative width as the phylogenetic scale is ascended from nonprimates to lemur, chimpanzee, and man (van't Hoog, 1920). On the basis of such comparative findings it was suggested (Ariëns, Kappers et al., 1936) that these short-axoned neurons, being interposed between long-axoned afferents and efferents and having merely local output, serve "higher associative" functions. In a study in which the Golgi impregnation technique was used, it was reported (Scheibel and Scheibel, 1963) that the granule cells in the cortex of kittens are the last cell types to develop their dendritic processes and that the conditionability of maturing animals may be dependent on the maturation of these elements.

What role can cell multiplication play in the mediation of higher associative functions in general or the storage of memory in particular? There are three major structural theories of neural storage of memory, which may be called intraneuronal, extraneuronal, and interneuronal theories (Altman, 1966a). Intraneuronal theories assume that the storage of individually acquired experience is dependent on intracellular, macromolecular coding mechanisms in a manner that may resemble the storage of hereditary information by DNA. Our studies do not bear on this theory. Extraneuronal theories assume that the structural changes underlying the memory trace are dependent on non-neural elements, such as various types of glia cells, particularly satellite cells. Our results indicating a facilitation of glial multiplication in the cortex of enriched animals may have some

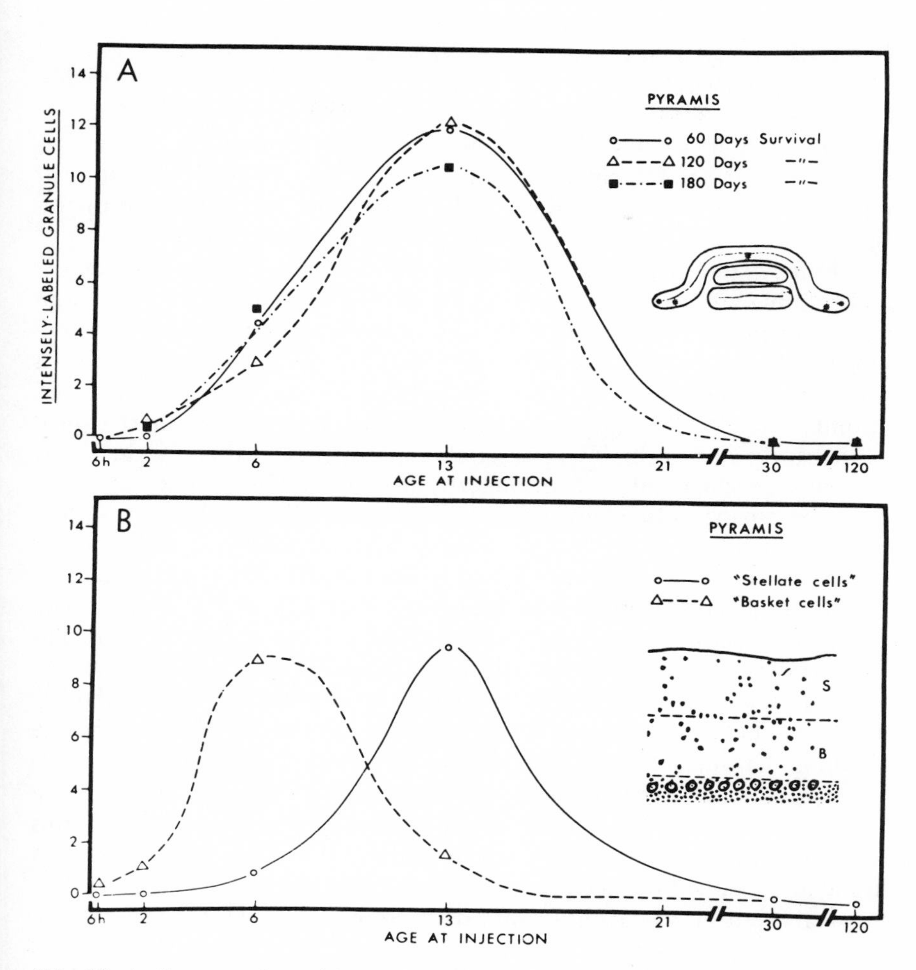

FIG. 13. A. Mean number of intensely-labeled cells in the internal granular layer of pyramis of the cerebellar vermis in pairs of rats aged 60, 120 and 180 days as a function of injection with thymidine-^{3}H at the ages indicated in the abcissa. Sampling sites (stars) schematically indicated (sample size 130 μ^2). Peak concentration in the animals injected at 13 days indicates that it is during this period that the highest number of granule cells are differentiating. B. Mean number of intensely-labeled in the upper half ("stellate cells") and in the lower half ("basket cells") of the molecular layer. The basket cells are apparently differentiating before the stellate cells.

bearing on this theory. Increased glial multiplication, through the role of glia cells in myelination, may be associated with recruitment of more conducting elements into newly formed neuronal circuits. Alternatively, since glia cells form the matrix in which nerve cells are embedded, the environs and hence properties of neurons may be radically altered by glial proliferation. Of course, rearing rats in an enriched environment represents a complex of variables, and factors other than increased opportunities for learning may be responsible for the obtained effect.

Finally, interneuronal theories assume that the storage of memory and learning are dependent on the formation of new synaptic connections among neuronal elements, due to such hypothesized factors as the outgrowth of new boutons. Another possibility, as we mentioned earlier, is that new synaptic connections are established among invariant input and output elements through the interposition of internuncial short-axoned neurons or granule cells. The postnatal origin of these cell types, their slow postnatal development, and an increase in their relative number as we ascend the phylogenetic scale, make it possible that these cell types are the plastic elements of the central nervous system and that their development is dependent on the organism's active commerce with the environment.

To test experimentally the hypothesis that behavioral variables can affect the structural maturation of the central nervous system, in particular the formation of short-axoned neurons, we studied with quantitative histological and autoradiographic techniques the possibility that infantile handling affects morphological development of the rat brain. Infantile handling has been shown to lead to pronounced physiological, endocrinological, and behavioral consequences in adult mice and rats (Denenberg, 1962; Levine, 1962). The specific question we have raised in this study was whether these effects are correlated in any way with alterations in the development of the brain.

Rats were handled for 15 min daily from two to 11 days of age, inclusive. In a pilot study, the handled rats and their unhandled controls were injected with thymidine-^{3}H on Day 11, and killed either 6 hr, 3 days, or 30 days after injection. The first result of the experiment was that the brains of the handled animals, contrary to expectation, were lighter in weight in all three age groups than the brains of the unhandled animals. This aspect of the study was reinvestigated on a larger series of animals with separate litters used for each survival period, and the brain weight difference between the two groups was confirmed for the 11- and 14-day-old rats; these differences, however, were no longer present at 41 days of age or at 101 days (90 days after cessation of handling). These findings led to the conclusion that handling results in a transient retardation of brain growth.

The quantitative histological and autoradiographic measurements were restricted to the brains obtained from the pilot study. Three classes of measurements were carried out: (1) determination of areal changes in the cerebellum and neocortex in an attempt to correlate these with gross brain weight differences in the two groups; (2) counting of labeled cells in the cerebellum, hippocampus, and neocortex to ascertain possible differences in the rate of acquisition of postnatally formed elements; and (3) analysis of possible differences in the composition of brain tissue by means of a technique of areal fraction analysis.

Planimetric areal measurements were carried out in matched histological sections of the cerebellum and neocortex. Areal determinations were, in general, in agreement with the results of the gross brain weight measurements, the sampled areas having been larger in the unhandled than in the handled animals. These results indicate that the gross brain weight measures, which in formalin-perfused brains include a considerable amount of fluid of extrinsic origin, reflect intrinsic differences in brain volume.

Whereas weight and areal measurements indicated that the brains of the handled animals were retarded in development, autoradiographic cell counts brought clear evidence of a higher rate in the acquisition of new cells in the cerebellum (Fig. 14), hippocampus (Fig. 15), and neocortex of handled, than of nonhandled animals during the period studied. The difference in the number of labeled cells was unambiguous 6 hr after in-

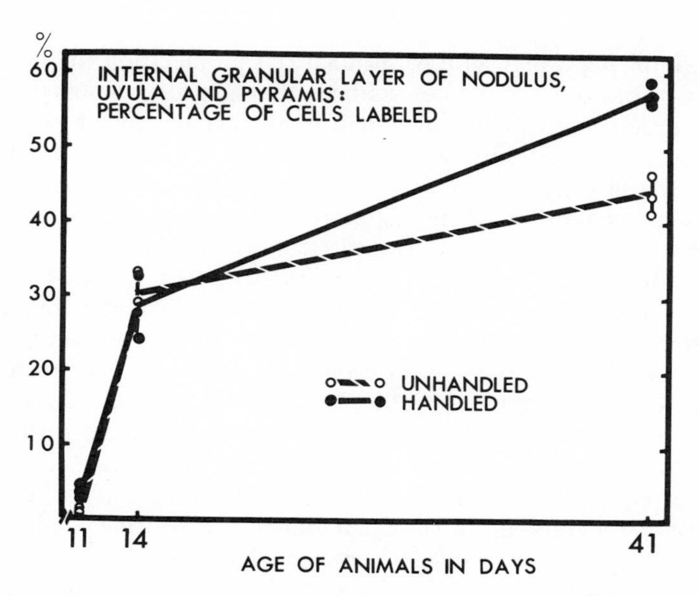

FIG. 14. Percentage of cells labeled in sampled regions of the internal granular layer. Each point refers to mean cell count in individual animal.

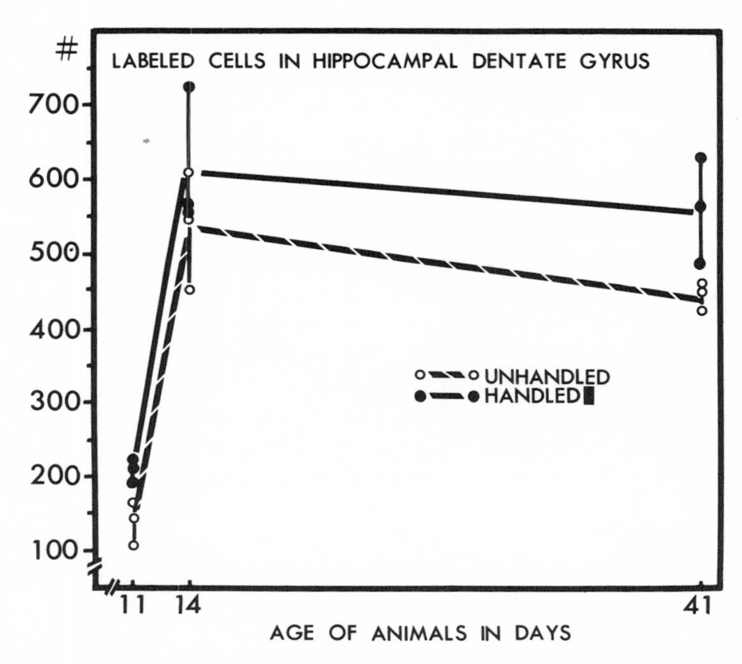

FIG. 15. Schematic illustration of hypothesis offered to account for higher rate of cell proliferation in handled rats (see text).

jection, implying a high rate of regional cell proliferation in the handled animals in the three structures studied. It was also pronounced 30 days after injection, suggesting a high rate in the acquisition of new cells by migration from such sites of production as the external granular layer of the cerebellum and the subependymal layer of the lateral ventricle.

A large proportion of the migratory cells in the cerebellar cortex and hippocampal dentate gyrus differentiates into microneurons. Thus these results indicate a higher rate in the acquisition of late-forming, short-axoned neurons in the handled animals than in unhandled controls. Our normative studies have established an inverse relationship between brain size and rate of cell proliferation during ontogenetic development in the rat. These two findings, a smaller brain at 11 and 14 days of age in handled rats as compared with nonhandled ones, combined with a higher rate in the acquisition of new neurons in the former, can be reconciled with our previous results if we assume that handling leads to a delay in the postnatal maturation of the brain. This interpretation implies that in the unhandled animals cell proliferation is declining by the latter half of the second week, but due to a delay in brain maturation cell proliferation is still brisk during this period in the handled animals (Fig. 16).

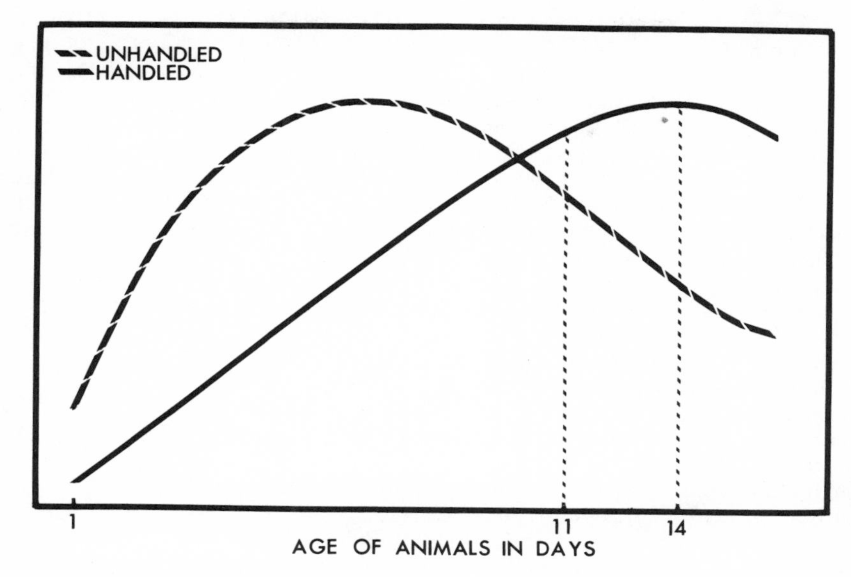

FIG. 16. Labeled cells in 5 200-μ-wide strips of the dentate gyrus of the hippocampus.

This last study indicates, as do some others, that the developing brain is particularly sensitive to environmental influences and that relatively minor perturbances can alter the structural organization of the growing nervous system. But the study also indicates that the consequences of environmental manipulations are complex ones and will not follow simple predictions. Evidently, a considerable amount of work remains to be done in this area and a variety of other techniques, in addition to the ones referred to here, will have to be employed if we are to obtain reliable and meaningful answers to the questions raised.

Acknowledgments

Several of the experiments described in this paper were carried out in collaboration with Mrs. Gopal D. Das and Robert B. Wallace, and Mr. William J. Anderson. These studies were supported by the National Institute of Mental health and the U.S. Atomic Energy Commission.

References

Altman, J. (1962a). Exp. Neurol., 5:302. ——— (1962b). Science, 135:1127. ——— (1963a). J. Histochem. Cytochem., 6:741. ——— (1963b). Nature (London), 199:777. ——— (1963c). Anat. Rec., 145:573. ——— (1966a). Organic Foundations of Animal Behavior. New York: Rinehart

& Winston. ——— (1966b). J. Comp. Neurol., 128:431. ——— (1969a). J. Comp. Neurol., 136:269. ——— (1969b). J. Comp. Neurol., 137:433. ——— and Altman, E. (1962). Exp. Neurol., 6:142. ——— and Das, G. D. (1964a). Anat. Rec., 148:535. ——— and Das, G. D. (1964b). Nature (London), 204:1161. ——— and Das, G. D. (1965a). J. Comp. Neurol., 124:319. ——— and Das, G. D. (1965b). Nature (London), 207:953. ——— and Das, G. D. (1966a). Physiol. Behav., 1:105. ——— and Das, G. D (1966b). J. Comp. Neurol., 126:337. ——— Das, G. D., and Anderson, W. J. (1968). Develop. Psychobiol., 1:10. ——— Das, G. D., and Chang, J. (1966). Physiol. Behav., 1:111.

Ariëns, Kappers, C. U., Huber, G. C., and Crosby, E. C. (1936). The Comparative Anatomy of the Nervous System of Vertebrates, Including Man. New York: Macmillan.

Bach, L. (1895). Acta Ophthal., 41:62.

Bech, K. (1957). Acta Ophthal., Suppl. 7:1.

Bélanger, L. F., and Leblond, C. P. (1946). Endocrinology, 39:8.

Bennett, E. L., Diamond, M. C., Krech, D., and Rosenzweig, M. R. (1964). Science, 146:610.

Birch-Hirschfield, A. (1900). Arch. Ophthal., 50:166.

Carlson, A. J. (1902). Amer. J. Anat., 2:341.

Das, G. D., and Altman, J. (1966). Physiol. Behav., 1:109.

Denenberg, V. (1962). In: The Behavior of Domestic Animals, Hafez, E. S. E., ed. Baltimore: Williams & Wilkins.

Diamond, M. C., Krech, D., and Rosenzweig, M. R. (1964). J. Comp. Neurol., 123:111.

Hughes, W. L., Bond, V. P., Brecher, G., Cronkite, E. P., Painter, R. B., Quastler, H., and Sherman, F. G. (1958). Proc. Nat. Acad. Sci. U.S.A., 44:476.

Hydén, H. (1962). In: Neurochemistry, 2 ed., Elliott, K. A. C., et al., eds. Springfield: Thomas.

Kershman, J. (1938). Arch. Neurol. Psychiat., 40:937.

Kopriwa, B. M., and Leblond, C. P. (1962). J. Histochem. Cytochem., 10:269.

Leblond, C. P., Messier, B., and Kopriwa, B. M. (1959). Lab. Invest., 8:296.

Levine, S. (1962). In: Experimental Foundations of Clinical Psychology, Bachrach, A. J., ed. New York: Basic Books.

Mann, G. J. (1895). J. Anat. Physiol., 29:100.

Scheibel, M. E., and Scheibel, A. B. (1963). EEG Clin. Neurophysiol., Suppl. 24:235.

Smart, I. (1961). J. Comp. Neurol., 116:325.

Taylor, J. H., Woods, P. S., and Hughes, W. L. (1957). Proc. Nat Acad. Sci. U.S.A., 43:122.

van't Hoog, E. G. (1920). J. Nerv. Ment. Dis., 51:313.

Wilzbach, K. E. (1957). J. Amer. Chem. Soc., 79:1013.

Chapter 15

Macromolecular Changes Within Neuron-neuroglia Unit during Behavioral Events

LEONID Z. PEVZNER

Pavlov Institute of Physiology of the Academy of Sciences of the USSR, Leningrad, USSR

A peculiar feature of nervous tissue is that its chief functional structure, nerve cells, represents as little as several percent of the whole mass of the tissue. Thus as soon as corresponding methodical achievements made it possible, investigations were carried out dealing with chemical composition (and its functional changes) within individual neurons. This work, now quite extensive, began with papers by Oliver Lowry (Lowry, 1941; Lowry and Bessey, 1946; Lowry et al., 1951) and Holger Hydén (Hydén, 1943, 1947, 1955). They and their collaborators, as well as a number of later authors, obtained an extensive collection of facts (see reviews by Hydén, 1960, 1962, 1964, 1967; Pevzner, 1963, 1966; Brodsky, 1966) which permitted several initial conclusions about biochemical mechanisms of fundamental states of the neuron: excitation and inhibition.

However, more and more data accumulated which indicated that study of the neuron alone is not sufficient for elucidating all the functional peculiarities of nervous tissue metabolism. Thus, more than a decade ago, Hydén (1959a; Hydén et al., 1958) suggested a fruitful conception about the existence of a biochemical relationship between the neuron and its satellite neuroglia, i.e., about the functioning of a single neuron-neuroglia unit.

Pathologists, morphologists, and physiologists have also made some indirect observations indicating intimate connections between the nerve and glial cells.

INDIRECT DATA IN FAVOR OF NEURON-NEUROGLIA METABOLIC UNIT

Neuroglial cells, unlike neurons, have preserved a capacity for cell division. So neuroglia acts as a less differentiated, more labile morphological structure of the nervous tissue as compared with neurons. Reactivity of glial cells under various pathological conditions is well known and has been carefully described in many handbooks and textbooks on pathomorphology of the nervous system (e.g., Penfield, 1932; Snesarev, 1946, 1961; Glees, 1961; Niculescu, 1963, etc.).

On the other hand, neuroglia can be more resistant than neurons to a number of harmful effects. Thus, hypoxic hypoxia (Jensen et al., 1948), clamping of main blood vessels of the brain (Van Harreveld and Stamm, 1954; Hornet et al., 1960), and tissue hypoxia due to administration of anoxic drugs (Lucas and Strangeways, 1963) induced prominent morphological changes in neuronal bodies with no marked reaction of neuroglial cells. Experimental C avitaminosis in guinea pigs was also characterized by a selective damage and even a death of the nerve cells only (Sulkin and Kuntz, 1950; Weatherford, 1961).

Great attention has been paid to the possible participation of glia in the barrier mechanisms of nervous tissue. Even the light-microscopic picture of the nerve cells surrounded by glial cells suggests a central strategic role of neuroglia as a particular cellular structure connecting the nerve cell bodies and capillary net of the nervous tissue and dividing them from each other (see Glees, 1955; Cammermeyer, 1960a; Polak, 1965). Cammermeyer (1960b) found a direct relationship between the degree of vascularization of brain regions and the number of glial cells.

Electronmicroscopy has provided a prominent advance in solving this problem. First, investigations of the nervous tissue at the ultrastructure level by Maynard and Pease (1955), Farquhar and Hartmann (1956, 1957), Wyckoff and Young (1956), Schultz et al. (1957), Luse (1956, 1960), and Hess (1958) showed that the nonstructural ground substance of the nervous tissue, described in light-microscopy, did not exist; the space around neuronal bodies is filled with bodies and processes of glial cells. As to the real intercellular space, it is represented only by narrow clefts which divide the membranes of nerve and glial cells and do not exceed 100–150° A (see DeRobertis and Gerschenfeld, 1961; Nakai, 1963; Kaplan, 1965; Manina, 1966). Calculation of the volume of these clefts by Hortsmann and Meves (1959) has shown as little as 5 percent of the total volume of the brain tissue.

In this connection it is worth mentioning that there are many papers on the determination of the extracellular space value by means of the substances that are assumed not to penetrate inside the cell (e.g., sucrose, inulin, chlorides, aminohyppuric acid, thiocyanate). The values calculated by different authors fluctuate between 2 to 3 percent (Barlow et al., 1961) and 30 to 50 percent (Streicher et al., 1964). Analysis of both earlier and more recent data on this point has enabled Dawson and Spaziani (1959) and Woodward et al. (1967) to conclude that the most realistic value of extracellular space of the brain tissue amounts, in all probability, to 12 to 15 percent.

This figure is about three times the value of the volume of intercellular clefts. Thus the question has arisen whether the protoplasm of glial cells is, at least partly, an analogue of extracellular brain space. Some authors are inclined to answer this question affirmatively. There are, for example, data that neuro glial cells, unlike neurons, are rich in Na and Cl, thus resembling, according to their ion composition, extracellular rather than intracellular fluid (Torack et al., 1959; Gerschenfeld et al., 1959; Katzman, 1961; Koch et al., 1962). Moreover, it has been shown by Gerschenfeld et al. (1959) that an administration to nephrectomized rats and rabbits of a great volume of isotonic salt solution as well as an incubation of brain pieces in isotonic, and particularly hypotonic, salt solution induces a marked edema of brain tissue. However, neither swelling of neuron bodies nor appearance around them of any increased extracellular space has been revealed by the electronmicroscope, the swelling being observed in all cases only in astrocyte bodies. Similar selective swelling of neuroglial cells which resulted from acute brain edema was subsequently confirmed repeatedly by many authors (Torack et al., 1959; Luse and Harris, 1960; DeRobertis and Gerschenfeld, 1961; Pappius et al., 1962; Rosomoff and Zugibe, 1963; Hirano et al., 1965; Van Harreveld et al., 1966). Many investigators (Quastel and Quastel, 1961; Coxon, 1964; Aleksandrovskaya, 1965; Vernadakis and Woodbury, 1965; Dimova et al., 1966; Lord et al., 1967) have supported DeRobertis' conception that neuroglia in many respects is in fact the brain-blood barrier (DeRobertis et al., 1960; DeRobertis and Gerschenfeld, 1961).

This point is not yet indisputable because the majority of the data referred to above are as a rule of qualitative and descriptive character. There are also some opposite data. Thus Lumsden (1955) and Pappius (1965) calculated that the swelling of the neuroglial cells could account only for a part of experimental edema of the brain tissue, both in vivo and in vitro. Of great interest are data by Lasansky (1965), who has shown that ferricyanide in vitro approaches toad retina neurons through intercellular clefts around astrocytes rather than through the astrocyte cytoplasm. Contrary to Hertz (1968), who calculated that the high diffusion rate found for ions and amino acids by no means can be accounted for by narrow intercellular clefts, Nicholls and Kuffler in a series of their papers have underlined the point of view that it is through the intercellular channels that the main flow of Na and K, as well as sucrose, proceeds in the brain tissue (Nicholls and Kuffler, 1964, 1965; Kuffler and Potter, 1964; Kuffler and (Nicholls, 1964, 1965, 1966; Kuffler et al., 1966; Kuffler, 1967).

This discussion, however, deals in fact only with some particular points of the problem but does not challenge at all the general conception of active participation of neuroglia in the specific neuron activity. This conception has been confirmed also by morphological data of another kind. Thus a prolonged preganglionic electric stimulation of cat and dog superior sympathetic ganglion has been shown by Kuntz and Sulkin (1947) to give rise to an increase in the number of both capsular and interstitial oligodendroglial cells, a part of these glial cells being divided amitotically. Moreover, Aleksandrovskaya et al. (1964) have found daily administration of aminazine to rats 7 to 14 days to induce a pronounced increase of the number of neuroglial cells in the cerebral cortex of these animals. Similar increase has been also revealed in the ventral horns of rat spinal cord within 20 and 40 min of swimming (Aleksandrovskaya et al., 1965); the authors have explained these data by a possible functionally dependent migration of neuroglia to the neuron. This explanation has been confirmed by our data (Pevzner, 1967, 1968; Pevzner and Litinskaya, 1968) that show an increase in the mean number of glial cell satellites per neuron in rat spinal cord ventral horns and cat superior sympathetic ganglia, as well as by data by Lodin et al. (1968) that indicates an accumulation of glial cells around cerebellar Purkinje neurons as a result of caloric stimulation.

Marked changes in the volume of glial cells have been observed by Kulenkamff and Wüstefeld (1954) and Matz (1966) in motor areas of spinal cord of animals under conditions of intensive muscular load. By means of a high-speed automatic analyzer of micro-objects, an increase of glial cell volume has been revealed in rat spinal cord after two-week daily injections of adrenaline (Pevzner and Litinskaya, 1968). Phase changes of glial cell volume, varying with the kind of muscular activity and functional pecularities of the spinal cord areas, have been shown by Haidarliu (1967a) in rats and by Brumberg (1969) in mice.

The possibility of neuroglia's participation in the functioning of the nervous system can be concluded also from the studies on nervous tissue cultures in vitro. Under these conditions neuroglial cells have been found to possess active pulsatory movements (Pomerat, 1952; Berg and Källen, 1959; Geiger, 1963); Geiger et al. (1960) described alterations of the pattern and rate of this pulsation that resulted from an addition to the culture of active agents such as adrenaline and eserine.

According to the data of neurophysiologists, neuroglial cells in tissue culture are characterized by a definite bioelectrical activity (Hild et al., 1958; Hild and Tasaki, 1962). Bioelectrical activity of neuroglia is not a feature peculiar only to glial cells in vitro. A number of slow processes in the electrical activity of the brain in vivo have been thought to be of glial origin (Birks et al., 1960; Sokoloff, 1962; Aladjalova and Koltsova, 1964; Laborit and Laborit, 1965). Considerable electrophysiological data, which can not be discussed in this chapter but are reviewed in details by Roitbak (1965, 1968) and by Kuffler and Nicholls (1966), indicate that neuroglia can influence such a peculiar feature of the neuron as its bioelectrical activity. Thus for the last few years an increasing number of authoritative neurophysiologists have expressed their opinion in favor of participation of neuroglia in the functioning of the neuron (Galambos, 1961, 1965; Roitbak, 1963, 1964, 1965, 1968; Hertz, 1965; Svaetichin et al., 1965; Tasaki, 1965).

These morphological and electrophysiological data have provided an indirect proof of the existence of a neuron-neuroglia single unit. Of great importance for solving the question is biochemical analysis of the neuron-neuroglia relationship. However, the possibilities of such analysis are limited by modern methods of biochemical investigation of individual cells of the nervous system. A detailed review of these methods—their main principles, advantages, and shortcomings—has been presented elsewhere (Pevzner, 1969a). Here only a brief comparison of five main methods will be provided.

Comparative Evaluation of the Methods of the Biochemical Analysis of Neuroglia

Topochemical Analysis. Topochemical analysis compares nervous tissue areas that differ in their cellular composition. Often compared, for example, are white and gray substances of the brain or individual layers of cerebral cortex (see Palladin, 1959; Kety and Elkes, 1961; Baranov and Pevzner, 1964; Friede, 1966; Pigareva, 1966). The main disadvantage of this method is the lack of exact information concerning the total mass and chemical composition of morphological structures other than neurons and neuroglia that are present in any sample of the nervous tissue: dendrites, axons, synaptic structures, blood vessels, connective tissue sheaths, etc. Meanwhile, dendrites, for example, represent metabolically very active structures (Lowry, 1955, 1957); axons are characterized both by intensive axonal flow of various substances and by active local synthesis (see Chapters 10 and 11, present volume). Thus the biochemical differences between the areas rich either in neurons or in neuroglial cells (e.g., cerebral cortex or corpus callosum, respectively) cannot be accounted for only in terms of the differences in metabolism between the nerve and glial cells.

Studies of Pathological Material. This involves analysis of tissue samples in which the relationship between neuronal and glial populations has been markedly changed. This is the case with areas of a selective neuronal degeneration (Koch et al., 1962; Utley, 1963, 1964; Talwar et al., 1966; Mihailović et al., 1968) or of proliferation and neuroglia (Dimova, 1966; Ruščak et al., 1967, 1968; Guth and Watson, 1968; Pappius, 1969). Biochemical investigation of glial tumors are used most frequently; metabolic features resulting from this analysis, particularly those which remain in spite of the changes in degree of malignancy, are thought to be characteristic for the neuroglia in general. A review of considerable data in the literature on brain tumor metabolism (Pevzner and Tomina, 1965), as well as our own experience on this point (Pevzner et al., 1964; Tomina and Pevzner, 1965), indicate that glial tumor cells can differ markedly from normal glial cells with respect to some metabolic features; these differences do not always progress parallel to the degree of dedifferentiation of tumor cells studied. The same holds true for other kinds of pathological material; great caution should be used when one considers proliferating glial cells or the glial cell satellites surrounding degenerating neurons as representatives of normal glia.

Enriched Fraction Method. This method entails isolation of sufficiently great masses of tissue fractions that selectively contain predominant nerve or glial cells. This isolation is done by a filtration of nervous tissue through special sieves (Roots and Johnston, 1964, 1965) and particularly by a combination of this filtration with a number of centrifugation stages in sucrose solution (Korey, 1957, 1958; Korey and Orchen, 1959). While the first attempts to obtain these enriched fractions failed to provide for a sufficient isolation of cell population, several subsequent schemes using concentration gradient in sucrose and its polymers have given both a greater yield of enriched fractions and a higher purity of cell populations desired (Rose, 1965, 1967; Satake and Abe, 1966; Freysz et al., 1967, 1968). In spite of a number of advantages of this method, both its main characteristics (yield and purity) are still rather far from ideal. The enriched neuroglial fraction has been shown by electronmicroscopy to contain myelin sheaths, blood cells, cellular debris, and other kinds of contamination while in the enriched neuronal fraction the neuron bodies deprived from axons, dendrites, and synaptic contacts are characterized by several signs of damage.

Microsurgery Method. Microsurgery involves mechanical isolation from the nervous tissue of individual neurons and clumps of neuroglia cells, followed by biochemical analysis. These investigations, because of an extremely low mass of samples, require modern ultramicromethods: modifications of Cartesian diver technique, spectrophotometry, microfluorimetry, etc. At present these techniques are widely used only in several laboratories: in the United States by Lowry (1953, 1957, 1962, 1966) and in Sweden by Hydén (1955, 1960, 1962, 1964, 1967), Edström (1953, 1956, 1958, 1964), and Giacobini (1962, 1964). The main disadvantage of this method is a mechanical disruption of the neuron-neuroglia unit, the degree of cell damage possibly being different for neurons and glial cells. This disadvantage should be particularly taken into consideration when studying ion permeability or some enzyme activities, the processes closely dependent on the state of external cellular membranes.

QUANTITATIVE CYTOCHEMISTRY IN SITU. This means chemical analysis of structures of histological slices by microscope. Absorbtion and emission cytospectrophotometry and cytointerferometry are examples of this method. It is undoubtedly to T. Caspersson's merit that the principles of cytospectrophotometry have been carried out (Caspersson, 1936, 1940, 1950, 1955). He was the first to analyze the conditions under which the photometry of heterogeneous objects could be performed under the microscope by using the laws of solution photometry. Detailed descriptions of the history of this method, of its main principles, error sources, etc., have been presented in a number of reviews (Caspersson, 1950, 1955; Leuchtenberger, 1954, 1958; Brodsky, 1956, 1966; Walker and Richards, 1959; Agroskin, 1962, 1968; Pevzner, 1963, 1966; Lodin, 1964; Swift, 1966; Sandritter et al., 1966). Later, ultraviolet and visible cytospectrophotometry were added by ultraviolet and visible cytofluorimetry (Brumberg, 1956; Rigler, 1966) and by cytointerferometry (Barer, 1956; Davies, 1958; Zakharyevsky and Kuznetsova, 1961; Brodsky, 1966). These methods allowed an analysis of the chemical composition of the cells without their mechanical disruption, with constant visual control of the objects studies. Disadvantages of cytochemistry *in situ* are the effect of fixation and the following histological treatment on the chemical composition of cells and the rather low chemical specificity of cytochemical determinations, and a greater error of individual measurements in cytochemistry than in biochemistry.

Even this short discussion shows all the modern methods of neuroglia analysis to possess both advantages and shortcomings. This should be remembered because in some cases discrepancies in data of various authors are associated mainly with the different methods used for the analysis of nerve and glial cells.

MACROMOLECULAR CHANGES WITHIN THE NEURON-NEUROGLIA UNIT DURING THE FUNCTIONAL STATES OF THE NERVOUS SYSTEM

Proteins

Dry weight as determined both by x-ray absorption method (Hydén, 1962, 1963, 1964) and by cytointerferometry (Pevzner, 1965, 1967, 1968) is of the same order in neurons and their glial cell satellites.

Freysz et al. (1967) found that in the enriched glial fraction of rat brain the concentration of protein nitrogen is somewhat higher than in the enriched neuronal fraction. However, when these authors calculated the corresponding chemical composition per cell, the dry weight and total protein in glial cells proved to be on average 3.5 and 2.8 times as much as that in neuron bodies (Freysz et al., 1968).

By means of histochemical stain reactions it has been shown by Portugalov (1959) and by Roskin (1959) that proteins of neurons contain a higher concentration of histidine, tryptophan, and arginine than do the proteins of glia. Content of protein sulfhydryl groups has been found to be higher in the glial cells than in the neurons, particularly in the phylogenetically old structures of the brain (Ellman and Sullivan, 1965). Saudargene (1969a), in our laboratory, has determined cytophotometrically the content of SH-groups per protein mass unit in rat spinal cord; according to her data the proteins of spinal ganglia neurons are 1.5 times richer than the proteins of glial cell satellites whereas in ventral horn motor neurons and their glial cells the concentration of SH-groups in cellular proteins is the same.

Historadiographically it has been shown by Gracheva (1964) and Roessman and Friede (1967) that the incorporation of labeled amino acids (glycine-^{14}C and leucine-^{3}H) into the nerve cells is quicker than into the glial ones. However, it should be taken into account that with the commonly used thickness of histological slices (5 to 30 μ) the whole glial cell body is as a rule located within the slice whereas the majority of nerve cells, particularly large ones, are dissected into several slices. Thus the external cellular membranes of glial cells are preserved in the slices while the external membranes of the nerve cell (in the slice occupying the middle part of the cell) are absent. It is quite natural that in these cases the conditions of the diffusion of low molecular precursors (as well as various products of histochemical, particularly enzymohystochemical, reactions) will be rather different for the neuron than for the glial cell. As an illustration of this point of view Leibson's data (1967) could be referred to; she found a discrepancy between high cholinesterase activity in cerebellum homogenates and quite low accumulation of the product of enzymohistochemical cholinesterase reaction in cerebellum slices; in the author's opinion this discrepancy was due to some barriers which were preserved in the slices but not in the homogenate; among these barriers the membranes of neuroglial cells possibly played a major role.

Using historadiography Kleihues and Schultze (1968) showed an increase of tryosine-^{3}H incorporation into white matter oligodendroglia cells in the areas of experimental brain edema.

Satake and Abe (1966) found the incorporation of ^{14}C labeled aminoacids into the proteins of enriched neuronal fraction proceeded more rapidly than into the proteins of brain total homogenate. Similar data were obtained by Blomstrand and Hamberger (1969), showing that leucine-^{3}H injection in rabbits resulted in a rate of label incorporation into enriched neuronal fraction approximately three times greater than into enriched glia fraction.

Rose (1969), by virtue of his in-vitro experiments with enriched fractions, reported the pool of free amino acids to be larger in neurons than in the glia. The incorporation of glutamate or pyruvate-^{14}C into amino acids proceeded in glial fraction more rapidly than in neuronal one. It is quite interesting that the Waelsch effect (quotient of radioactivity of glutamate higher than 1) was revealed only in the glial fraction. Taking into consideration the present state of enriched fraction techniques, it is difficult to judge whether the biochemical differences revealed reflect actual peculiarities of neuron and neuroglia metabolism or that these differences to a great degree are a consequence of some methodical factors —such as more pronounced damage of neuronal (as opposed to glial) external membranes, lower concentration of cell bodies in glial fraction, greater contamination of the glia enriched fraction with myelin (which is metabolically more inert material than cell bodies), etc.

Of a great interest are studies on an acidic S-100 protein which is predominantly present in the glial cells (Hydén and McEwen, 1966; McEwen and Hydén, 1966). This protein is thought to participate in biochemical mechanisms of learning (Hydén, 1969).

Data concerning functional changes of protein metabolism in the neuron-neuroglia unit are quite scanty. Thus Ford and Rhines (1967) compared glutamine-^{14}C incorporation into the proteins of rat spinal cord gray matter as well as into those isolated motor neurons under hyper- and hypothyroidism conditions. If the differences in protein metabolism between the whole gray matter and the separated motor neurons can be accounted for by protein changes in neuroglia, the data obtained indicate that the label incorporation is markedly increased in neurons and decreased in neuroglia as a result of hyperthyroidism while hypothyroidism induces an increase in glial cells with no changes in neurons. Therefore, the increased activity of the thyroid gland seems to decrease the protein synthesis in neuroglial cells while the hypofunction of the gland activated protein synthesis, the same regularity being observed by Ford and Rhines in other rat tissues examined. However, all these considerations do not take into account the nonglial protein of interneuronal space of the gray matter; this protein appears to be mainly of axonal origin. The question of functional changes of axonal proteins still remains to be answered.

By means of two-wave-length cytophotometry it has been shown by Saudargene (1969a) and Saudargene and Pevzner (1969) that acute corazol convulsions give rise to a decrease of total cytoplasmic protein content in rat spinal cord motor neurons and spinal ganglia neurons, while in neuroglia a similar decrease has been observed only in spinal cord ventral horns. The rest at the end of the seizures was characterized by a quicker normalization of protein content in glial cells than in the neurons.

Several Enzymes

Development of the methods of histochemical staining has resulted in the appearance of the great number of papers dealing with a demonstration of activity of many enzymes (particularly of hydrolases and dehydrogenases) within the morphological structures of the nervous tissue. Review of these data with respect to neuron-neuroglia unit has been presented elsewhere (Pevzner, 1969b). Here it should be stressed, however, that the usual enzymohistochemical analyses are sufficiently good to permit judgment as to the localization of the enzymes studied, but quite nonefficient as far as the attempt to compare activity level is concerned. As it has been stated above, the intensity of histoenzymatic reactions depends not only on the enzyme activity itself but also on a number of conditions of the reaction in the histological preparation. Moreover, because these data are as a rule merely qualitative, the conclusions stated concerning the predominance of the enzyme in a definite kind of cell or enzymatic changes accompanying some behavioral events should be considered with great caution. Thus the main object of this chapter will be to discuss the quantitative data obtained by microchemical determinations of enzyme activities.

As early as 1956 Lowry and his coworkers used original microchemical technique to show that the glutamate dehydrogenase activity in the glial capsule of spinal ganglia neurons was about half of the activity in the bodies of the neurons of spinal ganglia and spinal cord ventral horns. The activity of hexosephosphate isomerase was approximately the same (Lowry et al., 1956). Topochemical studies of Ammon horn layers did not reveal significant differences between the neuron and the neuroglia with respect to aldolase activity (Lowry et al., 1954).

Topochemical analysis led Pope and Hess (1957) and Robins et al. (1957) to a conclusion of predominance of hexosemonophosphate shunt (pentose cycle) within carbohydrate catabolism in the glial cells as compared with neurons. These biochemical data were subsequently confirmed by numerous histochemical staining reactions that showed a higher activity in glial than in nerve cells of glucose-6-phosphate dehydrogenase (Lazarus et al., 1962; Schiffer and Vesco, 1962; Abe et al., 1963; Friede et al., 1963; Smith, 1963; Kultas, 1965; Farkas-Bargeton, 1965; Carpenter, 1965; Robinson, 1966; Shimizu and Abe, 1966; Blunt et al., 1967; Matsuura, 1967; DeSibrik and O'Doherty, 1967), 6-phosphogluconate dehydrogenase (Schiffer and Vesco, 1962; Blunt et al., 1967; De Sibrik and O'Doherty, 1967), and transketolase (Collins, 1967).

Cytochrome oxidase activity—if judged by topochemical analysis of the gray and white matter (Abood et al., 1952; Ridge 1967)—is markedly higher in neurons than in the glial cells. Opposite conclusions result, however, when considerations are based on direct determinations of oxygen consumption by means of Cartesian diver technique. Activity of cytochrome oxidase had been shown to be two to three times higher in perineuronal glia clumps surrounding spinal cord motor neurons (Hydén et al., 1958) or Deiters' neurons of vestibular nucleus (Hamburger and Hydén, 1963) as compared with the corresponding neuronal bodies. There are some data concerning changes in the cytochrome oxidase activity within the neuron-neuroglia unit during behavioral events. The hypoxic hypoxia due to maintaining animals for 15 hr in an atmosphere of 8 percent oxygen, as well as the vestibular stimulation due to daily rotation of animals 25 min per day for seven days, have resulted in an increase of this enzyme activity in Deiters' neurons with no changes in perineuronal glial cells; the chemical stimulation due to tricyanoaminopropene administration has induced an activation of the cytochrome oxidase in Deiters' neurons and an inhibition in neuroglia (Hamberger and Hydén, 1963; Hamberger, 1963, 1964; Hydén, 1964). Similar changes have been revealed with respect to cytochrome oxidase activity in Deiters' neurons and glial clumps three and five days after the injection of a monoaminooxidase inhibitor (Hydén and Egyházi, 1968).

Succinate oxidation, on the other hand, has been shown to predominate in the neuroglia. According to Hydén and his co-workers the activity of succinoxidase is higher in the glial capsule around spinal cord motor neurons (Hydén et al., 1958), Deiters' neurons of vestibular nuclei (Hydén and Pigon, 1960; Hamberger and Hydén, 1963), neurons of hypoglossal nucleus (Hamberger and Sjöstrand, 1966), and neurons of formatio reticularis gigantocellular nucleus (Hamberger et al., 1966) than in corresponding neuron bodies.

It is this enzyme activity which has been studied to the greatest degree with respect to various behavioral events or pharacological agents. Thus artificial sleep deprivation induces an inhibition of succinoxidase in formatio reticularis neurons but an activation in perineuronal neuroglial cells (Hydén and Lange, 1964, 1965; Hydén, 1967). In these neurons, the enzyme activity was markedly increased during the natural sleep with no changes during the barbituate sleep, while in the neuroglia both kinds of sleep resulted in a decrease of the activity (Hamberger et al., 1966; Hydén, 1967). Intravenous urea infusion gave rise to an increase of succinoxidase activity; but in Deiters' neurons this increase was quite sharp while in the surrounding neuroglia it was rather slow (Hamberger and Lovtrop, 1964).

Rhythmic changes of intracranial pressure markedly activated the succinate oxidation in the Deiters' neurons without having any significant influence on their glia (Hamberger and Rinder, 1966). Regeneration of hypoglossal nucleus neurons, after the section of hypoglossal nerve, was accompanied by a slight increase of succinoxidase activity in the regenerating neurons and a much more pronounced increase (after the sixth day following the section, in particular) in their glial cell satellites (Hamberger and Sjöstrand, 1966). Intraperitoneal injection of γ-aminobutyric acid (GABA) decreased this enzyme activity in Deiters' neurons but increased it in the neuroglia of rabbit vestibular nucleus, whereas intravenous injection of hydroxylamine (inhibitor of GABA-transaminase) resulted in an opposite effect; intravenous injection of thiosemicarbazide induced an increase of the succinoxidase activity in the neurons with no changes in the glia (Aleksidze and Blomstrand, 1968).

The activity of lactate dehydrogenase is rather marked both in Deiters' neurons and in their glial capsule, but the isoenzyme pattern is quite peculiar: LDH-1 and LDH-2 are more pronounced in the microelectrophoregrams of the neurons while LDH-5 is somewhat more active in glial cells (Aleksidze and Haglid, 1970).

ATPase activity in Deiters' neurons has been found by Crummins and Hydén (1962) to be somewhat lower than in the clumps of neuroglial cells; it is interesting that pH optimum of this enzyme in the neuron is characterized by rather a large area around 8.0, whereas in the glial capsule it is limited strictly to 7.4.

Microchemical determination of glutamate-oxaloacetate transaminase has shown that its activity in the glial capsule of spinal ganglia neurons is six times higher than in the bodies of the ganglia neurons as well as of spinal cord motor neurons (Lowry et al., 1956).

In the enriched fraction of neuroglial cell nuclei the activity of NAD-pyrophosphorylase (ATP: NMN adenyltransferase) has been found by Kurokawa et al. (1966) to be one-tenth of that in the enriched fraction of neuronal nuclei.

Of a great interest are data by Giacobini (1961, 1962) concerning a most pronounced difference between Deiters' neurons enzyme proved to be 120 times more active in the glial clumps of the same volume as that of neuron body; when calculated per one cell, the activity in the glial cell was shown to be six times higher than in the nerve cell. This difference could even be used as a "Giacobini test" (Cremer et al., 1968) to reveal glial contamination of enriched neuronal fractions.

Since the first histochemical works done by Koelle (1950, 1962) and Hebb et al. (1963) a number of histochemical papers have appeared which show that the acetylcholinesterase is localized mainly in the neurons while the butyrylcholinesterase is in the glia (Cavanagh et al., 1954;

Giacobini, 1956; Gerebtzoff, 1959; Zaccheo and Viale, 1963; Kása et al., 1965; Friede, 1965, 1966; Martinez Rodriguez, 1965; Shanta et al., 1967; Matsuura, 1967). This difference in the cellular localization is preserved in the nervous tissue cultivated in vitro (Geiger and Stone, 1962; Geiger, 1963). Histochemical technique (as mentioned above) provides only relative data, the estimation of which should be done with a caution. However, these particular histochemical data agree sufficiently well with recent microchemical data of Pavlin (1969), who found that the rate of the hydrolysis of acetylcholine in the neurons of guinea pig nucleus reticularis ponti caudalis isolated by Hydén's technique is more than 1.5 times higher than in the glial clumps.

Nucleic Acids

DNA amount in the glial cell nuclei as determined both by ultraviolet (Pevzner et al., 1964; Pevzner, 1965b) and by visible cytospectrophotometry (Lapham, 1962; Lapham and Johnstone, 1963; Viola, 1963; Kushch and Yarygin, 1965) has been shown to be mainly diploid, this feature being peculiar to astrocytes as well as to oligodendrogliocytes (Pevzner et al., 1964; Pevzner, 1965a, b). Only a small part of glial cells is characterized by the tetraploid amount of nuclear DNA (Lapham and Johnstone, 1963, 1967; Pevzner et al., 1964; Svanidze, 1967). It is of interest that, according to Syanidze and Bershvili (1969), the polyploid glial cells in the brain cortex are localized as a rule around the bodies of the neurons with polyploid nuclei.

Confirmation of these data has been obtained also by the enriched fraction technique; calculation of DNA amount per cell has shown this amount to be practically the same in the neurons and in the glial cells (Freysz et al., 1968).

RNA amount in glial cells is, of course, much lower than in large nerve cells. According to Hydén and his coworkers (Hydén, 1960, 1962; Hydén and Pigon, 1960; Hydén and Lange, 1961), the giant Deiters' neuron of rabbit vestibular nucleus contains an average of 1,545 pg RNA whereas an approximately equal volume of glial clumps contains as little as 123 pg RNA. Thus the concentration of RNA in the glial clumps is about one-thirteenth that in the neuron body. Our cytophotometric measurements—by means both of ultraviolet (Pevzner, 1964, 1965a, 1966) and of visible cytophotometry (Saudargene and Pevzner, 1969; Saudargene, 1969b)—have indicated, however, RNA concentration in neuronal bodies to be not more than 1.5 times higher than in glial cell satellites. The clumps of glial cells obtained by Hydén's method of free-hand dissection are abundant in dendrites, myelin fragments, synaptic endings, and other structures which are characterized by a slight RNA content. Still greater contaminations of the structures poor in RNA are always present in the enriched glial fraction isolated by Korey's method, so that the calculation of RNA amount per glial cell gave the value 0.24 pg (Korey, 1957, 1958; Korey et al., 1958; Korey and Orchen, 1959) while according to our cytophotometric data the content of RNA per glial cell in various areas of the nervous system amount to 2 to 4 pg (Pevzner, 1965a, b; Brumberg, 1968a; Brumberg and Pevzner, 1968).

The base ratio in RNA of nerve and glial cells differ somewhat; these differences seem to be specific for each kind of neuron-neuroglia unit. Thus RNA of Deiters' neurons differ from that of their glial capsule in higher content of guanine and a lower amount of cytosine (Egyházi and Hydén, 1961). RNA of the neurons of hypoglossal nucleus contains more cytosine and less adenine and uracil than does RNA of surrounding glial cells (Daneholt and Brattgård, 1966). On the contrary, a higher content of adenine and uracil with lower content of guanine has been found in spinal cord motor neurons as compared with glial clumps (Slagel et al., 1966). Human globus pallidus (autopsy material) was characterized by practically identical base ratios of neuronal and glial RNA (Gomirato and Hydén, 1963; Hydén, 1964).

By means of Rose's enriched fraction method, Volpe and Giuditta (1967) separated neuronal and glial RNA. Their composition was rather similar, 4 S, 18 S and 28 S, but in the in-vivo incorporation of orotate-6-^{14}C was at first quicker into the glial 28 S RNA, after 3 to 6 hr into the neuronal RNA, and after 14 hr again into the glial RNA.

On separating the hypoglossal nucleus neurons and their glial capsules by Hydén's microdissection method, Daneholt and Brattgård (1966) in the 4-hr experiments with nucleotides-^{3}H found RNA turnover rate in the neuroglia to be about twice that in the neurons.

Microchemical data of Hydén and his coworkers have shown that an activation of the nervous system either by vestibular stimulation or by tricyanoaminopropene injection gives rise to an increase of RNA content in rabbit Deiters' neurons and to a decrease in the surrounding neuroglia (Hydén and Pignon, 1960; Hydén and Lange, 1961; Hydén, 1962, 1964). The critical question arose as to whether such a reciprocal relationship could deal with some artifacts due to the procedure of free-hand dissection or with some feature peculiar only to this particular kind of neuron-neuroglia unit. However, more recently our determinations, which were carried out by means of quite different methodical approaches (ultraviolet cytospectrophotometry of undisturbed, histologically fixed neuron-neuroglia unit), in different animals (cats, rats), and on different neurons (sympathetic ganglia neurons, spinal cord motor neurons), confirmed Hydén's data; under some conditions of the stimulation of the nervous system activity, we also observed an increase of RNA content in neurons and a decrease in their glial cell satellites (Pevzner, 1965a, 1966; Pevzner and Haidarliu, 1967).

This increase of neuronal RNA together with a decrease of glial RNA permitted Hydén (1959a,b; 1960) to suggest an interesting explanation about a direct transfer of glial RNA to the neuron body by pinocytosis. Although no direct proof of such a transfer has been shown at this time, there are several indirect data which favor this possibility. First of all, numerous observations indicate the possibility of macromolecular transfer by pinocytosis which actively proceeds in the membranes of both neurons (Tobias, 1960; Rosenbluth and Wissing, 1964) and neuroglia cells (Pomerat, 1952; Klatzo and Miquel, 1960; Geiger, 1963; Mizuno and Okamoto, 1964; Brightman, 1965; Nicholls and Wolfe, 1967). In addition, the reciprocal changes of RNA content in the neuron and neuroglia can take place very quickly; in our experiments, for example, the increase of RNA amount per cell in the motor neurons of rat spinal cord and the decrease in their glial cell satellites has been shown cytophotometrically as early as 5 min after the beginning of electric skin stimulation of animals (Pevzner and Haidarliu, 1967). In both cases in which we found these reciprocal changes in the nerve and glial cells (cat superior cervical sympathetic ganglion and rat spinal cord ventral horns), we did not detect any marked changes in RNA concentration in the whole tissue homogenate of the corresponding areas of the nervous system (Baranov and Pevzner, 1963; Haidarliu, 1967b). At last, very convincing data has been obtained by Hydén and Lange (1966); these workers determined RNA base ratio in Deiters' neurons as well as in their glial capsule before and after tricyanoaminopropene administration to rabbits and calculated that Δ RNA (i.e., the fraction which appeared in the neuron and disappeared in the glia) was characterized by identity of base ratios in two compartments of the neuron-neuroglia unit.

Recently, the increase of neuronal RNA and decrease of glial RNA content in rabbit vestibular nucleus has been shown by Hydén and Egyházi (1968) to take place 1 hr after the injection of monoaminoxidase inhibitor, tranylcyclopropylamine.

A number of cytophotometric determinations carried out in our laboratory has shown that the direction of RNA and protein changes in the neurons and their glial cell satellites depends on the functional state of the nervous system. Thus moderate activation of rat motor neurons by 3 to 4 hr swimming in water at 35° C was accompanied only by an increase of neuronal RNA while glial RNA amount per cell did not change during the swimming (Brumberg, 1968a, b; Brumberg and Pevzner, 1968; Pevzner, 1969b). Glial RNA levels has been found to be more stable, as compared with that of neurons, under the influence of hypoxia (Gazenko et al., 1968; Pevzner, 1969b) and antimetabolites such as 6-mercaptopurine and actinomycin B (Pevzner, 1969c). However, strong, acute excitation of the nervous system which gives rise to overexcitation, fatigue, and exhaustion of the nervous system induces a decrease of RNA content both in neurons and glial cells; such RNA decrease in the whole neuron-neuroglia unit has been observed as a result of prolonged electric skin stimulation (Pevzner and Haidarliu, 1967), long-term artificial hypodynamia (Brumberg and Pevzner, 1968), and acute cardiazol (metrazol) seizures (Saudargene and Pevzner, 1969). We consider it of great importance that after the termination of the action including this total decrease of RNA content, the restoration of initial RNA level proceeded in all cases more quickly in the neuroglia than in the neurons (Pevzner and Haidarliu, 1967; Brumberg and Pevzner, 1968; Pevzner, 1968, 1969b; Saudargene and Pevzner, 1969).

At the same time hormonal influences have acted more specifically upon neuroglia metabolism (Pevzner, 1964, 1967, 1968; Antonov and Pevzner, 1968; Doemin et al., 1968). Thus adrenaline (epinephrine) injections in rats and cats daily for two weeks gave rise to an increase of nucleic acid content in the glial satellites surrounding the neurons of the cervical sympathetic ganglion and ventral and lateral horns of spinal cord, whereas in the neurons RNA amount either decreased (spinal cord ventral horns) or did not change at all (Pevzner, 1964, 1965b, 1967). Adrenalectomy resulted in a decrease of RNA content in cerebellum Purkinje cells and spinal cord ventral horns with no changes in glial nucleic acid content. The decrease does not seem to be specific because it was not eliminated by daily injections of cortisol to adrenalectomized rats. But in the hypothalmic supraoptical nucleus (which is known to be more sensitive to hormones), there was a decrease of nucleic acid content only in the glial cells; this time the cortisol injections resulted in a complete normalization of the nucleic acid level.

Conclusion

The conception put forward by Hydén more than 10 years ago (Hydén, et al., 1958; Hydén, 1959a, b) that neuron and neuroglia are united in a single functional and biochemical unit is at present quite certain. However, the problem of elucidating the precise laws of activity of this unit in terms of metabolism is still rather far from being solved. The results of biochemical studies on neuroglia are numerous, but often contradictory. Such contradiction depends both on the use of a number of different methodological approaches and on the differences in metabolic features of neuroglia cells localized in various parts of the nervous system. Several biochemical differences between astroglia and oligodendroglia were reviewed by Laborit (1964, 1965), Friede (1965, 1966), and Pigareva (1966).

Based on the data presented in this chapter as well as on the histochemical data reviewed elsewhere (Pevzner, 1969b), the conclusion can be stated that, in general, the biochemical apparatus of the neuroglial cell is by no means less efficient than that of the neuron. Indeed, the activity of the majority of enzymes, the intensity of oxidative processes, the turnover of macromolecules, and so on are, as a rule, of similar order in the neuroglia and in the neuron.

Glial cells seem to be more resistant to hypoxia as compared with neurons (Hamberger and Hydén, 1963; Hydén, 1964; Pevzner, 1968, 1969b; Gazenko et al., 1968). This fact can be explained in at least two ways. On one hand, glia is characterized by close contact with the capillary network of the brain so that it should be provided with oxygen and energy-yielding substrates from the blood to a greater degree than the neuron. On the other hand, there are data about the predominance in the glia, rather than in the neurons, of lactate dehydrogenase isoenzymes more active under anaerobic conditions (Aleksidze and Haglid, 1970); it has been shown, although only histochemically, that adenyl cyclase activity is localized mainly in the glial cells (Shanta et al., 1966). If this aspect is confirmed by quantitative biochemical determinations, it will indicate the formation of glial glucose-6-phosphate predominantly through the glycogenolysis (i.e., without the use of hexokinase reaction which would require ATP).

There appear to be some differences in transmitter metabolism between the neuron and the glia. As mentioned above, the acetylcholinesterase is more peculiar to neurons whereas the nonspecific cholinesterase is more peculiar to glial cells (see Giacobini, 1956; Gerebtzoff, 1959; Friede, 1965, 1966). Manocha and Bourne (1968) histochemically demonstrated a lower activity of monoaminoxidase in the glial cells as compared with the neurons. This latter fact may be one of the reasons that would explain the difference between Deiters' neurons and their glia as revealed by the work of Hydén and Egyházi (1968): 1 hr after the injection in rabbits of tranylcypromine (an inhibitor of monoaminoxidase), the activity of cytochrome oxidase in the neurons increased markedly while in the glial clumps no statistically significant changes were observed. Several authors also indicate the formation of γ-aminobutyric acid to be localized in the neuron rather than in the neuroglia (Roberts, 1962; Hirsch and Robins, 1962; Pigareva, 1966).

Finally, neuroglial cells seem to be characterized by higher reparative synthesis of macromolecules after some injuries or stress activities of the nervous system. It should be borne in mind that the main mass of glial cell body consists of a nucleus in which the chief apparatus for the control of macromolecular synthesis is localized, whereas in the neuron body the nucleus is only a small part of the cell and a large bulk of the protein which is synthesized within the neuron body constantly leaves the nerve cell through the axonal flow of macromolecules. Perhaps the possibility of glial cells synthesizing nucleic acids and protein in the course of reparative processes depends not only on the close contact of neuroglia to capillary network but also on the predominance in the glia of hexosemonophosphate shunt, which provide for sources for biosynthesis as well as sufficient amount of reduced pyridine nucleotides (Pope and Hess, 1957; Robins et al., 1957; see also reviews by Laborit, 1964, 1965 and Friede, 1966).

It can be concluded that, with respect to macromolecule metabolism, neuron and neuroglia actually represent a single functional biochemical unit. Inside this unit, however, there exists both functional and metabolic "division of labor" between these cellular compartments. Neuronal metabolism appears to be more adapted to function-dependent fluctuations beginning with transmitter fluctuation, but it is more easily exhausted. Glial metabolism reacts to a lesser degree to specific transmitter alterations; we suggest that it is the neuron intermediates which are more specific control agents for neuroglia metabolism. Due to its contact with blood circulation and some biochemical pecularities discussed above, the neuroglia seems to be more resistant to the intensive activity of the nervous system as well as to the hypoxia and other stress factors. Thus it can not be excluded that at definite stages of stimulated activity of neuron-neuroglia unit a transfer of macromolecules takes place from the glial cell satellites to the neuron body. During the reparation period, after the cessation of the stress influence, it is neuroglia that first restores the initial level of its macromolecules and perhaps later promotes the normalization of neuronal macromolecules too. In other words, the neuroglia can be considered as a morphological substrate of homeostatic processes in the nervous system.

References

Abe, T., Yamada, Y., Hashimoto, P. H., and Shimizu, N. (1963). Med. J. Osaka Univ., 14:67.

Abood, L. G., Gerard, R. W., Banks, J., and Tschirgi, R. D. (1952). Amer. J. Physiol., 168:728.

Agroskin, L. S. (1962). Cytologiya, 4:585. ——— (1968). Cytologiya, 10:20.

Alajalova, N. A., and Koltsova, A. V. (1964). Bull. Exp. Biol. Med., 58:12, 9.

Aleksandrovskaya, M. M. (1965). In: Problems of Histo-Haematic Barriers. Stern, L. S., ed. Moscow: Nauka. ——— Geinisman, J. J., and Matz, V. N. (1965). Zh. Neuropath. Psychiat., 65:161. ——— Geinisman, J. J., and Samoilova, L. G. (1964). Bull. Exp. Biol. Med., 59:9, 80.

Aleksidze, N. G., and Blomstrand, C. (1968). Brain. Res. 11:717. ——— and Haglid, K. (1970). Proc. Acad. Sci., U.S.S.R., 190:972.

Antonov, L. M., and Pevzner, L. Z. (1968). Cytologiya, 10:599.
Baranov, M. N., and Pevzner, L. Z. (1963). Biochimiya, 28:958.——— and Pevzner, L. Z. (1964). Adv. Modern Biol., 58:246.
Barer, R. (1956). In: Pollister, A. N., ed. Physical Techniques in Biological Research, Vol. 3: Cells and Tissues. New York: Academic Press.
Barlow, C. F., Domek, N. S., Goldberg, M. A., and Roth, L. G. (1961). Arch. Neurol., 5:102.
Berg, O., and Källen, B. (1959). J. Neuropathol. Exp. Neurol., 18:458.
Birks, R., Katz, B., and Miledi, R. (1960). J. Physiol., 150:145.
Blomstrand, C., and Hamberger, A. (1969). J. Neurochem., 16:1401.
Blunt, M. J., Wendel-Smith, C. P., and Paisley, P. B. (1967). Nature (London), 215:523.
Brightman, M. N. (1965). Amer. J. Anat., 117:193.
Brodsky, V. J. (1956). Adv. Modern Biol., 42.87. ——— (1966). Cell Trophics. Moscow: Nauka.
Brumberg, E. M. (1956). Zh. Gen. Biol., 17:401.
Brumberg, V. A. (1968a). Proc. Acad. Sci., U.S.S.R., 182:228.——— (1968b). Cytologiya, 10:1193.——— (1969). Proc. Acad. Sci., U.S.S.R., 184:1231. ——— and Pevzner, L. Z. (1968). Cytologiya, 10:1452.
Cammermeyer, J. (1960a). Amer. J. Anat., 106:197.——— (1960b). Amer. J. Anat., 107:107.
Carpenter, S. (1965). Neurology, 15:328.
Caspersson, T. (1936). Skand. Arch. Physiol., 73:Suppl. 8.——— (1940). J. Roy. Microsc. Soc., 60:8.——— (1950). Cell Growth and Cell Function. New York: Norton.——— (1955). Experientia, 11:45.
Cavanagh, J. B., Thompson, R. H. S., and Webster, G. R. (1954). Quart. J. Exp. Physiol., 39:185.
Collins, G. H. (1967). Amer. J. Pathol., 50:791.
Coxon, R. V. (1964). In: Comparative Neurochemistry, Richter, D., ed. Oxford: Pergamon Press.
Cremer, J. E., Johnston, P. V., Roots, B. I., and Trevor, A. J. (1968). J. Neurochem., 15:1361.
Cummins, J., and Hydén, H. (1962). Biochim. Biophys. Acta, 60:271.
Daneholt, B., and Brattgård, S.-O. (1966). J. Neurochem., 13:913.
Davies, H. G. (1958). General Cytochemical Methods, Vol. 1. New York: Academic Press.
Dawson, H., and Spaziani, E. (1959). J. Physiol., 149:135.
DeRobertis, E., and Gerschenfeld, H. M. (1961). Int. Rev. Neurobiol., 3:1. ——— Gerschenfeld, H. M., and Wald, F. (1960). In: Structure and Function of the Cerebral Cortex, Tower, D. B., and Schade, J. P., eds. Amsterdam: Elsevier.
DeSibrik, I., and O'Doherty, D. (1967). Arch. Neurol., 16:628.
Dimova, R. (1966). Acta Neurol. Psychiat. Belg., 66:527.——— Duchesne, P. Y., and Csillik, B. (1966). C.R. Soc. Biol., 160:1325.
Doemin, N. N., Nechaeva, G. A., and Pevzner, L. Z. (1968). In: Hormones and Brain. Komissarenko, V. P., ed. Kiev: Naukova Dumka.
Edström, J. E. (1953). Biochim. Biophys. Acta, 12:361. ——— (1956). J. Neurochem., 1:159.——— (1958). J. Neurochem., 3:100.——— (1964). Methods in Cell Physiol., 1:417.
Egyházi, E., and Hydén, H. (1961). J. Biophys. Biochem. Cytol., 10:403.
Ellman, G. L., and Sullivan, C. V. (1965). Acta Neuroveget., 27:184.

Farkas-Bargeton, E. (1965). C.R. Acad. Sci., 261:5649.
Farquhar, M. G., and Hartmann, J. F. (1956). Anat. Rec., 124:288.
Farquhar, M. C., and Hartmann, J. F. (1957). J. Neuropathol. Exp. Neurol., 16:18.
Ford, D. H., and Rhines, R. (1967). Acta Neurol. Scand., 43:33.
Freysz, L., Bieth, R. Judes, C., Jacob, M., and Sensenbrenner, M. (1967). J. de Physiol., 59:239.——— Bieth, R., Judes, C., Sensenbrenner, M., Jacob, M., and Mandel, P. (1968). J. Neurochem., 15:307.
Friede, R. L. (1965). Progr. Brain Res., 15:35. ——— (1966). Topographic Brain Chemistry. New York: Academic Press.——— Fleming, L. M., and Knoller, M. (1963). J. Neurochem., 10:263.
Galambos, R. (1961). Proc. Nat. Acad. Sci. U.S.A., 47:129. ——— (1965). Progr. Brain Res., 15:267.
Gazenko, O. G., Doemin, N. N., Malkin, V. B., and Pevzner, L. Z. (1968). Proc. Acad. Sci., U.S.S.R., 179:997.
Geiger, R. S. (1963). Intern. Rev. Neurobiol., 5:1.——— Adachi, C., and Stewart, M. (1960). Feder. Proc., 19:281.——— and Stone, W. G. (1962). Acta Neurol. Scand., 38:Suppl. 1, 67.
Gerebtzoff, M. A. (1959). Cholinesterases. London: Pergamon Press.
Gerchenfeld, H. M., Wald, F., Zedunaisky, J. A., and DeRobertis, E. D. P. (1959). Neurology, 9:412.
Giacobini, E. (1956). Acta Physiol. Scand., 36:276.——— (1961). Science, 134:1524.——— (1962). J. Neurochem., 9:169.——— (1964). In: Morphological and Biochemical Correlates of Neural Activity, Cohen, M. M., and Sniker, R. S., eds. New York: Harper & Row.
Glees, P. (1955). Neuroglia, Morphology and Function. Springfield, Ill.: Thomas.——— (1961). Experimental Neurology. Oxford: Clarendon Press.
Gormirato, G., and Hydén, H. (1963). Brain, 86:773.
Gracheva, N. D. (1964). Cytologiva, 6:324.
Guth, L., and Watson, P. K. (1968). Exp. Neurol., 22:590.
Haidarliu, S. H. (1967a). Cytologiya, 9:644.——— (1967b). Biochimiya, 32:677.
Hamberger, A. (1963). Acta Physiol. Scand., 58:Suppl. 203. ——— (1964). Bol. Inst. Estud. Méd. Biol., 22:129.——— and Hydén, H. (1963). J. Cell Biol., 16:521.——— Hydén, H., and Lange, P. W. (1966). Science, 151: 1394.——— and Lovtrup, S. (1964). J. Neurochem., 11:687.——— and Rinder, L. (1966). J. Neuropathol. Exp. Neurol., 25:68.——— and Sjöstrand, J. (1966). Acta Physiol. Scand., 67:76.
Hebb, C. O., Silver, A., Swan, A. A. B., and Walsh, E. G. (1953). Quart. J. Exp. Physiol., 38:185.
Hertz, L. (1965). Nature (London), 206:1091. ——— (1968). J. Neurochem., 15:1.
Hess, A. (1958). J. Biophys. Biochem. Cytol., 4:731.
Hild, W., and Tasaki, J. (1962). J. Neurophysiol., 25:277. ——— Tasaki, J., and Chang, J. J. (1958). Experientia, 14:220.
Hirano, A., Zimmerman, H. M., and Levine, S. (1965). Amer. J. Pathol., 47:537.
Hirsch, H. E., and Robins, E. (1962). J. Neurochem., 9:63.
Hornet, Th., Appel, E., Nereanțiu, F., and Voinescu, S. (1960). St. Cercet. Neurol., 5:31.
Horstmann, E., and Meves, H. (1959). Z. Zellforsch, 49:569.
Hydén, H. (1943). Acta Physiol. Scand., 6:Suppl. 17.——— (1947). Symp. Soc.

Exp. Biol., 1:152. ——— (1955). In: Biochemistry of the Developing Nervous System, Waelsch, H., ed. New York: Academic Press. ——— (1959a). Nature (London), 184:433. ——— (1959b). Fourth Intern. Congr. Biochem., London, 3:64. ——— (1960). In: The Cell, Brachet, J., and Mirsky, A. F., eds. New York: Academic Press. ——— (1962). In: Neurochemistry, Elliott, K. A. C., Page, I. H., and Quastel, J. H., eds. Springfield, Ill.: Thomas. ——— (1963). In: The Effect of Use and Disuse on Neuromuscular Function. Gutmann, E., and Hnik, P., eds. Prague: Publishing House of the Czechoslovak Academy of Sciences. ——— (1964). In: Recent Advances in Biological Psychiatry, Wortis, J., ed. New York: Plenum Press. ——— ed. (1967). Neuron, Amsterdam: Elsevier. ——— (1969). Second Intern. Meeting of the Intern. Society for Neurochemistry. Milano, 10. ——— and Egyházi, E. (1968). Neurology, 18:732. ——— and Lange, P. 1961). Regional Neurochemistry. Oxford: Pergamon Press. ——— and Lange, P. W. (1964). Life Sci., 3:1215. ——— and Lange, P. W. (1965). Science, 149:654. ——— and Lange, P. W. (1966). Naturwissensch., 53:64. ——— Lovtrup, S., and Pigon, A. (1958). J. Neurochem., 2:304. ——— and McEwen, B. (1966). Proc. Nat. Acad. Sci. U.S.A., 55: 354. ——— and Pigon, A. (1960). J. Neurochem., 6:57.

Jensen, A. V., Becker, R. F., and Windle, W. F. (1948). Arch. Neurol. Psychiat., 60:221.

Kaplan, L. L. (1965). In: Morphology of the Pathways and Connections of the Central Nervous System. Iontov, A. S., and Mayorov, V. N., eds. Moscow-Leningrad: Nauka.

Kása, P., Joó, F., and Csillik, B. J. (1965), Neurochem., 12:31.

Katzman, R. (1961). Neurology, 11:27.

Kety, S. S., and Elkes, J. (1961). Regional Neurochemistry. Oxford: Pergamon Press.

Klatzo, J., and Miquel, J. (1960). J. Neuropathol. Exp. Neurol., 19:475.

Kleihues, P., and Schultze, B. (1968). Acta Neuropathol., 10, Suppl. 4:121.

Koch, A., Ranck, J. B., and Newman, B. L. (1962). Exp. Neurol., 6:186.

Koelle, G. B. (1950). J. Pharmacol. Exp. Ther., 100:157. ——— (1952). J. Pharmacol. Exp. Ther., 106:401.

Korey, S. R. (1957). In: Metabolism of the Nervous System, Lajtha, J., ed. London: Pergamon Press. ——— (1958). In: Biology of Neuroglia, Windle, S. F., ed. Springfield: Thomas. ——— and Orchen, M. (1959). J. Neurochem., 3:277. ——— Orchen, M., and Brotz, M. (1958). J. Neuropathol. Exp. Neurol., 17:430.

Kuffler, S. W. (1967). Proc. Roy. Soc., B 168:1. ——— and Nicholls, J. G. (1964). Arch. Exp. Pathol. Pharmakol., 248:216. ——— and Nicholls, J. G. (1965). Perspect. Biol. Med., 9:69. ——— and Nicholls, J. G. (1966). Ergebn. Physiol., 57:1. ——— Nicholls, J. G., and Orkand, R. K. (1966). J. Neurophysiol., 29:768. ——— and Potter, D. D. (1964). J. Neurophysiol., 27:290.

Kulenkamff, H., and Wüstenfeld, E. (1954). Z. Anat. Entwickl., 118:97.

Kultas, K. (1965). Folia Morphol., 13:43.

Kuntz, A., and Sulkin, W. M. (1947). J. Comp. Neurol., 86:467.

Kurokawa, M., Kto, T., and Inamura, H. (1966). Proc. Japan Academy, 42:1217.

Kushch, A. A., and Yarygin, V. N. (1965). Cytologiya, 7:228.

Laborit, H. (1964). Agressologie, 5:99. ——— (1965). Lés Régulations Metaboliques. Paris: Masson. ——— and Laborit, G. (1965). Agressologie, 6:639.

Lapham, L. W. (1962). Amer. J. Pathol., 41:1. ——— and Johnstone, M. A. (1963). Arch. Neurol., 9:194. ——— and Johnstone, M. A. (1964). J. Exp. Neuropathol. Neurol., 23:419.
Lasansky, A. (1965). Progr. Brain Res., 15:48.
Lazarus, S. S., Wallace, B. J., Edgar, G. W. F., and Wolk, B. W. (1962). J. Neurochem., 9:227.
Leibson, N. L. (1967). In: Evolutionary Neurophysiology and Neurochemistry. Kreps, E. M., ed. Leningrad: Nauka.
Leuchtenberger, C. (1954). Science, 120:1022. ——— (1958). In: General Cytochemical Methods, Danielli, J. F., ed. New York: Academic Press.
Lodin, Z. (1964). Ceskoslov. Fysiol., 13:126. ——— Hartman, J., Kazakhashvili, M., and Müller, J. (1968). In: Macromolecules and the Function of the Neuron, Lodin, S. F., and Rose, S., ed. Amsterdam: Excerpta Medica.
Lord, K. A., Gregory, G. E., and Burt, P. E. (1967). J. Exp. Biol., 46:153.
Lowry, O. H. (1941). J. Biol. Chem., 140:183. ——— (1953). J. Histochem. Cytochem., 1:420. ——— (1955). In: Biochemistry of the Developing Nervous System, Waelsch, H., ed. New York: Academic Press. ——— In: Metabolism of the Nervous System, Lajtha, J., ed. London: Pergamon Press. ——— (1962). Bull. New York Acad. Med., 38:789. ——— (1966). Feder. Proc., 25:846. ——— and Bessey, O. A. (1946). J. Biol. Chem., 163:633. ——— Roberts, N. R., and Chang, M.-L. W. (1956). J. Biol. Chem., 222:97. ——— Roberts, N. R., Leiner, K. Y., Wu, M.-L., Farr, A. L., and Albers, R. W. (1954). J. Biol. Chem., 207:39. ——— Rosebrough, N. J., Farr, A. L., and Randall, R. J. (1951). J. Biol. Chem., 193:265.
Lucas, B. G. B., and Strangeways, D. H. (1963). J. Pathol. Bacteriol., 86:273.
Lumsden, C. E. (1955). Excerpta Med., 8:832.
Luse, S. A. (1956). J. Biophys. Biochem. Cytol., 2:531. ——— (1960). Anat. Rec., 138:461. ——— and Harris, B. (1960). J. Neurosurg., 17:439.
Manina, A. A. (1966). In: Mechanisms of Activity of the Central Neuron. Biryukov, D. A., ed. Moscow-Leningrad: Nauka.
Manocha, S. L., and Bourne, G. H. (1968). J. Neurochem., 15:1033.
Martinez Rodriguez, R. (1965). Trab. Inst. Cajal. Investig., 57:29.
Matsuura, H. (1967). Mistochemie, 11:152.
Matz, V. N. (1966). Proc. Acad. Sci., U.S.S.R., 169:1474.
Maynard, E. A., and Pease, D. C. (1955). Anat. Rec., 121:440.
McEwen, B. S., and Hydén, H. (1966). J. Neurochem., 13:823.
Mihailović, Lj., Cupić, D., and Dekleva, N. (1968). Proc. Intern. Union Physiol. Sci., 6:883.
Mizuno, N., and Okamoto, M. (1964). Arch. Histol. Japon., 24:347.
Nakai, J. (1963). Morphology of Neuroglia. Tokyo: Igaku Shoin Ltd.
Nicholls, J. C., and Kuffler, S. W. (1964). J. Neurophysiol., 27:645. ——— and Kuffler, S. W. (1965). J. Neurophysiol., 28:519.
Nicholls, J. G., and Wolfe, D. E. (1967). J. Neurophysiol., 30:1574.
Niculescu, I. T. (1963). Pathomorphology of the Nervous System. Bucharest: Medical Publishers.
Palladin, A. V. (1959). Ukrain. Biochem. Zh., 31:765.
Pappius, H. M. (1965). Progr. Brain Res., 15:135. ——— (1969). In: Second Intern. Meeting of the Intern. Society for Neurochemistry. Pasletti, R., Fumagalli, R., and Galli, C., eds. Milano, 56. ——— Klatzo, J., and Elliott, K. A. C. (1962). Canad. J. Biochem. Physiol., 40:885.
Pavlin, R. (1969). In: Second Intern. Meeting of the Intern. Society for Neurochemistry. Pasletti, R., Fumagalli, R., and Gallis, C., eds. Milano, 314.

Penfield, W. (1932). Cytology and Cellular Pathology of the Nervous System. New York: Hoeber.
Pevzner, L. Z. (1963). Ukrain. Biochem. Zh., 35:448. ——— (1964). Proc. Acad. Sci., U.S.S.R., 156:1213. ——— (1965a). J. Neurochem., 12:993. ——— (1965b). In: Morphology of the Pathways and Connections of the Central Nervous System. Iontov, A. S., and Mayorov, V. N., eds. Moscow-Leningrad: Nauka. ——— (1966). In: Macromolecules and Behaviour., 1st ed., Gaito, J., ed. New York: Appleton-Century-Crofts. ——— (1967). In: Biochemistry and Function of the Nervous System. Kreps, E. M., ed. Leningrad: Nauka. ——— (1968). In: Macromolecules and the Function of the Neuron, Lodin, Z., and Rose, S., eds. Amsterdam: Excerpta Medica. ——— (1969a). Vopr. Med. Chem., 15:211. ——— (1969b). Adv. Modern Biol., Vol. 68. ——— (1969c). Cytologiya, 11:856. ——— and Litinskaya, L. L. (1968). Cytologiya, 10:812. ——— and Tomina, E. D. (1965). Vopr. Med. Chem., 11:3. ——— Tomina, E. D., and Chaika, T. V. (1964). Vopr. Med. Chem., 10:379. ——— and Haidarliu, S. H. (1967). Cytologiya, 9:840.
Pigareva, S. D. (1966). Zh. Neuropathol. Psychiat., 66:1716.
Polak, M. (1965). Progr. Brain Res., 15:12.
Pomerat, C. M. (1952). Texas Rep. Biol. Med., 10:885.
Pope, A., and Hess, H. (1957). In: Metabolism of the Nervous System, Lajtha, J., ed. New York: Pergamon Press.
Portugalov, V. V. (1959). In: Actual Problems of Modern Biochemistry. Orekhovich, V. N., ed. Moscow: Publishing House of the Academy of Science of the U.S.S.R.
Quastel, J. H., and Quastel, D. M. J. (1961). The Chemistry of Brain Metabolism in Health and Disease. Springfield: Thomas.
Ridge, J. W. (1967). Biochem. J., 102:617.
Rigler, R. (1966). Acta Physiol. Scand., 67:Suppl. 276.
Roberts, E. (1962). Res. Publ. Ass. Nerv. Ment. Dis., 15:288.
Robins, E., Smith, D. E., and Jen, M. K. (1957). Progr. Neurobiol., 2:205.
Robinson, N. (1966). Acta Neuropathol., 7:101.
Roessmann, U., and Friede, R. L. (1967). Exp. Neurol., 19:508.
Roitbak, A. I. (1963). Zh. Higher Nerv. Activity, 13:859. ——— (1964). In: Role of Gamma-Aminobutyric Acid in the Activity of the Nervous System. Yakovlev, N. N., ed. Leningrad: University Press. ——— (1965). In: Modern Problems of Physiology and Pathology of the Nervous System. Parin, V. V., ed. Moscow: Medicine ——— (1968). In: Integrative Activity of the Nervous System in Norm and Pathology. Anokhin, P. K., and Chernukh, A. M., eds. Moscow: Medicine.
Roots, B. J., and Johnston, P. V. (1964). J. Ultrastruct. Res., 10:350. ——— and Johnston, P. V. (1965). Biochem. J., 94:61.
Rose, S. P. R. (1965). Nature (London), 206:621. ——— (1967). Biochem. J., 102:33. ——— (1969). Second Intern. Meeting of the Intern. Society for Neurochemistry. Milano: Tamburini Editore.
Rosenbluth, J., and Wissing, S. L. (1964). J. Cell Biol., 23:307.
Roskin, G. I. (1959). Arch. Anat. Histol. Embriol., 36:3.
Rosomoff, H. L., and Zugibe, F. T. (1963). Arch. Neurol., 9:26.
Ruščák, M., Ruščákova, D., and Konîková, E. (1967). Biologia, 22:337. ——— Ruščákova, D., and Hager, H. (1968). Physiol. Bohemoslov., 17:113.
Sandritter, W., Kiefer, G., and Rick, W. (1966). In: Introduction to Quantitative Cytochemistry, Wied, G., ed. New York: Academic Press.

Satake, M., and Abe, S. (1966). J. Biochem., 59:72.
Saudargene, D. S. (1969a). Cytologiya, 11:1034. ——— (1969b). Cytologiya, 11:642. ——— and Pevzner, L. Z. (1969). Cytologiya, 11:1275.
Schiffer, D., and Vesco, C. (1962). Acta Neuropathol., 2:103.
Schultz, R. L., Maynard, E. A., and Pease, D. C. (1957). Amer. J. Anat., 100:369.
Shanta, T. R., Manocha, S. L., and Bourne, G. H. (1967). Histochemie, 10:234. ——— Woods, W. D., Waitzman, M. B., and Bourne, G. H. (1966). Histochemie, 7:177.
Shimizu, N., and Abe, T. (1966). Progr. Brain Res., 21A:197.
Slagel, D. E., Hartmann, H. A., and Edström, J. E. (1966). J. Neuropathol. Exp. Neurol., 25:244.
Smith, B. (1963). Brain, 86:89.
Snesarev, P. E. (1946). General Histopathology of Brain Injury. Moscow: Medgiz. ——— (1961). Selective Works. Moscow: Medgiz.
Sokolov, E. N. (1962). In: Principal Problems of Electrophysiology of the Central Nervous System. Makarchenko, A. F., ed. Kiev: Publishing House of the Academy of Science of Ukrainian S.S.R.
Streicher, E., Ferris, P. J., Prokop, J. D., and Klatzo, J. (1964). Arch. Neurol. 11:444.
Sulkin, N. M., and Kuntz, A. (1950). Anat. Rec. 108:255.
Svaetichin, G., Negishi, K., Fatenchand, R., Drujan, B. D., and Selvin de Tesla, A. (1965). Progr. Brain Res., 15:243.
Svanidze, I. K. (1967). Zh. Gen. Biol., 28:697. ——— and Berishvili, V. G. (1969). V All-Union Conference on Neurochemistry. Tbilisio.
Swift, H. (1966). J. Histochem. Cytochem., 14:842.
Talwar, G. P., Chopra, S. P., Goel, B. K., and D'Monte, B. (1966). J. Neurochem., 13:109.
Tasaki, J. (1965). Progr. Brain Res., 15:234.
Tobias, J. M. (1960). J. Gen. Physiol., 43:57.
Tomina, E. D., and Pevzner, L. Z. (1965). Bull. Exp. Biol. Med., 59:83.
Torack, R. M., Terry, R. D., and Zimmerman, H. M. (1959). Amer. J. Pathol., 35:1135.
Utley, J. D. (1963). Biochem. Pharmacol., 12:1228. ——— (1964). Biochem. Pharmacol., 13:1383.
Van Harreveld, A., Collewijn, H., and Malhotra, S. K. (1966). Amer. J. Physiol., 210:251. ——— and Stamm, J. S. (1954). Amer. J. Physiol., 178:117.
Vernadakis, A., and Woodbury, D. M. (1965). Arch. Neurol., 12:284.
Viola, M. P. (1963). Sperimentale, 113:317.
Volpe, P., and Giuditta, A. (1967). Brain Res., 6:228.
Walker, P. M. B., and Richars, B. M. (1959). In: The Cell, Brachet, J., and Mirsky, A. F., eds. New York: Academic Press.
Weatherford, T. (1961). J. Neuropathol. Exp. Neurol., 20:440.
Woodward, D. L., Reed, D. J., and Woodbury, D. M. (1967). Amer. J. Physiol., 212:367.
Wyckoff, R. W. G., and Young, J. Z. (1956). Proc. Roy. Soc., 144B:440.
Zaccheo, D., and Viale, G. (1963). Ann. d'Histochim., 8:259.
Zakharyevsky, A. N., and Kuznetsova, A. F. (1961). Cytologiya, 3:245.

Section IV

MODELS OF MEMORY

In this section two models of memory are presented: one is the viewpoint of the molecular biologist; the other, that of the neuropsychologist.

Chapter 16 (Bonner) presents a model that has great attraction for many scientists and which is consistent with considerable molecular biological data. It is suggested that DNA cistrons are repressed and that during learning events neurochemical reactions within the brain cause a depression of these gene sites.

The final chapter (Pribram) involves a synthesis of a wealth of information from broad areas of biology and psychology. The author presents an interesting set of speculations concerning macromolecules and behavior. One of his suggestions is that RNA molecules function as inducers to derepress the gene site.

Chapter 16

MOLECULAR BIOLOGICAL APPROACHES TO THE STUDY OF MEMORY

JAMES BONNER

California Institute of Technology
Pasadena, California

We have seen accumulate in recent time an impressive and ever-increasing body of information which indicates that learning and memory are associated with increases in RNA and protein content on the part of those neurons involved. If this association is in fact real, the next question will be: "What is the nature of the association and how might we imagine this synthesis in the neurons of RNA and of protein to be associated with the process of memory?" It is with this question that the present chapter is concerned.

The operation of the nervous system is, of course, basically electrical. Data acquisition, data processing, even short- and medium-term memories are all processes that are interfered with or abolished by random electrical discharge through the cortex (Brazier, 1960). These processes are not choice candidates for consideration in relation to gross chemical changes accompanying neural activity, but the long-term (or permanent) memory is a different matter. Given a sufficient time (an hour or so after the learning act), information becomes encoded in long-term memory, in a form resistant to electric shock, and resistant also to the cooling of the brain to a temperature sufficiently low that cortical electrical activity can no longer be detected (Andjus et al., 1956). The act by which the electrical display of short-term memory is converted to long-term permanent memory is clearly the obvious candidate for discussion in terms of chemical change. Let us then go to the encoding of information in permanent memory.

Information is encoded in the brain in a form much abstracted and symbolized from the original, and it is therefore convenient for our discussion to think of memory in terms of registers, in which information is displayed in binary form. This mode of thinking is almost certainly over-simplified, but it must do until some more sophisticated model comes along. Let us then imagine that we have, in the particular region of the cortex involved and before the learning process, an empty register, one that contains no information. As a result of the learning process the register is reset. Some information is displayed. The elements of the register are either individual neurons or more probably individual synaptic junctions between neurons. Resetting of an element of the register means that a change in electrical properties of that element have taken place. In the case of the short—and medium—term memories such change is reversible and in fact reversed by time, as well as by electrical activity. In the case of permanent memory the changes in electrical properties associated with the resetting of the unit of the register are permanent ones. The bulk of neuroanatomical evidence leans in the direction of indicating that learning and memory are not associated with the formation of new electrical junctions, that they are not associated with the soldering of the system. To be sure, not all neurophysiologists agree with this point of view (notably Weiss and Hiscoe, 1948). For the purposes of our present discussion we think rather in terms of changes in the chemical properties of the neurons involved. Halstead made the suggestion as long ago as 1951 that some kind of alteration in the nucleoprotein of the neuron is the neuronal chemical change responsible for memory. Halstead's suggestion has certainly been massively disregarded in the intervening years. It was a premature suggestion in the sense that the tools were not then available for the study of alterations in nucleoproteins, nor was there at hand understanding of the ways in which nucleoproteins are involved in cell economy. Today these tools and understanding exist. The nucleoproteins are related directly to the RNA metabolism of the cell. Let us then survey in more detail the experiments concerning the RNA content of neurons in relation to learning and memory. Consider for example the sensorily deprived or sensorily overprivileged rats studied by Hydén (1961). One set of rats are kept by themselves in dark, quiet boxes, so that they get no visual or auditory information. They are sensorily deprived. Other rats are put in a common box together; they may play with one another; they are taught to perform additional tasks that require visual or auditory discrimination. We then look at the visual and auditory cortices of the two kinds of rats. Structurally they are similar—that is, sensory deprivation has not altered the wiring diagram of the cortex in any obvious way. Compositionally, however, the two kinds of cortices are quite different. The sensorily deprived cortical areas are impoverished in both RNA and protein as compared with the similar cortical areas of rats which have led a culturally and sensorily rich life.

That increases in cortical content of RNA are associated with availability of sensory information, a fact first demonstrated by Hydén, does not rigorously show that such increases are, in fact, associated with memory. Other types of experiments do, however, succeed in doing just this —e.g., the experiments of Hydén and Egyházi (1963), in which rats were taught to climb and balance upon a wire in order to reach their food, and in which the RNA content of the vestibular geniculate was compared with that of rats who have neither performed nor learned such a task. The result is clear-cut: RNA content is increased in that part of the brain concerned with balance, when learning about balance is acquired. A similar conclusion flows from the experiments on handedness of Hydén and Egyházi (1964). This point is also elegantly demonstrated by the work of Morrell (1961), an experiment that concerned the way in which damage to a spot on one-half of the cortex (that is, an epileptic focus) results in the production of a mirror focus on the other, uninjured, side of the cortex. The undamaged side acquires information as to exactly where the lesion is on the damaged side; this learning is associated with an increase in RNA at the mirror focus, and there alone. (For further studies on this matter see Gaito, 1971.)

There are, then, good reasons to suspect that increases in RNA content are associated with those alterations in a neuron or neuronal network which constitute long term memory. What further can we say about RNA's role in memory? It has, of course, been suggested that information stored in memory is stored in the form of new RNA molecules which then contain the experiential data written out in the RNA code. This concept is behind Hydén's early emphasis on the fact that the RNA synthesized during learning possesses a nucleotide composition different from that of RNA made in the same cells during the course of nonlearning. No thought is so fantastic that the molecular biologist should not try thinking it for a while. However, the thought presently being considered becomes less and less appetizing as one thinks about it more. It requires that information encoded in electrical form be transcribed into base sequence form. It requires neurons to manufacture RNA of base sequences not specified by their DNA. This runs counter to all of our understanding of cell processes. In the normal cell all RNA is made by DNA-dependent RNA synthesis, and we know of no reason why the neurons of the cortex should be an exception to the general rule. Actinomycin D, a specific inhibitor of DNA-dependent RNA synthesis, when applied topically to the cortex is highly inhibitory to cortical RNA synthesis, as shown by Barondes and Jarvik (1964). The concept that RNA constitutes the memory molecule is then, not in its simple and naive sense, a persuasive one—as has been particularly cogently expressed by Gaito (1963). The basic problem upon which we center is what kind of biochemistry might conceivably relate electrical input to a permanently altered posture of a unit within our memory register. The unity of biochemical devices from lowest to highest cells and creatures encourages one to hold as a working hypothesis that the machinery of the neuron and of the brain is constructed and operated with the same logic so successfully used in the solution of other developmental and evolutionary life problems. Looked at in this light, our first thought will certainly be that electrical input to a neuron derepresses a particular gene whose function it is to be derepressed by this particular modality of effector, of derepressor substance. Derepression of such a previously repressed gene would result in increased RNA synthesis as is observed. The gene, once derepressed, may stay derepressed forevermore, thus causing a permanent change in posture of the unit, the encoding of a unit of permanent memory.

We can be somewhat more specific and more detailed in discussion of the derepression model of permanent memory. Consider a neuron and a single synapse. A signal is transmitted through the synapse and causes the neuron to fire. The problem is how, as a result of such an event, to permanently increase the probability that a signal from the same synapse will result in firing of the same neuron. We will assume that the DNA-dependent synthesis of RNA is somehow involved; we know, in fact, that the learning—and memory—induced RNA synthesis possesses the earmarks of derepression—for example, the increase in amount of such synthesis. We already know from neurophysiology of one way in which electrical signals are transformed into chemistry. We know a part, at least, of what happens at the synapse. An electrical impulse traveling down an axon reaches the end of the fiber. There it finds a small number of "bags" of transmitter substance; for example, acetylcholine. One or more bags are ruptured by the arriving impulse and the acetylcholine thus released is then free to diffuse across the synaptic junction and depolarize the membrane of the adjacent dendrite. In general the acetylcholine is then decomposed by cholinesterase in the dendrite. The derepressing substance of memory cannot be acetylcholine, since the experiments of Rosenzweig et al. (1960) have shown that alterations in acetylcholine and cholinesterase concentration in cortical tissues, as a result of learning, are small. One candidate model that one might envisage is that the dendrite also contains small bags, bags containing preformed (perhaps in early development) material of small molecular nature. That the postsynaptic dendrite is in fact full of such bags is known from the electronmicroscopic studies of Gray and Young (1964). These bags, according to the model, are to be broken by the act of transmission across the synapse. The substance thus released would be, in the first place, perhaps a sensitizor of the dendrite for synaptic conduction, and in the second place an effector substance for the derepression of genes to make messenger RNA, to make the enzymes, to make more of the substance itself. Since a single cortical neuron may have a hundred or more synaptic inputs, it is possible that a hundred or more different effector substances and their relevant genes might operate in a single neuron, each specific to a particular synapse. But, let us not make our model overspecific, because to do so would almost certainly render it incorrect. To summarize its principal features, we may say that the learning-induced chemical alteration in the neuron that results from permanent memory and which appears to be associated with the synthesis of new RNA would, according to this model, be derepression of a previously repressed gene or genes for the ultimate manufacture of substances which render more sensitive (or less sensitive) to conduction a particular synapse or synapses. Our model possesses the virtue that it makes predictions which are subject to experimental test, and in fact to simple tests. It predicts in the first place, of course, that the act of storage in permanent memory should be actinomycin D inhibitable. A number of experiments with actinomycin D have been conducted, but the results are variable (see review by Gaito, 1971). Our model predicts in the second place that the RNA made during the learning process should be a gene product and be therefore hybridizable with the genomal DNA. It is in contrast in this respect to the memory-molecule view of cortical RNA which would envisage the learning-induced RNA as made by non-DNA-dependent RNA synthesis. According to this latter view, the learning-induced RNA should of course be not hybridizable with the genomal DNA. And finally, our model predicts that the learning-induced RNA, since it is made by transcription of genes previously repressed, should be different from the RNA made in the same cell in the absence of learning. This last prediction gets at the very heart and core of the question of whether storage in permanent memory is associated with derepression of genes. An extraordinarily desirable experiment along this line would be that of hybridization competition experiments with use of pulse-labeled RNA from the cortices of sensorily deprived rats and of sensorily affluent rats. If learning and memory are associated with gene derepression, the pulse-labeled RNA of sensorily affluent rats should contain new species, distinguishable by hybridization and different from the species present in the pulse-labeled RNA of sensorily deprived rats. The outcome of this experiment would tell us with certainty and rigor whether derepression is at work in the memory process. Experiments have been conducted with competition hybridization procedures for shock avoidance trained rats with positive results, but further work is required before definite conclusions can be offered (Machlus and Gaito, 1969). If the three predictions of the model outlined above are not sustained by experiment, we are of course no worse off in our understanding of memory. If, on the other hand, some of our predictions are fulfilled, then the entire armament of molecular biology and of classical enzymology may be trained upon one of the oldest, hardest, and most slippery of biological problems: namely, the nature of memory and, in a very real sense, the nature of human nature.

References

Andjus, R. K., Knopfelmacher, F., Russell, R. W., and Smith, U. (1956). Quart. J. Exp. Psychol., 8:15.
Barondes, S., and Jarvik, M. (1964). J. Neurochem., 11:187.
Brazier, M. A. D. (1960). Electroenceph. Clin. Neurophysiol., Suppl. 13.
De Robertis, E., Salganicoff, L., Zieher, L. M., and DeLores Arnaiz (1963). Science, 140:300.
Gaito, J. (1963). Psychol. Rev., 70:471. ——— (1971). DNA Complex and Adaptive Behavior. Englewood Cliffs, N.J.: Prentice-Hall.
Halstead, W. (1951). In: Cerebral Mechanisms in Behavior, Jeffress, L. A., ed. New York: Wiley.
Hydén, H. (1961). Sci. Amer., 205:62. ——— and Egyházi, E. (1963). Proc. Nat. Acad. Sci. U.S.A., 49:620. ——— and Egyházi, E. (1964). Proc. Nat. Acad. Sci. U.S.A., 52:1030.
Machlus, B., and Gaito, J. (1969). Nature (London), 222:573.
Morrell, F. (1961). Physiol. Rev., 41:443.
Rosenzweig, M. R., Krech, D., and Bennett, E. L. (1960). Psychol. Bull., 57:476.
Weiss, P., and Hiscoe, H. B. (1948). J. Exp. Zool., 107:315.

Chapter 17

Some Dimensions of Remembering: Steps Toward a Neuropsychological Model of Memory

KARL H. PRIBRAM

Stanford University, Stanford, California

Interest in the relation of macromolecules to behavior centers on the memory process. It is memory that allows an organism to act on the basis of occurrences removed in time—past and future. It is memory also that allows an organism to act appropriately to present circumstances, for without memory these events constitute nothing more than William James' "buzzing, blooming confusion." The guiding assumption is that this "memory" is effected by macromolecular change in protoplasm, especially in brain tissue. A good part of the search has been for *the* memory macromolecule; the contents of the present volume attest to the success attained by this approach.

Yet, psychology and neurology and even molecular biology stand to lose much if this continues to be the main approach to the problem. Memory is not of-a-piece; it is multidimensional. Psychologists have long been aware of the differences between recognition and recall, between long—and short—term memory span, and similar dichotomies. And neurologists have been concerned not only with memory storage but also with the mechanism of retrieval. As a rule, macromolecular processes have been dismissed by these disciplines as important only at the most reductive level—i.e., usually macromolecules have been relegated to the task of long-term storage, *period*. This makes sense only if the sole dimension of memory recognized is that of duration.

There are, however, complexities in the process of remembering which are not easily resolved by this time honored—I am tempted to say hoary—approach. Perhaps the most obvious regards one already mentioned; appropriate reactions to present circumstances. For example, recognition involves not only a memory mechanism of such short duration that it is practically instantaneous (How many faces can one recognize in a second?) but also a memory store which is practically unlimited in duration ("You look just the same as when I last saw you—no, it couldn't be twenty years, could it?").

A fresh look at memory seems in order. In the following account I have drawn freely from both old and new knowledge in experimental psychology, neurology, computer and information sciences, as well as from classical molecular biology, for suggestions about the dimensions of remembering and a model of memory. Much of what I have to say is speculative, but the speculation rests solidly on data ordinarily ignored in discussions of memory. The hope is that in this context memory molecules will become properly plural and some of the old pros and cons will give way to new questions for experimentalists.

Experiencing Experience

Look at a friend, then look at his neighbor, and immediately you experience the difference. In the auditory mode, such transient, rapidly paced recognition—of phrases in music, of phonemic combinations of speech, and so forth—are commonplace. Ordinary views of the memory mechanism have considerable difficulty handling the immediacy, precision and apparent multidimensionality of the evanescent experience. Here a unique process must be in operation. What could it look like; how might it work?

Habit, Habituation, and Awareness

Let me begin by detailing a paradox concerning experience on the one hand and behavior on the other. There are influences on behavior of which we are not aware. In fact, instrumental behavior and awareness are often opposed—the more efficient a performance, the less aware we become. This antagonism is epitomized by Sherrington (1947): "Between reflex action and mind there seems to be actual opposition. Reflex action and mind seem almost mutually exclusive—the more reflex the reflex, the less does mind accompany it." Thus a range of problems is ignored if the focus of inquiry on memory is purely behavioristic.

The reciprocal relationship between experiencing and behavior is perhaps best illuminated by the psychological processes of habit and habituation. If we are repeatedly in the same situation, in an invariant environment, two things happen. One is that if we have consistently to perform a similar task in that environment, the task becomes fairly automatic, i.e., we become more efficient. We say the organism (in this case, ourselves) has learned to perform the task; he has formed *habits* regarding it. At the same time the subject habituates, by which we mean that he no longer notices the events constant to this particular task in this environment. His verbal reports of introspection, his failure to move his head and eyes in the direction of the stimulus—electrophysiological measures such as galvanic skin response, plethysmography, and EEG—all attest to the disappearance of orienting with repetition of unvarying input in an unvarying situation. However, habituation is *not* an indication of some loss of sensitivity on the part of the nervous system but rather the development of a neural model of the environment, an expectancy, a type of memory mechanism against which inputs are constantly matched. The nervous system is thus continually tuned *by* inputs to process further inputs (Sokolov, 1960).

It is hardly necessary to state that habitual performance of the organism is also due to neural activity. In the case of expectancy there appears to be a diminution of input processing with repetition; in the case of performance, enhanced efficiency of output processing apparently occurs. So the question is: What is the difference between the two kinds of neural activity that makes awareness inversely related to habit and habituation?

Nerve impulses on the one hand, and the slow potentials that occur at synaptic junctions and in dendrites on the other, are available as two kinds of neural processes that could function reciprocally. A simple hypothesis would state that the more efficient the processing of arrival patterns of nerve impulses into departure patterns, the shorter the duration of the design formed by the slow potential junctional microstructure. Once habit and habituation have occurred behavior becomes "reflex." Meanwhile, the more or less persistent designs of slow potential patterns that become constituted are coordinate with awareness. If this view is accepted, it carries with it a corollary, viz., that nerve impulse patterns per se and the behavior they generate are unavailable to immediate awareness. Thus even the production of speech is "unconscious" at the moment the words are spoken. My hypothesis, therefore, is an old fashioned one: that we experience in awareness *some* of the events going on in the brain but not *all*.

In short, nerve impulses arriving at junctions generate a slow potential microstructure. The design of this microstructure interacts with that already present by virtue of the spontaneous activity of the nervous system and its previous "experience." The interaction is enhanced by inhibitory processes and the whole procedure produces interference effects. The interference patterns act as analogue crosscorrelation devices to produce new figures from which the patterns of departure of nerve impulses are initiated. The rapidly paced changes in awareness could well reflect the duration of the correlation process.

What evidence is there to suggest that the junctional electrical activities of the central nervous system are involved in awareness? Kamiya (1968) and others have shown, by using instrumental conditioning techniques, that people can readily be taught to discriminate whether their brains are producing alpha rhythms or not, despite the fact that they have difficulty in labeling the difference in the states of awareness they perceive. Subjects who have been able to label alpha rhythm state claim that it is one of pleasantly relaxed awareness. More experiments of this kind are now being carried out in order to find ways to shorten the long educational process currently entailed in Zen, Yogi, and western psychotherapeutic procedures aimed at identifying and achieving pleasant states.

More specific are some recent experiments of Libet's (1966) that have explored a well-known phenomenon. Since the demonstrations in the 1880s by Fritsch and Hitzig (1969) that electrical stimulation of parts of man's brain results in movement, neurosurgeons have explored the brain's entire surface to determine what reactions such stimulations will produce in their patients. For instance, Foerster (1936) mapped regions in the postcentral gyrus which give rise to awareness of one or another part of the body. Thus sensations of tingling, of positioning, etc., can be produced in the absence of any observable changes in the body part experienced by the patient. Libet has shown that the awareness produced by stimulation is not immediate: a minimum of a half second and sometimes a period as long as 5 seconds elapses before the patient experiences anything. It appears that the electrical stimulation must set up some state in the brain tissue and only when that state has been attained does the patient become aware.

The Hologram

But in order for recognition to be effected some more permanent alteration of substrate must act to influence the configuration of arrival patterns. If one looks at EEG records coming off an EEG machine for a number of hours during the day, and then goes home to try to sleep, what happens? The day's records go by in review; but note—they go by *in reverse*. This is known as the "waterfall effect."

Obviously, some neural change has taken place to allow the record to be re-viewed but also obvious is the fact that the re-viewing takes place from a different vantage point than did the original viewing. The record must therefore have "stereo"-like properties that allow it to be examined now from this, now from that, standpoint. This re-viewing from various vantages must not lose its identity relative to the entire record: a familiar face gains, rather than loses, its familiarity and recognizable identity by being viewed from different angles.

The proposal was made above that interactions among the patterns of excitation which fall on receptor surfaces become, after transmission over pathways organized in a parallel fashion, encoded by virtue of interference patterns of horizontally interacting processes in the slow potential activities of neuronal aggregates to form temporary microstructures whose design is dependent more on the functional organization of neural junctions than on neurons per se as units.

Recently important new advances have been made in the study of interference effects. These advances are the results of a new photographic process which produces images by way of a record called a hologram, which shows some startling similarities to the perceptual process.

Most of us are familiar with the image-generating aspects of physical optical systems. A camera records on photographic film placed at the image plane a copy of the light intensities reflected from the objects within the camera's visual field. Each point on the film stores information which arrives from a corresponding point in the visual field, and thus the film's record "looks like" the visual field. What have been studied more recently are the properties of records made on film which is placed somewhere in front of the image plane on an optical system. When properly exposed by a coherent light source, such a film record constitutes an *optical filter* in which information from each point of the visual field is stored throughout the filter itself. These filters display a number of remarkable characteristics.

As we have all experienced, when a film does not lie exactly in the image plane of a camera, the image becomes blurred, boundaries become less sharp, contrast less marked. In the case of an optical filter, the information is so distributed that there is no resemblance whatever between the stored image and the visual field itself. Thus the optical filter does not visually resemble the original object—rather, it is a record of the wave patterns emitted or reflected from an object. "Such a record can be thought of as 'freezing" of the wave pattern; the pattern remains frozen until such time as one chooses to reactivate the process, whereupon the waves are 'read out' of the recording medium" (Leith and Upatnieks, 1965). Thus, when transilluminated by a coherent light source, an optical filter reconstructs the wavefronts of light which were present when the exposure was made. As a result, a virtual image of the visual field can be seen by looking towards the filter. This virtual image appears exactly as the visual scene did during the exposure, complete and in three dimensions. In essence, all the information describing the visual field (and from which an image of the visual field can be reconstructed) is contained in the filter.

> As the observer changes his viewing position the perspective of the picture changes, just as if the observer were viewing the original scene.

> Parallax effects are evident between near and far objects in the scene: if an object in the foreground lies in front of something else, the observer can move his head and look around the obstructing object, thereby seeing the previously hidden object. . . . In short, the reconstruction has all the visual properties of the original scene and we know of no visual properties of the original scene and we know of no visual test one can make to distinguish the two (Leith and Upatnieks, 1965).

Even before the practical demonstration of the use of optical filters in the reconstruction of images, Gabor (1948; 1949; 1951) had mathematically described another way of producing images from photographic records. Gabor began with the intent of increasing the resolution of electronmicrophotographs. He proposed that a coherent background wave be allowed to interfere with the waves refracted by the tissue. (Reflection from an opaque object would serve as well.) The resulting interference pattern would store both amplitude and spatial phase (neighborhood interaction) information which could then, in a second step, be used to reconstruct, when transilluminated with a coherent light source, an image of the original tissue. Gabor christened his technique holography and the photographic record a hologram because it contains all of the information to reconstruct the whole image.

Gabor holograms can be composed in two ways. A wave form is divided by a beam splitter (e.g., a half-silvered mirror) so that one part can serve as a reference, the other reflected off the object to be photographed. The reference alone can then be used to reconstruct an image. Or, each part of the divided beam can be reflected off a different object. When this is done and one of the objects is used (as a reference) at the time of image reconstruction, the other appears as a "ghost" image. In this instance, the hologram can be used as a mechanism for associative storage of information.

The formal similarity between Gabor's refraction and reflection holograms and the various types of optical filters gradually became evident. The basic similarity between them lies in the fact that the resultant coding of information in each is a linear transformation of the pattern of light—not only in terms of the intensity (as in an ordinary photographic process) but also in terms of neighborhood interactions (spatial phase). The most intensively studied holograms have been those in which these phase relationships can be expressed mathematically as Fourier transforms. These equations are a special form of convolutional integrals which have the property that the identical equation convolves and deconvolves. Thus any process represented by the spatial Fourier transform can encode and subsequently decode simply by recurring at some second stage.

Holograms of whatever sort have some interesting properties in common which make them potentially important in understanding brain function. First, the information is replicated and distributed throughout the hologram. This makes the record resistant to damage. Each part of the hologram, no matter how small, can reproduce the entire image; thus the hologram can be broken into small fragments, each of which can be used to construct a complete image. As the pieces become smaller, resolution is lost. On the other hand, as successively larger parts of the hologram are used for reconstruction, the depth of field of the image decreases—i.e., the focus becomes narrowed. Thus an optimum size for a particular use can be ascertained. These curious properties derive from the fact that the hologram is a representation of an image "defocused" in an orderly manner so that information becomes replicated and distributed.

Second, the hologram has a fantastic capacity usefully (i.e., retrievably) to store information. This capacity stems from the fact that when incorporated in a suitable retrieval system, information can be immediately located and accurately reconstructed. The density of information storage is limited only by the wave length of the coherent light and the grain size of the film used. When holograms are produced in solids, many different patterns can be simultaneously stored. Each image is stored throughout the solid, yet each image is individually retrievable. Alternatively,

> several images can be superimposed on a single plate on successive exposures, and each image can be recovered without being affected by other images. This is done by using a different spatial-frequency carrier for each picture. . . . The gating carriers can be different frequencies . . . and there is still another degree of freedom, that of angle (Leith and Upatnicks, 1965).

At the time of this writing, some ten billion bits have been usefully stored holographically in a cubic centimeter. As Van Heerden and others have pointed out, if we should store during a lifetime as little as one bit per second, the human brain requires approximately 3×10^{10} elementary binary operations *per second:* "If that sort of thing was going on it was incomprehensible . . . However, once confronted with this paradox, it gradually became clear . . . that optical storage and processing of information can provide a way of accomplishing this 'impossible' operation" (Van Heerden, 1968, pp. 28–29).

A final point about physical holograms. Optical systems are not the only ones that can be subjected to the holographic process. Now that the mathematical relationships have been specified, computer programs have been constructed that "simulate" optical information processes. One such program represents the intensity of an input by the size of a disc; spatial phase relationships are represented by the angular direction of a slit within that disc. Holograms are thus not dependent on the physical presence of "waves" even though they are most readily described by the equations of wave mechanics.

This independence of holography from physical wave production is an important consideration in approaching the problem of a neutral holographic process. There is some considerable doubt whether "brain waves" as presently recorded form the substrate of any meaningful interference pattern organization for information processing—although they may be indicative that some such process is taking place. The wave forms recorded for the most part have a long time constant and can therefore be carriers of only very small amounts of information—even in the form of spatially interfering holographic patterns. The hypothesis proposed in the next section, therefore, emphasizes the role of junctional slow-potential microstructures in brain function. These microstructures can be described either in statistical, quantal terms or in the wave-mechanics language of convolutional integrals and Fourier transforms. The microstructures do not change their characteristics because a choice is made as to description. Each language, each descriptive form, has its own advantages. With respect to the problems of perception, especially the questions of image formation and the fantastic capacity of recognition memory, holographic description has no peer. So why not try out its application to brain processes?

A Neural Holographic Process

The essence of the holographic concept is that images are reconstructed when representations in the form of distributed information systems are appropriately engaged. In fact, as noted, one derivative of the holographic process comes from a consideration of optical filtering mechanisms. Holography in this frame of reference is conceived as an instantaneous analogue cross correlation performed by matched filters. In the brain correlation can take place at various levels. In more peripheral stations correlation would occur between successive configurations produced by receptor excitation, the residuals left by adaptation through self-inhibition forming a buffer memory register to be updated by current input. At more central stations correlation would entail a more complex interaction: at any moment input would be correlated not only with the configuration of excitation existing at any locus but also with patterns arriving from other stations.

According to the holographic hypothesis the mechanism of these correlations is not by way of some disembodied "floating field" nor even by disembodied "wave forms." Consider instead the construction of more or less temporary organizations of cortical columns (or in other neural locations, other aggregates of cell assemblies) by the arrival of impulses at neuronal junctions which activate horizontal cell inhibitory interactions.

When such arrival patterns converge from at least two sources their designs would produce interference patterns. Assume that these interference patterns are made up on classical postsynaptic potentials coordinate with awareness, as suggested above. Assume also that this microstructure of slow potentials is correctly delineated by the equations that describe the holographic process, which is also composed of interference patterns. The conclusion would follow that information representing the input is distributed over the entire extent of the neutral pattern just as it is over the entire extent of the physical holographic pattern. This does not mean, of course, that input information becomes distributed willy nilly over the entire depth and surface of the brain. Only those limited regions where reasonably stable junctional designs are initiated by the input, partake of the distribution. Furthermore, for any effect beyond the duration of a particular input more localized enduring memory mechanisms must be invoked. However, these mechanisms can be engaged in loci distributed in neural space once information has become dispersed. Addressing the more permanent store demands merely the repetition of the pattern (or essential parts thereof) which originally initiated storage. This content addressability so readily accomplished by the holographic process does away with the need for keeping track of "where" information is stored.

What are some of the possibilities for making the junctional microstructure endure? Some more lasting property of protoplasm must be invoked to account for storage which can be of varying duration. Profound temporary interactions do occur between inputs separated by hours (as in the McCulloch effect, in which exposures to a set of colored bands influence subsequent observations of color) or in some individuals for days (as in the rare person who shows true eidetic capacities). And, of course, the longer range interactions that account for recognition and recall must also be accounted for. Conformational changes in macromolecules such as lipids or proteins and even longer-lasting anisotropic orderings of macromolecular structure lend themselves to speculation in the following terms. Successive junctional microstructures formed in a region of cortical cytoplasm may produce, when similar in configuration, a cumulative residual effect by inducing ordering into previously disordered macromolecular chains or fibrils, or by increasing an existing order, so that the region thereafter responds more easily to a repetition of the same excitation. Early results of experiments performed on retinal tissue examined with the electronmicroscope show that such changes in molecular conformation can occur with excitation (Sjöstrand, 1969). Similar but as yet unconfirmed suggestions have been made by Whyte (1954) and by Halstead (1951). The former investigation suggests that:

> this cumulative medium- and long-range ordering of some of the macromolecular chains throughout a particular volume of cortical cytoplasm

> is a kind of growth process of a pattern determined not by heredity but by activity, and involving the development not of a differentiated tissue but of an element of ordering in the molecular arrangement of an extended mass of cytoplasm. Here we are concerned with the differentiation of particular vector directions, possibly parallel to the cortical surface in particular cortical layers. The templates of memory are not single localized molecular structures, but extended components of long-range order set at various angles to one another. . . . [However] the ordering will correspond only to the statistically dominant pattern of activity, or simplest overall pattern common to the successive activity patterns. Moreover this tendency to select the dominant pattern will be reinforced by the fact that the simplest overall patterns will be the most stable, since their parts will mutually support one another. The random protein structures may thus act as a structural sieve taking a stable impress at first only of the simplest, most unified, and statistically dominant component in all the patterns of activity of a given general form. In general [then] the development of the modification proceeds from a grossly simplified to a less simplified and more accurate record. This process of the development of a hierarchically organized modification corresponds to Coghill's "progressive individuation" of behavior patterns during ontogeny, and may hold the clue to the self-coordinating capacity of cortical process (Whyte, 1954).

Nor can this be all there is to registering the wave forms. Conformational changes in macromolecules are apt to be reversible. A more permanent record probably demands such mechanisms as the tuning of "averaging circuits" in cortical columns described below and growth induced by the changes in membrane permeability consequent to and dependent on these macromolecular alterations. The "filter," "sieve," or "screen" of holographic patterns is composed not only of the lattice of membrane macromolecules making up the synaptodendritic net but also of a facilitation of all tendencies toward image formation and the initiation of certain departure patterns of nerve impulses.

How then can we approach the problem of changes in protein confirmation as a basis for memory? Sensitization akin to the development of immunities have been proposed. And some initial experimental efforts have been directed toward this view (Mihailović and Janković, 1961). Another lead comes from some incidental observations made during the course of experiments carried out for initially different purposes. In my laboratory we have had occasion to cause epileptic seizures in monkeys by implanting aluminum hydroxide cream in their cortex (Pribram, 1951; Kraft et al., 1960; Stamm and Knight, 1963; Stamm and Pribram, 1960, 1961; Stamm et al., 1958; Stamm and Warren, 1961). Such implantations cause havoc in the learning process. Yet even a major convulsive episode will leave the immediate performance of a learned task unimpaired in these animals. Only 24 to 48 hr *after* such seizures does performance deteriorate—and this in the absence of further seizures. Also, the deterioration is temporary, lasting only about 48 hr. In short: some process takes this many hours to build up sufficiently to challenge the otherwise dominant neural pattern established by learning. And the challenge is temporary; apparently total recrudescence of the learned pattern is reestablished shortly. Organic chemists must have available many macromolecules with similar peculiar characteristics. Are protein conformations subject to such temporary deformations and is the time course of such alterations consonant with that observed in these experiments?

Arranging Memories

As I have already indicated, there are many memory processes in which permanent and impermanent features mingle in a variety of ways: memorization of telephone numbers in a strange city, the use of experience in a related-but-novel situation, the schedules which guide us through our daily tasks and pleasures, and the recrudescence of extinguished performances when the conditions of extinction are lifted. These all are memory processes in which more or less temporary arrangements are produced by more permanently stored mechanisms.

Memory and Circuitry

Does the suggestion of a protein conformation mechanism for memory storage dispose, then, of the "neural" or "synaptic growth," or "strengthening" hypothesis? Not necessarily. As I pointed out on another occasion (Pribram, 1963), the electroconvulsive shock experiments have provided evidence that consolidation of the memory trace is at least a twofold process. Immediately after an experience—or 5 sec afterwards, or even up to 1 hr afterwards—all traces of the experience can be wiped out. This suggests, as already noted, that the protein conformation change mechanism is disruptible during this period. After this, more permanent changes gradually take place. But concomitant with the protein conformation change, alterations in the design of the neural circuitry must also take place. Otherwise, retrieval through the generation of appropriate arrival patterns becomes impossible.

Thus another aspect of brain function needs to be called into account: namely, some change in neural connectivity that accompanies the protein changes. A problem arises here mainly because the brain's nerve cells do not divide. However, they can grow new branches. This has been dramatically demonstrated in a study (Rose et al., 1961) of the effects on brain of high-energy radiations produced by a cyclotron. Remarkably minute and sharply demarcated laminar destruction (often limited to a single cell layer, and this is not necessarily the most superficial one) were produced in rabbit cerebral cortex when high-energy beams were stopped short by the soft tissue. The course of destruction and restitution was then studied histologically. Intact nerve cells were seen to send branches into the injured area; these branches became progressively more organized until, from all that could be observed through a microscope or measured electrically, the tissue had been repaired.

The organization of the branches of nerve cells could well be guided by the glia that pervasively surround these branches. Such directive influences are known to be essential, for example, in the regeneration of peripheral nerves. Schwann cells, close relatives of glia, form a column into which the budding fibers must grow if they are not to get tangled in a matted mess of their own making.

The operative assumption is that glial cell division is somehow spurred by those same activities recounted above as being important to memory storage. Data to support this assumption are presented below. The resulting patterns of the glial bed would form the matrix into which nerve cell fiber growth occurs. Thus guided, fiber growth is directed by its own excitation—with the whole mechanism based, however, on the long-lasting intervention of glia. This "arranging" mechanism would account for the later "interfering" effects obtained in the consolidation experiments and in the spontaneous "restitution" as well. The growing nerve cell fiber is ameboid and can temporarily retract its tip, which is made up of a helical winding of small globular protein molecules. After the convulsive "insult" is over, first tentative, and then more vigorous probings are found to be resumed in some "random-walk" fashion by the nerve fiber tip (as has been suggested regarding normal growth by von Foerster, 1948). The glial substrate, assumed to be undamaged in this experiment, will perform its guiding function to effect the apparent restitution. Support for the glially guided "growth" hypothesis comes from the work of Krech et al. (1960). These investigators found that the cerebral cortex of rats actually becomes thicker as a function of experience—thickening of visual cortex with visual experience and of somatic cortex with somatic experience were demonstrated. The increased cortical volume was not due to an increase in the number of neurons; rather, glia and fibers were responsible for the change.

The glially guided neural growth hypothesis, in addition to accounting for the late-interference effect data, has another attractive feature. The electrochemical memory storage process per se has no built-in mechanism which satisfactorily explains retrieval. A neural network structured through growth glially guided by experience could, on the other hand, serve retrieval much as do the "feelers" on the magnetic memory core of a computer. The patterns of electrical signals that activate particular network configurations would then correspond to the lists or programs fed to a computer, as well as to the schemata proposed by Bartlett (1961) to account for the results of his studies on memory in man.

Dismembering and Remembering

According to the view developed thus far, inputs are both isomorphically recorded as protein-conformation changes and coded into programs through neural growth. These programs, when properly activated, reconstruct the appropriate protein conformation, i.e., the "memories." Three observations in addition to the facts of recognition given earlier support the isomorphic recording of input items. One is the occurrence of eidetic imagery; another is the phenomenon of hypnotic regression (Gebhard, 1961); and a final one is the evocation of "memories" by electrical brain stimulation. There are limitations to all of them. The evidence for verisimilitude in hypnotic regression has been questioned. Evocation of "memories" by electrical brain stimulation occurs only in epileptic (i.e., scarred) cortex and is subject to influences of environmental set (Mahl et al., 1962). The occurrence of eidetic imagery in the adult is extremely rare and—curiously, considering the interest such a phenomenon must arouse in psychologists—studies on eidetics are relatively few (e.g., Haber and Haber, 1964; Haber, 1969; Luria, 1968; Stromeyer, 1970). The evidence is thus overwhelmingly in favor of the suggestion that, in addition to some memory storage record, memory processing depends heavily on programs. Bartlett (1961) amply documented the view that schemata are stored in the head.

In many ways this clarifies the memory problem considerably. If storage were only isomorphic to experience, one should be able to locate and find direct correspondences between all of the stored items and the world "out there." In a schematic or programmed memory no such isomorphic relation would have to obtain. The difference is essentially that between, say, a dictionary and a typewriter, between a trigonometry table in a handbook of physics and chemistry and a calculating machine. For example, if I take a simple adding machine and add to it the capability to multiply, I am putting a new memory mechanism into it. If I look into the machine I will find a change and that change may be the addition of a set of registers. Yet I will never find any specific "product" by opening the machine. "Products" are obtained when the machine is presented with inputs which "signal" that a product is required, inputs anisomorphic to the "products" themselves. This seems self-evident and is often forgotten in our more erudite arguments about memory.

Much confusion would be resolved if we adhered to the notion—deceptively simple, yet immensely significant—that "remembering" is the opposite of "dismembering." Even our language reflects that remembering is a putting together, a reconstruction. Once accepted, the conclusion this leads to is remarkable; namely, *it means that a good deal of what we call the memory storage problem is a hoax.* Most "memory" is stored in our libraries and in our jobs and homes as inputs to our brain machines. The human organism is thus signaled to remember what he is programmed to remember. The very word "remember," as I have suggested, reflects this process of reconstruction from parts as by a machine. A word of caution is appropriate at this point, however; in no sense do I want to imply that man is "nothing but a machine." Man as often as not goes to the library in search of the appropriate signals; he plans and controls, as often as not, the significant—i.e., signifying—aspects of his home and job. The point is that, in these respects as well as in the assembly of appropriate routines and programs, remembering is an *active,* not a passive process.

The Numbers Game

Once we dispose of the hoax that isomorphic coding and recording of all inputs is the sole necessity for a "proper" memory mechanism, we can also get rid of the "numbers game" that is constantly being played when memory is discussed. Bits of information are thus seen as irrelevancies—every book an author writes can be "stored" in his typewriter, which possesses fewer than 50 symbols on its registers. One can raise the objection that the brain must be more complicated than a typewriter—and I agree; but the number of states that it can register involves an experimental rather than a logical or psychological debate. An alphabet of only 26 letters does an heroic job.

I have repeated these things, which by now are almost truisms, because I find that in our discussions and our literature we do *not* hold these facts in mind. Over and over, the argument revolves only around storage of particulars. There need *not* be 10^{10} units for storage; there need *not* be an RNA change specific to a Y maze but not to a T maze. The rules of the numbers game hold only if one elects to play it. Only if the model one holds is one based *exclusively* on item storage—the storage of inputs in some isomorphic manner—is this kind of argument valid. And the evidence is overwhelming that there is more to memory than bit-by-bit storage.

Hierarchy

Implicit, then, is the idea that our memory machinery is capable of hierarchical organization—that all small units and probably some larger combinations of the memory machinery are permanent and undamage-able, but that at least some of the larger units can be flexibly combined through programming operations initiated either by the input or by even larger permanent units. Also implicit is the suggestion that a particular memory unit or state can serve in a variety of combinations and thus participate in the production of a variety of re-membrances.

"Hierarchy" here implies several things: first, on any occasion I know all-of-a-piece whether I have anything at all relevant in memory to express; second, the mechanics of expression demand that I produce only one memory at a time. This limitation on output is the "keystone in the construction of the individual," as Sherrington (1947) so beautifully stated it. Thus serial ordering accomplished by an hierarchy of processes prior to output is yet another dimension essential to remembering (see, e.g., Hart, 1965).

A sophisticated statement by Werner (1969)—based in part on his own extensive research and that of Hartline et al. (1956), of Mountcastle et al. (1963), Poggio and Mountcastle (1960), and of Hubel and Wiesel (1968), and in part on computer programming formulations presented by Miller et al. (1960)—describes a process for somesthetic perception which applies with only minor modifications to the other input systems. Werner discerns a basic columnar structure in the brain cortex in which each neuron of the column displays a receptive field which "is the finest indivisible unit" of representation of the input. Columns of neurons tend to display identical or at least similar receptive fields and thus make up one level of representation. Columns are in turn combined into more complex structures by directionally sensitive units which serve as pointers connecting the activities of the columns. These pointers, depending on the preferred direction of response, structure the electrical activities of the columns into various relationships to one another; if pointers with more than one direction are available, blocks of columns become connected to form "ring structures." Werner compares his cortical columnar structures to the list structures out of which computer programs are constructed. Each list contains items that point to other lists. Thus complex interactions (list structures) can be programmed by this simple device. In fact, Spinelli (1970) has designed a program (called Occam) to stimulate a feature analyzer based on this cortical structure. This program can be tuned by the presentation of patterns of nerve impulses or wave forms to respond subsequently when certain features of the wave form are repeated.

A somewhat simplified version of Werner's and of Spinelli's feature analyzer is composed as follows. A cortical column is conceived to consist of input and operator neurons, and of interneurons and test cells. An input to a neural unit of the column that displays a receptive field is distributed to interneurons which in turn connect to an operator neuron. The interneurons are tunable—i.e., they adapt and habituate; they have memory. Each interneuron thus acts as does a bin in a computer that averages the patterns of input to which it is exposed. Only when a pattern is repeated does structured summation occur—nonrepetitive patterns simply raise the baseline and average out. Thus the operator neuron, sensitive solely to *patterns* of excitation, is activated only when input patterns are repeated. The entire process is sharpened by feeding the output from the operator neuron back onto the input cell via a test neuron that compares the pattern of neural activity in the input and operator neurons. When match is adequate, the test cell produces an exit signal, otherwise the tuning process continues. Thus each cortical column comes to constitute an engram by virtue of its specific sensitivity to one pattern of neural activity, a "list" of interresponse times of a firing neuron or the wave form that describes the envelope of the firing pattern.

Each cortical column is conceived of as being connected with others via horizontal cells and their basal dendrites which are responsible for inhibitory interactions. Whenever these horizontal cells are activated in an unsymmetrical fashion, as they are by directional sensitive inputs, a temporary structure constructed of several columns is put together. These extended structures, dependent as they are on hyperpolarization rather than on nerve impulse transmission, are composed therefore by processes taking place at neural junctions and constitute temporary neural states.

We now have good evidence that the so-called association areas of the cerebral cortex exert control over the input systems, control which is in many respects similar to that exercised when a zoom lens is extended and retracted (Pribram, 1969). This function would have the effect of changing the number (and perhaps the complexity) of cortical columns that can be contained in a temporary structure.

The logic of the input systems can thus be conceived constituting a feature filter on input, a screen that is being continually tuned by that input. One of the characteristics of the filter is, therefore, that it constitutes a self-adapting system whose parameters of adaptation are controlled by its own past history and by the operations performed on it by other neural mechanisms.

The Temporal Code

This flexible rearrangement of hierarchically organized memories demands that some important attribute of neuronal function is sufficiently flexible to be temporarily but effectively alterable. This attribute might well be the temporal code with which the nerve discharges, or to which it is sensitive (see, e.g., Hydén, 1961; Landauer, 1964).

Direct experimental evidence for any such flexibility in the temporal code with which neurons fire has hitherto been sparse. Almost the sole evidence that the brain is at all capable of altering its rhythms as a function of experience comes from the pioneering studies of John and Killam (1959). These investigators flashed light to their subject at certain frequencies (e.g., 30/sec) and recorded from various locations in the brain.

In brief, their experiments demonstrate that at the beginning of training the electrical activity of a wide variety of brain structures appears to be synchronous with a repetitive stimulus. After learning has occurred and performance is at criterion, the electrical activity synchronous with the stimulus can be recorded only from the appropriate projection system. In the earlier phases of learning, the electrical activity of many of the core areas of the brain stem and forebrain show such synchrony: the reticular formation, hippocampus, and amygdala are only a few of the structures involved. The synchronous rhythms drop out progressively and the dropping out is correlated with progressively better performance on the part of the animal.

An additional observation which may point the way toward which future efforts for evidence of temporal coding may be directed comes from my laboratory (Spinelli and Pribram, unpublished data). Small wire electrodes were implanted in the lateral geniculate body and in the striate cortex of monkeys. Those in the geniculate were so placed that electrical stimulation would encompass a large portion of the entire nucleus, and continuous stimulation with 5-volt biphasic pulses, occurring at approximately 8/sec, was applied. Bipolar recordings were made from the cortical electrode placements. Most of them reflected more or less accurately the rhythm of stimulation imposed on the geniculate station of the visual system. From some of the cortical placements, however, arrythmic recordings were consistently obtained; they sounded like a complex tap dance when transduced by a loudspeaker. The brain cortex apparently has a remarkable power to alter a rhythm imposed at an adjacent station.

The Temporal Hold

The above-noted observation leads us immediately to the question: How, then, are rearrangements among temporal codes accomplished? As yet no biological mechanism has been proposed to effect flexibility; nor will I attempt to propose one in detail here. But the imposition of local D.C. potentials on brain tissue is effective not only in altering the firing pattern of nerve cells but also in maintaining—i.e., temporarily storing—this change (Chow, 1964; Chow and Dewson, 1964; Dewson et al., 1964).

Further, lesions of the limbic forebrain and of the anterior frontal isocortex impair just the type of task which demands the flexible rearrangement of memory processes. I have elsewhere (Pribram et al., 1964) suggested that this deficit is due to a failure of the regulation of the "temporal hold" imposed by an input on a particular matrix of registers. This "temporal hold" is assumed to be accomplished through an operation similar to that which gives rise to a temporary dominant focus in the experiments of Ukhtomski (1962), Rusinov (1956), and Morrell (1961). Without regulation by such a hold mechanism, the organism fluctuates inordinately among possible temporal codes and thus produces only a jumble of arrival patterns. In such circumstances even the temporary conditions necessary to the registration of interference patterns as holograms cannot be achieved. Support for some sort of temporal hold process emanating from the frontolimbic portions of the brain comes from electrical recordings made in man:

> When conditional and imperative stimuli are presented in this way a remarkable change appears in the frontal brain response; a negative potential appears immediately after the conditional response and endures until the imperative response, when it declines rapidly to zero or becomes positive. This has been described as the "Contingent-Negative Variation" or Expectancy Wave (Walter, et al., 1964). In conditions such as those described, the E-wave is the most constant and stable of all electro-cerebral phenomena in normal adults. It does not depend on the character of the intrinsic normal rhythms and is as large and extensive with the eyes open as shut. In children, however, and in mentally disturbed patients, the E-wave is often elusive and variable; above all, it is extremely sensitive to social influences.
>
> As already mentioned the E-wave arises always and only during sensory-motor association, but both the sensation and the motion may be of quite a subtle nature. In the simplest case the presentation of a conditioned stimulus in any modality, followed by an imperative stimulus in another modality, evokes an E-wave following the primary conditional responses and lasting until the moment when the imperative response would have occurred.
>
> The striking feature of the E-wave is that it appears, as it were, to submerge the imperative response, and terminates very abruptly at the instant when the latter would have subsided. The typical sawtooth waveform of this phenomenon is remarkably like that of the time-base of an oscilloscope, rising steadily toward a maximum value over a time determined by the established stimulus interval, and dropping suddenly to zero. The duration of the E-wave as studied systematically so far is several seconds, but in some subjects the potential difference seems to be sustained much longer during "extinction" trials when there is no imperative stimulus to act as a "fly-back" trigger. Sometimes there is even a suggestion of a staircase or "Treppen" effect when conditional stimuli are presented at intervals of a few seconds without reinforcement to subjects with a very slow rate of extinction. Since the E-wave presumably represents depolarization of the apical dendritic plexus, the possibility of "recruitment" in such a mechanism would be interesting to study in more detail. The subjects who have shown signs of this effect are highly

suggestible and easily hypnotized (Black and Walter, 1963); the capacity to maintain a high and even cumulative level of expectancy may be typical of this disposition, and may depend on some idiosyncrasy of the electrochemical relations in the superficial cortical levels (Walter, 1964).

And so we are back to the problem of matching temporary changes in protein conformation against some more lasting arrangement of memory structures so that, on occasion, some permanent neural organization can be achieved.

RNA and Reinforcement

RNA and Behavior

Despite the difficulty in ridding ourselves of conceptual shackles, progress *is* being made by leaps and bounds. Hydén's work has often been criticized by both psychologists and biochemists; yet the picture he began to draw for us is nevertheless taking form. The RNA changes he reports may indeed be occurring—but not necessarily as evidence of item storage on evanescent messenger molecules, but rather as evidence of derepression of genomes. Bonner's theory (see Chapter 16, present volume) and Hydén's (1961) evidence are in accord.

But greater difficulties are posed by such phenomena as cannibalism and the injection of "knowledgeable" RNA. Here is a focus of discrepancy—here is the point where experimentation must take the offensive and attack. The evidence must be firmed-up; new directions must be taken to decipher the relationship. But, again, the problem comes into better focus if RNA is not considered *the* memory molecule. Rather, the question might be put: Just what *is* the relation between RNA and derepression? An increase in RNA can signal to the experimenter merely that derepression has occurred. Could it also be that RNA in some way can *initiate*—i.e., *induce*—derepression? There is good evidence from embryology that this may actually be so. The process of induction in the embryo has many similarities to the process of reinforcement which establishes the memory trace in the adult.

Inductors in Review

In essence, induction is a "chemical conversation"—as Bonner has called it—between the intrinsic determining mechanisms of the morphogenetic field (or its already-independent differentiated parts) and the extrinsic organizing properties which guide its flowering. An early experiment, the classic example, is that of the determination of the lens by the eye vesicle. Contact between this vesicle with the overlying epidermis stimulates the latter to form a lens in the region of contact. If the eye vesicle is removed the epidermis fails to differentiate a lens. This experiment raised a whole set of problems which generated a direction of research in experimental embryology bearing a striking resemblance to current explorations in experimental psychology and ethology (see, e.g., Hamburger and Levi-Montalcini, 1950).

The first and logical assumption was that the inductor acted merely as a trigger; that, in the classical example, the head skin is already "predisposed" to form a lens and that it requires only a signal to start. Two lines of evidence disproved this concept of induction. First, the optic vesicle was shown by transplantation to induce a lens in skin other than head skin—for example, flank skin. Second, the area of head skin which normally forms a lens was shown by other transplantation experiments to be *polypotential* and therefore definitely not "predetermined" for lens formation *only*. If the region of the head epidermis which normally forms the lens is combined with an ear induction, for example, it will respond with ear formation; if combined with a nose inductor, it will form a nose.

These facts do not deny, however, that the reacting system must be "ready" or "competent," i.e., in the proper state of responsiveness, to allow induction to become effective. For example, tissue which is already "launched," as it were, toward a different destination, will fail entirely to respond.

Another point is that inductors are not species-specific. An inductor can be effective on tissues which belong to a different species, genus, or even order. The suggestion is, therefore, that inductors are made up of chemicals common to many organisms (more of this in a moment). These chemicals apparently determine the overall character of the induced structure while the hereditary equipment of the cells of this structure determines its detailed form. For example, when the flank skin of a frog embryo was induced to form head structures by salamander tissue into which it was transplanted, the embryo had a salamander head with the horny jaws and other features of the frog.

A long series of chemical experiments has currently culminated in the view that the ribonucleic acids (RNAs) are most likely, and perhaps uniquely responsible for the inductive effect (see Niu, 1959), though ribonucleoproteins and steroids have not been entirely ruled out. For the most part RNAase destroys the inductive effect, although the problem remains that RNAase has other effects on the induced tissue which may disrupt its differentiation. More direct evidence, however, comes from demonstrations of the inductive effect of RNA extracted from different organs. Not only has this been accomplished, but RNA isolated from different sources was shown to be capable of inducing the recipient tissue to differentiate into different specific structures. These experiments suggest that there are *many* species of RNA in an organism and that each has a specific function.

With the use of ^{14}C-labeled RNA, another problem has been tackled. Evidence has been established to show that it is possible for RNA molecules actually to move from the microsomes of the inductor tissue into the cells of the tissue induced, most likely by a process of pinocytosis.

To sum up: embryogenesis is dependent not only on the inherited and inherent properties of the genetic constitution of the organism; rather, these properties are also evoked and organized by the inductive capacity of the milieu in which the cells grow. The inductive capacity is itself specific, but in a somewhat different sense than is the genetic potential. The *genetic capability* is individual-, species- (and genus- and order-) specific. Hereditary factors proscribe commonalities with the past and future while assuring variation within any single generation. Inductors, on the other hand, are nonspecific with respect to individuals, species, and so forth. They are relatively simple chemicals—RNAs—common to all living organisms. Inductors thus provide the existential commonality which allows the possibility of modification of whole generations according to the exigencies of the time.

Induction and Reinforcement

The superficial descriptive similarity between induction as studied in embryological tissue and reinforcement as studied in conditioning situations is easily drawn. (1) Inductors evoke and organize the genetic potential of the organism. Reinforcers evoke and organize the behavioral capacities of organisms. (2) Inductors are relatively specific as to the character they evoke but are generally nonspecific relative to individuals and tissues. Reinforcers are relatively specific in the behaviors they condition but are generally nonspecific relative to individuals and tasks. (3) Inductors determine the broad outlines of the induced character; details are specified by the action of the substrate. Reinforcers determine the solution of the problem set; details of the behavioral repertoire used to achieve the solution are idiosyncratic to the organism. (4) Inductors do not just trigger development; they are more than just evanescent stimuli. (5) Inductors must be in contact with their substrate in order to be effective. Contiguity is a demonstrated requirement for reinforcement to take place. (6) Mere contact, though necessary, is insufficient to produce an inductive effect; the induced tissue must be ready, must be competent to react. Mere contiguity, though necessary, is insufficient to produce reinforcement; shaping, deprivation, readiness, context, expectation, attention, hypothesis—these are only some of the terms used to describe the factors which comprise the competence of the organism without which reinforcement cannot become effective. (7) Induction usually proceeds by a two-way interaction—or, as stated earlier, by way of a chemical conversation. Reinforcement is most effective in the operant situation where the consequences of the organism's own actions are utilized as the guides to its subsequent behavior.

But when this much has been said, the question still remains as to whether these descriptive similarities point to homologous mechanisms. My hypothesis states that they do. What evidence is there in support? What neural processes become operative during conditioning?

What is required is an anatomical pathway that functions at multiple locations in the brain to release the reinforcing "inductor," the chemical that can induce the recipient tissue to differentiate. Kety (1970) in a provocative synthesis of recent neurochemical research, makes the following suggestion. Scheibel and Scheibel (1967) have described a system of neurons (some of which lie in the brain-stem reticular formation) which function as nonspecific afferents to the cerebral cortex. The axons of these cells are characteristically long and at their terminations produce climbing fibers which twine around the apical dendrites of pyramidal cells with a loose axodendritic coupling in contrast to the well defined synapses that characterize specific afferents. Fuxe et al. (1968) claim to have shown that the nonspecific couplings are largely aminergic, that the axons and their cells of origin contain nor-epinephrine. The local release of nor-epinephrine can be responsible for instigating the induction process by stimulating the secretion of RNA and facilitating protein synthesis. Evidence is accumulating that cyclic adenine monophosphate may be the mediator of this process. Kety remarks that "it is interesting that the stimulation of protein kinase by cyclic adenine monophosphate can be markedly potentiated by magnesium or potassium ions and inhibited by calcium which suggests means whereby an effect of adrenergic stimulation could be differentially exerted on recently active and inactive synapses."

But such experiments tell us only that certain neural processes are possibly equivalent to those set up when reinforcement is manipulated extrinsically. Somewhat closer to demonstrating the mechanism with which we are concerned are experiments in which a temporary dominant focus is produced in the brain. The classical example is the chemical stimulation of the exposed motor cortex of a dog that has been conditioned to lift his left forepaw. When strychnine is placed on the cortical area that controls the right hind limb, the dog will lift the right hind limb instead of the left forepaw when given the usual signal. Once the chemical is removed, the dog reverts to its former behavior unless the stimulation has been often repeated. It is plausible to conclude that field-like configurations of such temporary dominant foci as these are produced during conditioning and the function to organize subsequent neural, and therefore behavioral, activity. But this, although relevant, is another story and has already been alluded to.

A more chronic and therefore more easily studied change in neural discharge can be obtained by making epileptogenic lesions in cortex with implantations of aluminum hydroxide cream or by locally freezing the cortical tissue.

In my laboratory different areas of the brain cortex of monkeys have been treated with aluminum hydroxide cream to produce local irritations manifested by altered electrical activity (abnormal slow waves and spike discharges). Such irritative lesions, while they do not interfere with a monkey's capacity to remember the solution to problems repeatedly solved prior to the irritation, do slow their original learning of these problems some fivefold (Henry and Pribram, 1954, Kraft et al., 1960; Pribram, 1951; Stamm et al., 1958; Stamm and Pribram, 1960, 1961; Stamm and Warren, 1961). Moreover, problem-solving in general is not affected; the defect is specific for those solutions to tasks which cannot be remembered when that particular part of the brain has been removed. Furthermore, the impairment is restricted to the early part of the learning process, the part before there is actual demonstration that learning is occurring. Thus the irritative lesions do not block consolidation but do delay its manifestation. Could it be that a single engram restricted to one neural locus is insufficient to be manifest? There is a good deal of evidence from human learning experiments that considerable rehearsal must take place in order that an experience be remembered (Trabasso and Bower, 1968). What appears to occur during rehearsal is a distribution of the rehearsed material so that it becomes linked to a larger assortment of previously stored experience. The results of the irritative lesion experiments can thus be interpreted as showing that the process of reduplication and distribution of the engram has been retarded. A test of this interpretation would come from a comparison of learning by irritative-lesioned monkeys under spaced and massed trial conditions.

Histological analysis of the tissue treated with aluminum hydroxide shows tangles of nerve fibers much as those described in peripheral nerves when growth is not properly guided by an adequate Schwann cell column. Could it be that oligodendroglia are selectively killed off by the treatment, allowing the disordered growth to occur? Chemical analysis of the tissue implanted with aluminum hydroxide cream is, of course, impractical. But an ingenious experiment designed to answer this question has been achieved (Morrell, 1960). An irritative lesion made in one cerebral hemisphere produces, after some months, a "mirror focus" of altered electrical activity in the contralateral cortex by way of the interhemispheric connections through the corpus callosum. This "mirror focus" has not been directly damaged chemically, yet it possesses all of the epileptogenic properties of the irritative lesion. The RNA in this mirror focus has been shown to be considerably altered when compared to that found in normal brain tissue. Once more, RNA production by nerve cells has become involved in experiments undertaken to study the memory mechanism.

Thus there is every reason to believe memory induction, just as embryological induction, to be a multistage process which takes time to run off. Each stage in such a process would be expected to show its own vulnerabilities, vulnerabilities that can be demonstrated by appropriate techniques applied at the critical period. We have learned much about critical periods in embryogenesis and behavior development. A rich field of exploration and experimentation lies ahead in determining the nature of critical, i.e., sensitive, periods in mnemogenesis.

Conclusion

The thoughts expressed in this chapter have centered on the process we call remembering, but the interwoven complexities of the psychological mechanism have led me into a discussion of awareness, of temporal coding in the nervous system, of the "holding" functions I attribute to the frontal and limbic forebrain, and of a molecular mechanism of reinforcement. I have not given all of the evidence available in support of the proposals made, nor have I given evidence to jeopardize the views presented. I have chosen this course deliberately—for, aside from pressing the fruitful course of laboratory exploration now in progress—I feel a great need to re-view and restructure my image of the problem. I fear that the present views of the memory problem will soon—or perhaps already have begun to—lead prematurely to a dead end and thereby permit the experimental challenge to wither away unanswered. Already I am tired of hearing that RNA *really* doesn't have anything to do with learning—i.e., not *real* learning—because it has not been known to store the "association" necessary to learning. And how many times have we seen the memory problem reduced to information storage? I can image as many bits of succulent and poignant detail about a loved one as you will give me time and an interested ear. And in my imagination I can do this while possessing the routine of my daily affairs, with hardly a perceptible effect on my behavior. Just where *are* the questions about "short-term," "intermediate," and "long-term" memory processes leading? Are there more types than these (e.g., very, very short), or are we dealing with a continuum? So goes the argument, which unfortunately misses the point that memory has structure; that in order to process nonsense syllables, man must know language; that to "forget" the irrelevant, the relevant must be properly available.

This need to restructure my thinking has thus produced this chapter and the classification used here, the division of memory mechanisms into "evanescent," "arranging," and "rein-forcing" processes which I believe to be pertinent to a range of problems in biology and psychology. In it *novelties* are emphasized at the expense of tried truths: the neural hologram, rearrangements among neural configurations, an aminergic-RNA-mechanism of reinforcement by induction. Here are some new possibilities which may finally enable us to realize that memory mechanisms are no more monolithic in their structure than are macromolecules.

Acknowledgments

The author thanks Dr. Walter Tubbs for editorial assistance, and the United States Public Health Service for Research Career Award MH 15,214 which made the present manuscript possible.

References

Bartlett, F. C. (1961). Remembering: A Study in Experimental and Social Psychology. Cambridge: The University Press.

Chow, K. L. (1964). Neurosychologia, 2:175. ——— and Dewson, J. H., III (1964). Neuropsychologia, 2:153.

Dewson, J. H., III, Chow, K. L., and Engel, J., Jr. (1964). Neuropsychologia, 2:167.

Foerster, S. von (1948). Das Gedächtnis. Vienna: Franz Deuticke.

Foerster, O. (1936). In: Handbuch der Neurologie, Bumke, O., and Foerster, O., eds. Berlin: Springer.

Fritsch, G., and Hitzig, E. (1969). In: Brain and Behavior, vol. 2, Perception and Action, Pribram, K. H., ed. London: Penguin Books.

Fuxe, E. Hamberger, B., and Hokfelt, T. (1968). Brain Res., 8:125.

Gabor, D. (1948). Nature, 161:177. ——— (1949). Proc. Roy. Soc. (London), A197:454. ——— (1951). Proc. Phys. Soc. (London), B64:449.

Gebhard, J. W. (1961). Amer. J. Clin. Hypn., 3:139.

Haber, R. N., ed. (1969). Information Processing Approaches to Visual Perception. New York: Holt, Rinehart and Winston. ——— and Haber, E. G. (1964). Percept. & Motor Skills, 19:131.

Halstead, W. C. (1948). Res. Publ. Ass. Nerv. Ment. Dis., 27:59.

Hamburger, V., and Levi-Montalcini, R. (1950). In: Genetic Neurology: Problems of the Development, Growth and Regeneration of the Nervous System and of its Functions, Weiss, P., ed. Chicago: Univ. of Chicago Press.

Hart, J. T. (1965). Recall, Recognition, and the Memory-Monitoring Process. Ph.D. Dissertation. Stanford University.

Hartline, H. K., Wagner, H. G., and Ratliff, F. (1956). J. Gen. Physiol., 39:651.

Heerden, P. J. van (1968). The Foundation of Empirical Knowledge. The Netherlands: N.V. Uitgeverij Wistik-Wassenaar.
Henry, C. E., and Pribram, K. H. (1954). Electroenceph. Clin. Neurophysiol., 6:693.
Hubel, D. H., and Wiesel, T. N. (1968). J. Physiol., 195:215.
Hydén, H. (1961). In: Man and Civilization, Farber, S. M., and Wilson, R. H. L., eds. New York: McGraw-Hill.
John, E. R., and Killam, K. F. (1959). J. Pharmacol. Exp. Ther., 125:252.
Kamiya, J. (1968). Psychology Today, 1:56.
Kety, S. S. (1970). An hypothesis of learning based upon an action of biogenic amines on synaptic protein synthesis. Personal communication.
Kraft, M. S., Obrist, W. D., and Pribram, K. H. (1960). J. Comp. Physiol. Psychol., 53:17.
Krech, D., Rosenzweig, M., and Bennett, E. L. (1960). J. Comp. Physiol. Psychol., 53:509.
Landauer, T. (1964). Psychol. Rev., 71:167.
Leith, E. N., and Upatnieks, J. (1965). Sci. Amer., 212:34.
Libet, B. (1966). In: Brain and Conscious Experience, Eccles, J. C., ed. New York: Springer.
Luria, A. R. (1968). The Mind of a Mnemonist. New York: Basic Books.
Mahl, G., Rothenberg, A., Delgado, J., and Hamlin, H. (1962). Psychological responses in the human to intracerebral electrical stimulation. Presented at Annual Meeting of the American Psychosomatic Society, 1962.
Mihailović, L. J., and Janković, D. B. (1961). Nature (London), 192:665.
Miller, G. A., Galanter, E. H., and Pribram, K. H. (1960). Plans and the Structure of Behavior. New York: Holt, Rinehart & Winston.
Morrell, F. (1960). Epilepsia, 1:538. ——— (1961). In: Symposium on Brain Mechanisms and Learning, Delafresnaye, J. F., Fessard, A., and Konorski, J., eds. Oxford: Blackwell.
Mountcastle, V. B., Poggio, G. F., and Werner, G. (1963). J. Neurophysiol., 26:807.
Niu, M. C. (1959). In: Evolution of Nervous Control from Primitive Organisms to Man. Washington: Amer. Assoc. Adv. Sci., Publ. No. 52.
Poggio, G. F., and Mountcastle, V. B. (1960). Bull. Johns Hopkins Hospital, 106:283.
Pribram, K. H. (1951). Surg. Forum, 36:315. ——— (1969). Sci. Amer., 220:73. ——— Ahumada, A., Hartog, J., and Roos, L. (1964). In: The Frontal Granular Cortex and Behavior, Warren, J. M., and Akert, K., eds. New York: McGraw-Hill.
Rose, J. E., Malis, L. I., and Baker, C. P. (1961). In: Sensory Communication, Rosenblith, W. A., ed. New York: Wiley.
Rusinov, U.S. (1956). 20th Internat. Physiol. Cong. (Brussels), 785 (Abstr.).
Scheibel, M. E., and Scheibel, A. B. (1967). Brain Res., 6:60.
Sherrington, C. (1947). The Integrative Action of the Nervous System. New Haven: Yale University Press.
Sjöstrand, F. S. (1969). In: The Future of the Brain Sciences, Bogoch, S., ed. New York: Plenum Press.
Sokolov, E. N. (1960). In: The Central Nervous System and Behavior, Brazier, ed. New York: Josiah Macy, Jr. Foundation.
Spinelli, D. N. (1970). In: The Biology of Memory, Pribram, K. H., and Broadbent, D. New York: Academic Press.

Stamm, J. S. (1964). J. Neurophysiol., 24:414. ——— and Knight, M. (1963). J. Comp. Physiol. Psychol., 56:254. ——— and Pribram, K. H. (1960). J. Neurophysiol., 23:552. ——— and Pribram, K. H. (1961). J. Comp. Physiol. Psychol., 54:614. ——— Pribram, K. H., and Obrist, W. (1958). Electroenceph. Clin. Neurophysiol., 10:766. ——— and Warren, A. (1961). Epilepsia, 2:229.

Stromeyer, C. F., III (1971). Nature (London), in press.

Trabasso, T., and Bower, G. H. (1968). Attention in Learning Theory and Research. New York: John Wiley.

Ukhtomski, A. A. (1927). Psychol. Abstr., 2388.

Walter, W. G. (1964). Arch. Psychiat. Nervenkr., 206:309.

Werner, G. (1969). The Topology of the Body Representation in the Somatic Afferent Pathway (Neurosciences, vol. II). New York: Rockefeller Univ. Press.

Whyte, L. L. (1954). Brain, 77:158.

Index

Page numbers in italic type refer to tables.